Planning and Estimating Dam Construction

ALBERT D. PARKER, P.E.
Heavy Construction Consultant
Formerly, Chief Heavy Construction Engineer
Kaiser Engineers, Oakland, California

McGRAW-HILL BOOK COMPANY

New York St. Louis San Francisco Düsseldorf Johannesburg
Kuala Lumpur London Mexico Montreal New Delhi
Panama Rio de Janeiro Singapore Sydney Toronto

PLANNING AND ESTIMATING DAM CONSTRUCTION

Library of Congress Catalog Card Number 77-141922

07-048488-0

1234567890MAMM754321

This book was set in Caledonia by Textbook Services, Inc., and printed and bound by The Maple Press Company. The editors were William G. Salo, Jr., and Carolyn E. Nagy. The designer was Naomi Auerbach. George E. Oechsner supervised production.

Contents

Preface

This book is intended to serve as a reference for heavy construction estimators, as an estimating guide for estimators who only occasionally estimate dam construction, and as an introduction to dam estimating for young engineers who are just entering the estimating profession. It is based on the knowledge I have accumulated during my 34 years of experience in construction. The new and interesting construction problems which I have frequently encountered and my associations with experienced and competent engineers have made these years exciting and gratifying. I have found dam-construction estimating particularly stimulating when unique conditions have made it necessary to develop new construction techniques and apply new types of construction equipment.

The procedure in preparing cost estimates for dam construction consists of reviewing and evaluating all job conditions and specification constraints so that the most economical construction system can be selected and a work schedule can be prepared. The system and schedule are then utilized in the determination of the required quantities and types of construction equipment and plant facilities. Supervisory and office forces required for work supervision are also determinable by utilizing the construction schedule in connection with work quantities. In-depth planning is then done so that the cost components of all bid-item work can be listed. The estimate is then completed by costing all this construction planning. Work planning is simplified if precedence diagraming and resource allocation is used to assist in work scheduling. Estimate costing is simplified when computers are used to eliminate routine calculations.

Dam estimating is more interesting and intriguing than any other type of construction estimating because many different construction systems, types of construction equipment, and construction techniques can be used in their construction. Because dams can be constructed in so many different ways, the selection of the most economical construction system and type of equipment often taxes the ingenuity of the estimator.

The first eight chapters of this book provide a background in dam construction and pre-estimate planning to enable the estimator to envision all the construction operations and to select the proper construction system and equipment for a particular project. In the last four chapters descriptions are given on estimating procedures, and the application of these procedures is illustrated by the inclusion of a sample estimate for a concrete dam. The formation and operation of joint ventures and the establishment of bid prices is described. Bidding procedures are illustrated by the computation of bid prices for the sample dam estimate. I have tried to be clear and concise in describing construction methods, estimating procedures, and bid preparation. If reference is made to the descriptions of the estimating procedures when the sample estimate is reviewed, the overall concept of dam estimating may be more readily apparent.

The descriptions included in this book are broadly general in order to furnish an overall concept of the procedures used in planning and estimating dam construction. Emphasis is on the constructor's approach to dam construction rather than on the dam design. If more detailed descriptions on any subject are desired, such descriptions can be found in other publications such as those listed as references at the end of each chapter.

I wish to express my appreciation to Kaiser Engineers for the valuable experience I gained during thirty years in their employment; to Geraldine Randall for her editorial assistance in the preparation of this book; to coworkers Jerry Smith, Robert Snyder, and Mel Loomis for their assistance; and to my wife, Helen, who gave me invaluable help and encouragement.

ALBERT D. PARKER

CHAPTER ONE

Introduction

Rapid population growth has increased the demand for additional dam construction to provide sufficient water storage for domestic and industrial consumption, land irrigation, power generation, and flood control. This book is intended to fill the need for a reference book on dam construction, and specifically to provide information on construction planning, scheduling, and cost estimating. Much of the information presented is also relevant to the construction of other large concrete- or earth-moving projects.

Traditionally, water-storage reservoirs have been provided by constructing dams across deep-river canyons or across natural basins that have sufficient runoff from their watersheds to fill their reservoirs. Dams in such locations will always be required, but due to the complexity of our civilization, there is also a need for dams constructed on natural storage basins that have insufficient runoff to fill their reservoirs. These reservoirs must be filled with water pumped or conveyed from other watersheds, or from reservoirs formed by the construction of dams on other watersheds. One use of this type of dam is to store water for large population areas that are lacking in natural water sources. Examples of this type of dam are San Luis Dam and Cedar Springs Dam in California, which store water for the southern California area.

Another use for dams built across dry-storage basins is to provide reservoirs for pumped storage power projects. A typical installation of this type consists of two dams forming reservoirs at different elevations with a penstock conveying the water between the two. During daily periods of excess power (midnight to early morning) electric pumps lift the water from the lower reservoir to the higher. During daily periods when there is a high demand for power (late afternoon and early evening), the flow of water is reversed and the pumping station functions as a power-generating station. Castaic Dam and Pyramid Dam in California will function in this manner. Water is conveyed between these two reservoirs through Angeles Tunnel and, from the tunnel portal, through a penstock to the pump-generator station located on the edge of Castaic reservoir. This pumping power-generating plant will have an installed capacity of 1,200,000 kw.

DAM TYPES

Dams may be any combination of concrete, earth, or rock fill.* Before the advent of large earth-moving equipment, most dams were concrete-gravity-arch dams or hydraulically placed earth-fill dams. The main disadvantage of the latter is that the stability of the hydraulically placed embankment is always questionable. Today, compacted-fill and rock-fill dams have replaced hydraulic-fill dams. These require a much smaller volume of materials than similar hydraulic-fill embankments. Futhermore, unit placement cost is low with modern dirt-moving equipment. These two factors have reduced the total cost of earth- or rock-fill dams so that they are, at most damsites, more economical than concrete dams. However, concrete dams are still used at sites where suitable borrow materials are unavailable, or when the damsites are located in narrow canyons where river diversion creates special problems. Combination dams are used when the site is broad enough to allow the spillway and power-intake section to be built of concrete and the remainder of the dam to be constructed of earth or rock fill.

There are two factors which may reverse the cost trend in favor of earth-fill dams: increasing first costs and operating costs for earth-

*In rare instances, timber-crib dams are constructed, but because of their very limited application, they will not be covered in this book.

moving equipment as the size of this equipment continues to increase and the economy of using concrete thin-arch dams in narrow canyons, or buttress-type concrete dams on flat damsites.

PURPOSE OF A DAM ESTIMATE

The primary purpose of a dam estimate is to furnish a logical cost and time basis for a bid for dam construction, with a bid price that will allow the contractor to make a reasonable profit. Dam construction is a competitive field, and it is therefore necessary for the contractor's organization to prepare an accurate estimate of the cost of doing the work. If the estimated cost is too low and the contractor is the successful bidder, he will lose money on the project. Conversely, if the estimated cost is much higher than the actual cost, the contractor will have very little chance of being the low bidder and hence will not get the job. Therefore, the estimate must be based on the use of the most economical construction system, on performing work with construction plants and equipment units that have been selected to operate efficiently as component parts of the construction system, on proper work scheduling, and on accurate costing of all work items. The contractor places a great amount of confidence in the skill and integrity of his estimator, since proper bidding is essential to his success. Estimating is less important outside the construction industry; the contractor is the only manufacturer who must price his product before it is produced.

If the contractor is the low bidder on the construction project, the estimate has several secondary uses. All the adjustments made before bidding are carried back into the body of the estimate and it is then published as a "budget" estimate. This budget estimate provides the job management with a list of the plant and equipment required to construct the job, is used as a guide in determining the methods of constructing the work, and furnishes the basis for job control since the estimated cost can be compared with actual cost and estimated production can be compared to actual production. This comparison points out the cost items that are critical and indicates when work falls behind the schedule and therefore should receive attention by construction management and job supervision. Cost-accounting procedures and cost accounts are established in accordance with the estimate's format and the cost items and divisions used in the estimate. If there are changes in the work or in job conditions, or changes that affect job completion which are sufficiently extensive to warrant presenting a claim to the owner, the budget estimate is used as a basis for its preparation.

If the contractor is not the successful bidder, he should compare his estimate with the bid submitted by the successful bidder. This comparison, with field inspections of the low bidder's work progress, may indicate improvements in construction planning that can be incorporated into future estimates.

HEAVY-CONSTRUCTION ESTIMATING METHOD

The heavy-construction estimating method is used for dams, tunnels, powerhouses, and similar projects. This estimating method treats each construction project as a separate entity, and the construction-cost computations are based on the use of the construction equipment units, the plant facilities, the labor rates, the labor productivity, the working conditions, the work schedule, the work sequence, the subcontract prices, and the permanent material and equipment prices that apply to each project.

This estimating method requires that the estimator take an overview of the entire project. The specification work constraints, the type of project, its location, any special requirement such as stream diversion, the topography, the work scheduling, and the work sequence are all studied and evaluated. A construction system is selected on the basis of these factors as well as considerations of economy and the required completion date. A construction schedule is then prepared indicating the time available for each construction operation. Selection of construction equipment of appropriate type, size, and capacity completes the construction system. The work required for each work item is planned so that the construction equipment is used to maximum capacity. The number of men, hours of equipment operation, quantities of consumable supplies and repair parts, the subcontracted work, and the required permanent materials and equipment are listed for each work item. The listings are then costed using the labor rates, the supply cost, the price of repair parts, the subcontract prices, and the materials and equipment quotes that apply to each project. The individual work-item costs are then summed to arrive at the total cost for direct labor, consumable supplies, subcontract work, and permanent materials and equipment.

Plant and equipment depreciation chargeable to the project is estimated by preparing a detailed list of all required construction equipment units and plant facilities. Acquisition plus erection cost, less resale value at the time of project completion, represents the depreciation for each unit.

Indirect expenses for supervisory costs and labor, bond premiums, taxes, and insurance are also estimated. These costs vary with work volume, time required for construction, and the amount of the contract. Escalation is estimated by determining the labor and material expenditures that will be made in each yearly time period, estimating the percentage that the labor and material cost will increase during each period, and then computing the total increases.

The total estimated cost for the project is the sum of the costs of labor, consumable supplies, subcontract work, permanent material and equipment, equipment depreciation, indirect expense, and escalation.

UNIT-COST ESTIMATING METHOD

The unit-cost estimating method is used to estimate the cost of all but heavy-construction projects and requires much less effort than the heavy-construction estimating method. Work quantities are established for each bid item and then multiplied by the applicable unit cost. The sum of these extensions represents the prime contractor's on-site construction cost. To this figure is added the payments to subcontractors and the cost of procuring permanent equipment, materials, and facilities, to arrive at the total estimated cost. This procedure varies in minor ways among contractors. Some contractors use unit costs that include all on-site costs. Other contractors use unit costs that exclude equipment depreciation and/or indirect expense, and these costs are then computed separately and added to the extended total.

The unit costs used in an estimate are costs that have been developed by each contractor from his experience in similar work. These costs can be readily adjusted for changes in labor productivity or in labor rates. However, there is no accurate method of adjusting these unit costs to reflect the use of different types of construction equipment.

In preparing an estimate using the unit-cost method, the major work is taking off the quantities of work. Estimating skill is required to make proper use of, and adjustments to, the unit costs. Computer manufacturers have developed programs for taking off quantities, for storing and using unit costs, and for making and totaling the cost extensions. The use of the computer for this purpose will greatly lighten the estimating work load for contractors who use this estimating system.

COMPARISON OF ESTIMATING METHODS AND THEIR APPLICATIONS

A heavy-construction estimate is prepared using the plant and equipment, the labor rates, and the labor productivity applicable to the specific project. A unit-cost estimate is prepared using work performance on similar projects. Adjustments must be made for changes in labor rates and for labor productivity, but there is no way to make accurate adjustments for changes in equipment use. Because of the savings in estimating effort, contractors use the unit-cost method for construction projects where unit costs do not vary from project to project as a result of changes in sizes or types of construction equipment. However, the heavy-construction method must be used to produce estimates for projects where the costs do vary, such as for tunnels, dams, and powerhouses.

Projects of these types are constructed most economically using large amounts of construction equipment. Furthermore, different types of equipment are required to construct similar projects, since the physical characteristics at the job locations will always be different. On such projects, construction cost for the major work items is primarily composed of the cost of operating, maintaining, and depreciating the construction equipment. Changes in equipment can result in large differences in the unit costs of the major work items and in the total construction cost.

Heavy-construction estimating for dams can be simplified by using the unit-cost method for estimating the cost of many minor work items that are constructed by hand labor, such as the construction of built-in-place wood forms, the placement of reinforcing steel, hand excavation, electrical work, plumbing, and architectural work. Unit costs for these items will vary between dam projects only because of differences in labor rates and labor productivity.

The unit-cost estimating method can be used to prepare road-construction cost estimates since contractors in this field maintain their own fleets of construction equipment which are used on each of their road projects at standard depreciation rates. Since all projects use the same basic equipment and are charged the same depreciation rates, unit costs will not vary among projects because of equipment usage.

The unit-cost method is also suitable for estimating the cost of industrial, commercial, and building construction. For such projects, the prime contractor's on-site cost is a minor portion of the total construction cost; the greater portion of the construction cost is for permanent materials, equipment, facilities, and subcontracted specialty work. The prime contractor's construction equipment will be basically the same on all jobs, and in any case represents such a small portion of his on-site cost that a change in equipment types would not result in changes to unit costs.

COMPUTER ESTIMATES

Computer preparation of estimates can eliminate many tedious calculations and is an excellent way of preparing dam estimates provided the computer's limitations are recognized. The computer does exactly as it is instructed; it cannot be used as a substitute for estimating ability. In computer jargon *GIGO* still controls—that is, garbage into the computer, garbage out of the computer. A great amount of planning ability is needed to instruct the computer on all the necessary details involved in preparing a dam estimate. The overall planning discussed earlier must still be done. The construction system must be determined, plant and equipment components selected, production rates determined, hourly operating and maintenance costs established, crew sizes determined, and supply consumption estimated.

After this planning is completed, data can be fed into the computer and it will produce a typed estimate. This estimate should be checked to determine if it was properly programed. As an example of inaccurate programing, overbreak concrete has been programed to reduce instead of increase the pay quantity unit cost. Proper programing requires both an experienced dam estimator to establish the instructions for the computer and a good programer to see that these instructions are properly translated into computer language.

The program used should include equipment hours, plant hours, man-hours, fuel, oil and gas consumption, and repair parts consumption. The best programs output typed estimates in the same basic form that is used for calculator estimates. Such programs have been developed by certain consultants and contractors but are not available from computer manufacturers. The manufacturers' programs use the unit-cost method and have its same limitations. Most of the inaccurate heavy-construction computer estimates have been produced by personnel who understood computer programing but who were not experienced in heavy-construction estimating. The computer cannot supply estimating knowledge; it supplies only rapid computations based on adequate instructions.

Heavy-construction firms would do well to investigate computer estimating since it can greatly reduce the amount of manual work done in the estimating department. Computers can be gradually introduced into a construction company by using them to prepare payrolls, maintain equipment records, record costs, and make detailed schedules from precedence diagrams. They then can be introduced into the estimating department to make quantity takeoffs of variable arch dams and to compute cycle times for haul trucks, scrapers, and cableways. As the estimating personnel become acquainted with the computer, they will recognize the advantages of using it to prepare the complete estimate.

A good first project with the computer for an estimating department is to use it to prepare the budget estimate for a project that was successfully bid with a longhand estimate. The estimate data can be the computer input; an estimate will be produced that will furnish a large amount of detail helpful for job control. Many estimators who have both computer and calculator estimating experience think this is the right application for computer estimates. They believe that a bid estimate can be turned out more rapidly by the longhand method, but that if the bid is successful, a computer estimate should be run so that the details of the estimate will be more readily available.

CALCULATOR ESTIMATES

Calculator or longhand estimates have been used in the heavy-construction industry since its infancy. If they are prepared in the right format, they provide great flexibility. Estimate costing of part of the dam project can be performed while the remaining work is being planned. At any stage of estimate preparation, it is relatively simple to make changes to completed work when new work concepts are conceived. With a computer, planning must be complete before the estimate is costed. With calculator estimates, there will be more chance of arithmetical errors than with computer estimates; however, programing errors will not occur. Arithmetical errors can be reduced by using an estimating format that provides checks for all major arithmetical computations.

Since calculator estimates more readily illustrate estimating procedures, this type of estimate is used for illustrative purposes in Chap. 11.

CONSTRUCTION-SYSTEM SELECTION

A construction system comprises the construction time, manpower, construction equipment, construction plants, and supporting facilities as an integrated unit for performing all the work required to construct a project. Successful dam estimating requires the selection of construction systems that can produce the work in the most economical manner.

Because of its high productivity and despite its high purchase cost, construction equipment is more economical to use on heavy-construction projects than hand labor with today's high labor cost. As a result, the cost connected with operation and depreciation of the construction equipment forms the largest portion of a dam's total construction cost. Different equipment units will differ in their operating, maintenance, and depreciation costs and in their rates of production. Construction cost is lowest when construction is performed with equipment having the most favorable relationship between cost and productivity. However, the construction equipment that can be used on any dam project is limited to the types that will function together as an integrated construction system. Thus the construction system must be determined prior to the selection of construction equipment.

Main Construction System

The main construction system is the coordinated arrangement of the subsystems into one complete operating unit. The system must have sufficient capacity so that crash scheduling will not be required. It must be capable of constructing the major items of work with maximum efficiency, and yet with enough flexibility to supply the minor demands placed on the system during the construction of the minor work items.

The main construction system is used as the control of the composition and production capacity of each subsystem so that a balanced operation will result without any activity gaps. Main-system review will indicate whether any equipment supplied for one subsystem can also be utilized in another subsystem. For example, excavating equipment supplied to the excavation subsystem can often be utilized later in the concrete subsystem for either aggregate-pit or rock-quarry excavation. This review will also reveal whether all equipment supplied to the subsystem is in balance with the work to be accomplished. Often large expenditures for equipment are planned for the construction of a minor amount of work. Systems review may show that this work should either be done with rented equipment or be subcontracted.

Subsystems

The specification constraints, construction schedule, work sequence, method of diversion, and the topography of the damsite must be studied and evaluated since they control the selection of construction subsystems to be used to construct the dam project. That is, the subsystems selected must meet the work constraints and comply with the work conditions, and also must be economical and capable of functioning efficiently together as components of the main system.

The number of subsystems required depends on the type of dam to be constructed. For a concrete dam, an excavating and a concreting subsystem are minimum requirements. Subsystem selection is a complex problem because so many different types of work are involved in dam construction. Work may include site clearing, tunneling, river diversion, common excavation, rock excavation, drilling and grouting, concrete construction, rock quarrying, rock embankments, earth-fill embankments, powerhouse construction, permanent-material installations, and equipment erection. The types of work that must be performed for the construction of any dam project vary with the purpose of the project, the type of dam, the damsite characteristics, and the method of stream diversion. Because the work required for dam-construction projects may differ so widely, contractors have developed a large number of subsystems. Each can be used economically for some dams but will not necessarily be efficient in the construction of similar dams with different working conditions.

Limitations on Construction-system Development

Work constraints limit the contractor's development of new construction systems. Most of these work constraints are delineated in the specifications or established by labor contracts and are therefore totally binding upon the contractor. Many specifications are so out of date that they contain work constraints that do not apply to modern construction practices and are, in fact, a great hindrance to the development of better construction systems. The systems that have developed for dam construction are the most suitable to use with present specification and labor constraints. These are continually becoming more expensive and complex as additional work constraints are added to the specifications or included in labor contracts. All system improvements to date have been the result of contractors' efforts in developing new systems, or of their modification of systems to include new work concepts developed by other industries.

SCHEDULING

The construction schedule is utilized as the coordinating tool in the systems approach to dam estimating. It shows the time available for constructing each phase of the project. By using the schedule in conjunction with work quantities, the desired production rates and work sequence can be established, and from this, the required construction equipment and facilities determined. The construction schedule is also used for resource leveling—that is, the scheduling of the construction of noncritical work items at time periods when the construction equipment is not used to full capacity on critical items. Without resource leveling, additional plant and equipment may be required for these noncritical work items. To make the construction schedule as useful as possible as a coordinating tool, it should be prepared in a form that illustrates the flow and sequence of all operations and yet is simple enough to be readily understandable.

Advantages of Arrow Diagraming

Four types of construction schedules are used for dam construction: the bar chart and three types of arrow diagraming. In the past, the bar chart has been widely used because it is simple and readily understandable. Recently it has been modernized by some contractors to show work sequence, with triangles indicating the original schedule, the revised schedule, and the actual schedule. Bar-chart schedules do not show work flow and the interrelation among activities, nor do they highlight essential work. Because arrow diagraming provides methods for illustrating these points, it has largely replaced the bar chart in the heavy-construction industry.

Arrow diagraming enables the construction planners to study the effect that the performance and completion of one operation will have on others. Use of an arrow diagram to determine the critical path in dam construction is a minor consideration since the critical path for dam construction always follows the same path: mobilization of plant and equipment, plant and equipment erection, construction of the diversion facilities, dam excavation, dam placement, and the placement of gates and/or other mechanical items. Even if the critical path is known, the arrow diagram is helpful in the determination of the minimum construction time since it indicates whether there can be any overlapping of the activities that are part of the critical path. It also indicates when the construction of noncritical work items can be scheduled, thereby facilitating resource leveling. Furthermore, arrow diagraming can assist in scheduling the receipt of permanent materials and equipment so that on-site construction will not be delayed. Another advantage of the arrow diagram is that it can be readily revised as construction is performed, reflecting any change in work performance or in job planning.

Comparison of Arrow-diagraming Methods

The three types of arrow diagraming presently in use are PERT (Program Evaluation and Review Technique), CPM (Critical Path Method), and precedence diagraming. With the PERT system, arrows are labeled with events and connect to numbered nodes so that each event is described by the number in the node at the start of the event and the number in the node at the end of the event. PERT uses three estimated times for each activity: the optimistic time, the pessimistic time, and the most likely time. The CPM method is very similar, the difference being that the arrows are labeled with activities and only one estimated time is used. Both methods of scheduling must be done in such a manner that in order to properly show work sequence, it is necessary to insert dummy activities. The preparation of these arrow diagrams is so complex that errors frequently creep into any complicated network, usually because of the absence of some required dummy activities.

Both of these methods of diagraming are so complicated that special knowledge is required for their preparation; therefore, this function is often performed by specialists or consultants who may or may not have had construction experience and whose only job contact is from occasional visits. Many consultants specialize in the preparation of PERT or CPM arrow diagrams for contractors who must submit schedules of this type to the owner, but who do not want to be bothered with their preparation. Since these construction schedules have not been prepared by the job-construction personnel, the construction forces may classify them as consultant-prepared theoretical diagrams, ignore them, and not keep them up to date. When this happens, their main usefulness is destroyed. Another disadvantage of a consultant-prepared schedule is that it is often so complex with so many activities and dummy activities that only an arrow-diagraming specialist can comprehend its ramifications.

The dummy activities required by CPM and PERT are often confusing to job personnel, who prefer a schedule that shows only required activities. Furthermore, a change or modification in these diagrams often requires additional arrows, necessitating a renumbering of the arrow nodes; thus, an activity may not retain the same number for the life of the project. To summarize, PERT and CPM schedules are often ponderous to use and very difficult to follow in determining the overall job direction and construction-system requirements. For those interested in PERT or CPM diagraming, there is considerable published data on these methods.[1]

Precedence Diagraming

Precedence diagraming is a simplified form of arrow diagraming. It is quite similar to a flow chart and does not require the insertion of dummy activities. It has all the advantages of PERT and CPM, and none of the disadvantages.[2] It is not as stylized so the diagraming can be done in many ways. Activities are placed in circles or squares and given an activity number which does not have to be changed during the life of the project. The squares can also carry a cost-code number, earliest starting dates, latest starting dates, time required to complete the activities, and the calendar date of the start or completion of any operation as illustrated in Figure 1*a*.

The squares can be plotted on a time scale and connected with arrows showing the sequence of operations. These arrow connections are illustrated by Figure 1*b*, which shows that foundation

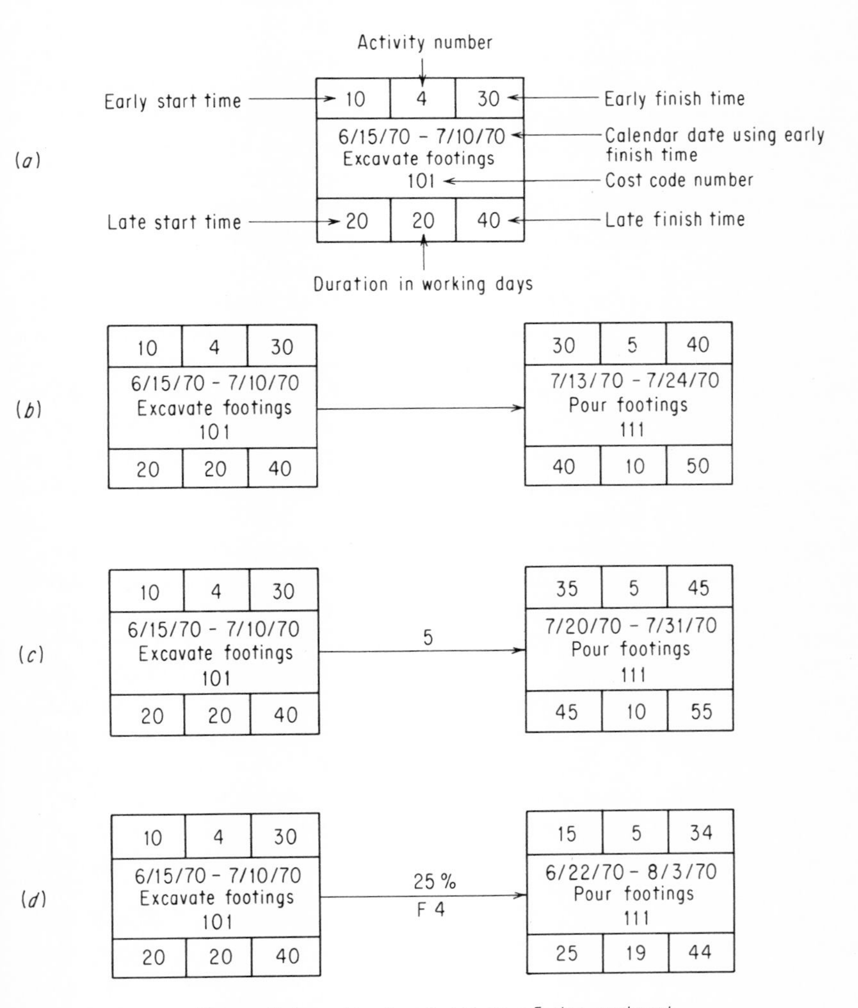

Fig. 1. Precedence diagraming.

concrete cannot be placed until all the foundations are excavated. Restrictions and constraints occurring between activities can be shown. For example, Figure 1*c* indicates that concrete cannot be placed in the footings until 5 days after all the excavation is completed. Figure 1*d* indicates that foundation pouring can start after 25 percent of the foundation has been excavated but cannot be completed until 4 days after all footings have been excavated. With these simple designations, a complete arrow diagram can be constructed. Many other diagraming and designation methods are used for precedence diagraming, depending on the owner's requirements or the contractor's practice. One indicates that the next operation can start before the first one is complete if the connecting arrow starts along the bottom of the first square.

ESTIMATING ABILITY

To plan a dam estimate and develop the most efficient construction system, the estimator must have knowledge of dam construction, the types of equipment and facilities required to accomplish the construction, the crew sizes required for each operation, their established wage rates, and the time required to complete each construction phase. He must understand construction-plant layout, aggregate manufacturing, refrigeration, mixing-plant operation, form layout, concrete-placement methods, dam-foundation excavation requirements, methods of diversion, economical dirt-moving procedures, quarry operations, embankment compaction, equipment installation, labor regulations, equipment maintenance, camp construction, logistic support, overhead cost, insurance coverage and rates, safety practices, subcontract administration, bond rates, escalation, cash requirements, and accounting procedures.

It is difficult for an estimator to visualize all the problems that will be encountered in construction unless he has had construction-field experience. This experience can be supplemented by job inspections, the perusal of past job records, the study of job reports published in trade magazines and the reading of construction books. Construction experience is invaluable in determining labor production for operations where machines are not the basic production units but production is the result of labor working with hand tools. The amount of work that construction labor will produce with hand-held tools is the hardest part of an estimate to prepare since it is dependent on the efficiency of the laboring force and the planning, knowledge, and leadership of the supervisory force. Productivity of this laboring force is not mathematically determinable but must be based on the judgment and experience of the estimator.

In contrast to this, the labor cost of operations where production is accomplished by utilizing equipment and plants can be computed mathematically. These types of operations are machine excavation, truck haulage, embankment construction, aggregate production, mixing and cooling plant production, cableway or crane operation, etc. The number of men required for each unit of equipment and for each plant is dependent on their size and type. The time required for constructing any facilities is dependent on equipment and plant production rates which are primarily a matter of determining the production capacities of machines. Therefore, labor-cost estimating for these types of operations is based on equipment production rates and varies only in a secondary manner with the ability and efficiency of the workmen.

The time available for the preparation of a dam estimate is so limited that it is imperative that the estimator have the ability to schedule his time so he can spend as much time as possible on the major work items and only a minimum amount of time on the minor items whose total cost will be relatively minor.

ESTIMATING DATA

Dam estimating is such a complex subject and requires such a wide range of construction knowledge that estimators should maintain reference files of estimating data. Catalog information and other data secured from equipment manufacturers will provide information for the selection of the types of construction equipment and of construction plants. Files should be maintained on the available plant and equipment, its production capacities, and on repair parts and supply consumption. If any complicated facility is required, such as an aggregate plant, refrigeration plant, mix plant, or a cableway, the manufacturer's representatives will provide design assistance and assist in plant selection so that the proper facilities are provided to accomplish the work requirements. Manufacturers will also supply quotations on equipment and construction plants to enable the estimator to properly price these facilities.

Copies of the latest labor agreements are important to the estimator as they furnish information on craft jurisdiction, working rules, working hours, wage rates, and fringe benefits. In order to determine the payroll burden, files should be kept current on insurance quotes and on state and federal payroll taxes.

A file of costs from construction projects and a file of completed estimates will provide information on construction methods and equipment selection and will assist in determining the equipment's consumption of fuel, oil, and grease; maintenance cost; labor productivity; and the like. These files furnish invaluable assistance as a check against estimated unit costs; if differences occur, they should be analyzed to determine if they are a result of estimating errors, or of the use of different construction systems, changes in labor rates and labor productivity, or differences in efficiency of the supervisory personnel that will be used on the project.

A list of interested subcontractors and material suppliers should be maintained. These should be reliable contractors who have a good record of work performance. The estimator can then rely on these firms to quote on permanent materials and subcontracts, which will relieve him of this responsibility.

A file of articles from trade magazines describing the construction of dam projects may furnish solutions to many of the problems the estimator encounters.

DAM-ESTIMATING PROCEDURES

Dam estimating is so complex that it is necessary to develop the construction background before it can be explained. The broad scope has been presented in this chapter. In Chaps. 2 to 8, background information on construction scheduling, construction-system and equipment selection, construction planning, and cost estimating is presented. This consists of defining the types of work required for the dam construction, explaining how each type is performed, describing the available construction equipment and plant facilities, explaining how their work capacity can be computed, and explaining the controls for construction-system selection. All the information presented in the first eight chapters is a prerequisite to understanding the information presented in Chap. 9.

Chapter 9 explains how the construction planning of a dam located within the United States is performed and describes the preparation of an estimate of its construction cost.

Chapter 10 explains how the planning and estimating procedures presented in the previous chapters must be modified to make them applicable to a dam located outside the United States.

In Chap. 11, a sample estimate for a fictitious concrete dam is presented, and in the final chapter the preparation of a competitive bid for the same fictitious concrete dam is described and illustrated.

REFERENCES

1. *Control of Construction Projects through IBM Data Processing*, IBM Data Processing Application, IBM Technical Publications Department, White Plains, N.Y.
2. John Fondahl, Professor of Civil Engineering, Stanford University, Let's Scrap the Arrow Diagram, *Western Construction*, August, 1968.

CHAPTER TWO

Diversion, Cofferdams, Dewatering

The method of stream diversion must be determined prior to selection of the construction system since it controls when the construction equipment and plant can be erected, the method and sequence of dam excavation, and the method and sequence of material placement in the dam. Since it is part of the critical path of a dam's construction schedule, it controls the time required for construction. Selection of the diversion method is therefore one of the first tasks that should be undertaken in planning dam construction.

The method of stream diversion must be one that will not endanger life or cause excessive property damage when the diversion facilities are subject to maximum flood flows. If the method chosen restricts the flow of the water past the damsite, it must either prevent overtopping of the diversion facilities, or it must permit controlled overtopping. For economic reasons it must be the simplest method that will divert the water past the damsite without delaying construction, with a minimum exposure to overtopping. To determine whether diversion facilities should be designed for controlled overtopping or with sufficient capacity to prevent overtopping, the overtopping cost (taking into account how frequently overtopping could occur) should be compared with the cost of providing increased diversion-facility capacity. Based on this study, the determination of whether the diversion facilities should be designed with sufficient capacity to handle flood flows with a frequency of 1 year in 5, 1 year in 10, 1 year in 20, and so on, is simply a matter of construction economics. It is usually more economical to risk occasional overtopping than to provide sufficient diversion-facility capacity to handle all flood flows.

Overtopping may result in increases in the direct cost, indirect cost, plant and equipment cost, and escalation cost of the project. The direct cost will increase if the diversion facilities are damaged as a result of the overtopping; if flood waters must be removed from the construction areas; if debris must be removed from the construction area; or if formwork, scaffolds, bridges, or other temporary or permanent works must be repaired or replaced. If overtopping results in an extension of the construction period, the indirect cost will increase since most indirect costs are proportional to the time required for construction. Plant and equipment cost increases will be the result of damage to the construction equipment, or of construction delays which, by increasing the time required on the job, reduce the salvage value of the equipment. Construction delays can also result in escalation cost increases since higher wage rates and higher supply prices may come into effect. The extent of the delay and the increased cost due to overtopping is dependent upon the design of the diversion facilities and the severity of the flood flow.

Dam-construction contracts vary with respect to the division of responsibility for the selection and design of diversion facilities. On some jobs this is done by the owner; other contracts allow the contractor to select and design the diversion facilities; and in some cases the job of designing and selecting is a joint effort by both parties. Similarly, depending on the contract documents, the cost exposure to overtopping may be entirely the owner's responsibility, the responsibility of the contractor, or the responsibility may be divided. Insurance covering the cost of overtopping can be obtained by either the owner or the contractor. If there is a force majeure clause in the contract,* the contractor can recover from the owner the cost of any overtopping of the diversion facilities by unprecedented floods. In this case, partial relief may also be available under the Disaster Relief Act from the national government for either the contractor or owner.

*Provides relief for the contractor for conditions over which he has no control.

Even if the diversion method is selected and designed by the owner, the contractor's estimator must completely review the diversion system so that he can plan and estimate the cost of its construction, its effect on the other work, the time required for its construction, the chances of overtopping, and the cost to the contractor if overtopping occurs. When the contractor is to select and design the diversion method, which is a large factor in the total construction cost, the low bidder is often the contractor who has selected and designed the most efficient method with the minimum exposure to overtopping. If it is a joint effort, the contractor must thoroughly review the owner's diversion method in order to properly select and design his portion of the facilities and determine his exposure to overtopping expense, as well as to determine the effect of the diversion method on the construction time and cost of the total project.

The remainder of this chapter discusses the stream-flow characteristics that should be compiled for diversion planning, methods of diversion and the types of damsites for which they are suitable, cofferdam construction, methods of restricting subsurface water inflows, and foundation dewatering.

STREAM-FLOW CHARACTERISTICS

Before attempting to select the diversion method, the volume of runoff that will pass by the damsite and the anticipated volume of flood flows with varying frequencies should be determined and plotted. The basis for this study is the historical record of stream flow at or near the damsite, and the longer the period for which records have been kept, the more accurate will be the results. Graphs should be prepared showing the volume and the calendar months when flows could occur for floods rated as 5-, 10-, 15-, 20-, 50-, and 100-year floods. (This rating, of course, does not mean that these floods will occur in exactly this frequency, but is rather a measure of exposure.) Graphs should also be prepared showing the maximum daily flow, the average daily flow, and the minimum daily flow.

If stream-flow records are not available or have been kept for only a short period of time, the design of the diversion facilities must be based on the area of the watershed and the amount of anticipated precipitation, correlated with the stream-flow records that are available. Though this is a logical computation, it cannot replace stream-flow records, and insufficient data can result in errors in determining the diversion facilities and dam storage capacities. Dams have been built where, because of the inadequacy of the records which determined the design, there has never been sufficient runoff to fill their reservoirs.

For more detailed information on how to compute and tabulate stream flow, books specializing in this field should be reviewed.[1]

HYDRAULIC-MODEL STUDIES

The best method of stream diversion to use at any damsite is dependent on the stream-flow characteristics, the water velocities that will occur during diversion, the type of foundation material at the damsite, the type of dam that is to be built, the topography at the damsite, the methods and equipment with which the dam material will be placed, the sequence of placement of the materials in the dam, the time required for construction of the dam, and the cost that would result if overtopping of the diversion facilities occurred.

If large flows and high velocities are involved, diversion-design calculations should be substantiated by hydraulic-model studies. Such studies have been used to verify and improve diversion methods planned for many dams, including those on the lower Columbia River and the Guri Dam in Venezuela. They assist in determining and developing solutions for inadequacies in the diversion plan. Specific points that may be checked by hydraulic-model studies are the water velocities that will occur along the cofferdam sides or during cofferdam closure, and the size of rocks that will be required to prevent cofferdam erosion and effect closure. If diversion channels controlled by diversion dikes are used, the proper location of these dikes can be determined. The hydraulic-model studies performed for Guri Dam were used for this purpose in addition to aiding in the determination of the best diversion method. The study for McNary Dam on the Columbia River indicated that the best method of cofferdam closure was the gradual raising of the closure dike with concrete tetrahedrons.

DIVERSION THROUGH THE DAM

Diversion through the dam is the most economical method when dams are to be constructed on large rivers where flood flows are of such volume that the construction of diversion tunnels, diversion flumes, or diversion channels is a major undertaking. Irrespective of stream size, this diversion method has wide application at sites where the water channel is wide enough to permit one-half the chan-

Photo 1 First-stage diversion at Amistad Dam, Texas. (C. S. Johnson Division of Koehring Company.)

nel to be enclosed in a cofferdam and still leave sufficient waterway in the remaining half of the channel to handle all flood flows. Dams that have been constructed with this type of diversion include those built on the lower Columbia River and the lower Snake River.

When this diversion method is used for a dam of concrete construction, diversion is handled in three stages. First, a cofferdam enclosing one-half the water channel is constructed at the damsite. Then the cofferdammed area is unwatered, the enclosed dam foundation area excavated, and the foundation surface prepared to receive concrete. Concrete is poured in the enclosed dam monoliths until they reach an elevation that will be above the impounded water surface to be encountered during second-stage diversion. Typical construction during first-stage diversion is shown in Photograph 1 of Amistad Dam, Texas.

If the water is to pass over low blocks during second-stage diversion, alternate blocks are left low because alternate low- and high-block concrete placement is required to furnish clearance for the transverse forms, and because a high block is required on each side of a low block to support the bulkhead which restricts the water flow while concrete is being raised in the low block. The number of blocks to leave low, and the elevation of these blocks with respect to the upstream pool can be computed using the broad-crested weir formula:[1]

$$Q = 3.087\, L\, (h + h_v)^{3/2} C$$

C = compensation factor for end restrictions and other losses. For this application, its value is approximately 0.85.
Q = maximum theoretical discharge, cfs
L = width of block, ft
h = depth of water on block, ft
h_v = velocity head of approach, ft

If during second-stage diversion, the water is to pass through diversion conduits instead of over low dam blocks, these conduits and provisions for future closing of the conduits against a head of water must be incorporated into the dam. Photograph 2 shows the conduit technique in use during the construction of Guri Dam in Venezuela. The river was exceptionally large for handling in this manner.

The number, size, and elevation of the diversion conduits can be computed using the orifice-flow formula. Since this formula involves a factor that varies with the shape of the conduit entrance and would require many sketches to explain, it is beyond the scope of this book. The interested reader is referred to hydraulic textbooks that cover this matter in detail.[1]

Photo 2 Second-stage diversion through diversion conduits, Guri Dam, Venezuela.

While concrete is being placed inside the first-stage cofferdam, the dam foundation excavation and preparation can be performed on the dam abutments so that these areas will be ready for concrete placement when the conversion is being made from first- to second-stage diversion.

After the dam concrete in the first-stage cofferdammed areas has been raised above the rock foundation, the stream dividing leg of the second-stage cofferdam is erected on top of any of the upstream and downstream concrete aprons and connected to the dam concrete. This portion of the second-stage cofferdam is erected at this time and located so that upon final completion of the second-stage cofferdam, the dam concrete that was placed during first-stage diversion will extend underneath the second-stage cofferdam and will connect to the concrete placed during second-stage diversion.

Second-stage diversion starts at the beginning of the first low-water period that occurs after the concrete within the first-stage cofferdam has reached an elevation that will be safe from unplanned overtopping. The first-stage cofferdam is removed, then the dividing leg of the second-stage cofferdam is connected to the opposite banks of the river by upstream and downstream cofferdam legs. This second-stage cofferdam will enclose the unexcavated portion of the river channel and when it is completed, water will be impounded upstream of the cofferdam until it reaches an elevation that will furnish sufficient head to pass the water through conduits in the dam or over low blocks. The second-stage cofferdammed area is unwatered; the enclosed dam-foundation area is excavated and prepared for concrete placement; and concrete placement is started. During construction of the second-stage cofferdam, dam-concrete placement can continue in the high blocks in the first-stage diversion area and in the blocks in both dam abutments.

After the concrete in the dam monoliths, enclosed by the second-stage diversion, reaches the height of the upstream leg of the second-stage cofferdam and any required grouting is completed, third-stage diversion can commence.

Third-stage diversion begins with the removal of all cofferdams, allowing the water to pond against the dam and to flow either over low blocks in the dam or through diversion openings constructed for this purpose. If the low-block system is used for third-stage diversion, concrete is poured in these blocks by cofferdamming off one low block with an upstream bulkhead and then placing concrete in this block until the level of the concrete is above the upstream water surface. The bulkhead is moved from block to block, and as each block is raised, the water surface behind the dam will also gradually rise until both the concrete and the upstream water surface reach an elevation that will permit the river flow to be handled by the permanent river outlets, or, if there are no outlets or the outlets are of insufficient capacity, then until the water flows over the permanent spillway. When this elevation is reached, the bulkhead will no longer be required and concrete can be placed in a normal manner. If the diversion-conduit method is used for third-stage diversion, the conduits are usually left open until the dam is topped out. The openings must then be closed at the upstream ends with gates capable of resisting the head of water in the reservoir. The diversion conduits are then plugged with concrete, the concrete is cooled, and contact grouting is performed.

If high-pressure gates are used to close the openings, a safeguard to ensure closure of the openings is to install stop log slots ahead of the gates so that stop logs can be dropped in these slots if the gates become inoperative.

The time between bid advertising and bid submittal is usually too short to allow the estimator to make detailed designs of these closure gates, so he must secure this information from manufacturers. Closure-gate design is covered in detail in other books.[1]

Unless special provisions are made for overtopping, diversion through the dam has limited application for earth- and rock-fill dams as the flow of water over the unprotected dam embankment will cause rapid erosion. If overtopping of an earth- or rock-fill dam is anticipated, a spillway section should be constructed before the start of the flood season on the top and on the downstream slope of the fill to resist erosion. Large rocks, pavement, or timber sheeting are used for spillway paving. If the dam is of rock-fill construction, water should not be impounded against the rock fill unless the impervious core has been placed to the same elevation as the rock fill, since the passage of water through the rock fill will transport rock away from the downstream face and weaken the dam.

With earth- and rock-fill dams, diversion through the dam has its greatest application at damsites where, during the low-flow season, the water in the stream will decline to a trickle or to nothing. Then a portion of the dam is omitted to provide a channel for flood-flow passage and the remainder of the embankment is placed to a height that will be safe from overtopping when the channel-flow passage is plugged. This height can be computed by evaluating the storage capacity of the reservoir upstream from the dam, the elevation of water-release openings, and the elevation of the spillway crest. After placement of this part of the embankment and at the beginning of a low-water season, the water is piped or flumed through the dam, the dam foundation is excavated in the former channel, and fill is placed in the embankment opening until the embankment reaches the bottom of the pipe or flume level. The pipe or flume is removed and fill placement in the embankment opening must then proceed at a rate that will prevent the water in the dam reservoir from reaching the top of the fill. When this section of the dam embankment is brought up to the height of the remaining embankment, fill is placed across the entire dam until it is topped out.

Diversion through the dam can be used to advantage in the construction of combination dams containing earth or rock fill and concrete spillway and/or powerhouse sections. First-stage construction consists of cofferdamming the concrete section, placing concrete in this section, and providing facilities in the concrete section for the passage of water during second-stage diversion. During second-stage diversion, the water is passed through the concrete section, the earth- or rock-fill section is cofferdammed, and the embankment is completed. During the third-stage diversion, the concrete section is completed. Many combination dams are constructed with the concrete section located on one abutment, which simplifies the construction of the first-stage cofferdam. Examples of this are Wells and Wanapum Dams constructed on the Columbia River.

When stream diversion is through the dam, the type of cofferdam that is used varies with the volume of water, the area available for cofferdam construction, the water velocities adjacent to the cofferdam, and the water velocities which will occur. When the velocities are moderate despite the decreased waterway, earth- or rock-fill cofferdams can be safely used. If the river channel is narrow and space is at a premium, or if high velocities occur adjacent to the dividing leg of the first-stage cofferdam, the best method of constructing the dividing leg of the cofferdam is with steel-sheet-pile cells or with timber cribs. If the upstream and downstream shore connecting legs of the cofferdams are exposed to high water velocity near the dividing leg, the sheet-pile cells or cribs are carried back a short distance on each leg. If high velocities occur when a cofferdam is being closed, then closure must be made with rocks large enough to resist the transporting action of the water.

Other closure methods may be required if velocities are such that closure cannot be made with large rocks. Cofferdams have been closed by raising their surface with precast tetrahedrons. Cofferdams used when high water velocities are encountered across the total area have been constructed by building timber cribs on shore with the bottoms tailored to fit the streambed. These are then launched, floated into place, and sunk in position. This type of cofferdam was used to construct Bonneville Dam on the lower Columbia River. In other cases, erosion-resisting abutments are constructed on each side of the closure opening; temporary cofferdam closure to stop the flow through the opening is made with a trussed bulkhead floated into place between these abutments, allowing the cofferdam to be completed behind the bulkhead. This system was used in closing the first-stage cofferdam at Long Sault Dam on the St. Lawrence River. When site conditions present definite problems, unique closure methods have been devised, such as closing the channel by blasting in the river banks or by casting a concrete bulkhead on the bank and moving it into position by an explosive charge.

Construction of the dividing leg of the second-stage cofferdam does not present any difficulties, as it usually is constructed in the dry within the cofferdammed area during first-stage diversion. The cofferdam is located on top of any downstream apron concrete and must be tied into the upstream and downstream sides of the dam concrete. Due to space restrictions and because part of the cofferdam is located on the concrete downstream apron, it is usually of timber-crib or sheet-pile cell construction. The downstream section of this cofferdam may be subject to fast, flowing water as the water is discharged from the diversion openings or over low blocks in the dam, which may result in a negative force on the cofferdam, causing it to slide toward the water. If it is built where these conditions will occur, the cofferdam must be designed to resist this action.

DIVERSION THROUGH TUNNELS

This method of diversion is used at damsites located in narrow, steep-walled canyons where space is not available for other types of diversion. It is also used at damsites when the diversion tunnel can be utilized as part of the permanent facilities. Dams that have diversion tunnels converted to permanent use are Hungry Horse Dam, where the diversion tunnel became part of the glory-hole spillway; Hoover Dam, where some of the diversion tunnels were used to convey water to the powerhouse and the others were used as part of the spillway; Oroville Dam, where diversion tunnels were used to convey water to the powerhouse; and Glen Canyon Dam, where diversion tunnels were converted to spillway use.

This diversion method requires the construction of a diversion tunnel or tunnels in the dam abutments to convey the water past the damsite. Whether the diversion tunnels are lined with concrete or left with bare rock surfaces depends on the characteristics of the rock. If the tunnel must be supported with steel sets, the tunnel is usually lined in the supported section to protect the supports from erosion action. The upstream portal of the diversion tunnel must be located a sufficient distance upstream of the damsite so that the upstream cofferdam can be positioned between the tunnel portal and the upstream edge of the dam excavation. To prevent damage to the dam foundation, the diversion tunnel should be located so that there will be a minimum distance of 100 feet of solid rock between the bottom of the dam and the closest part of the diversion tunnel or tunnels. The designed capacity of a diversion tunnel is a function of the waterway area and the head of water on the tunnel entrance; this head determines the velocity of flow in the tunnel. In designing the system, the cross-sectional area of the tunnels must be economically balanced with the height of the upstream water, which in turn determines the height of the upstream cofferdam. For the upstream cofferdam, a minimum freeboard height of 5 ft should be allowed.

A preliminary determination of the diameter of a diversion tunnel and the height of the upstream cofferdam required to handle a given stream flow can be made utilizing Figure 2.[1] This chart applies to tunnels that are submerged at their inlet and have a free flow from their outlet. If a water velocity is assumed, for example 25 cfs, the diameter of the diversion tunnel, the velocity head, and the friction loss per 100 ft of tunnel can be taken from the chart. Then the difference in feet between the elevation of water impounded by the upstream cofferdam and a theoretical point that is 0.8 of the tunnel diameter above the invert at the tunnel's downstream portal is equal to the total head. The total head consists of the friction loss in the tunnel, the velocity head (both of which were secured from Figure 2), and the entrance head. The entrance head for a square-tunnel opening is equal to one-half the velocity head, and for a bellmouth tunnel opening it is equal to one-fourth the velocity head. If lower upstream cofferdam heights are desired, the above-described use of the chart can be reversed.

The upstream cofferdam can be constructed as a small earth- or rock-fill dam with an impervious core, or the fill can be faced with an impervious upstream blanket when there is ample space available for its construction. If the main dam is either earth or rock fill, its design often incorporates the upstream cofferdam as part of the permanent dam. When earth- or rock-fill cofferdams are used, stream closure velocities should be checked to determine if final closure of the cofferdam can be accomplished with earth or rock. When cofferdam space is at a premium or when water velocities during cofferdam closure are excessive, either a timber-crib cofferdam or a steel-sheet-pile cofferdam is used.

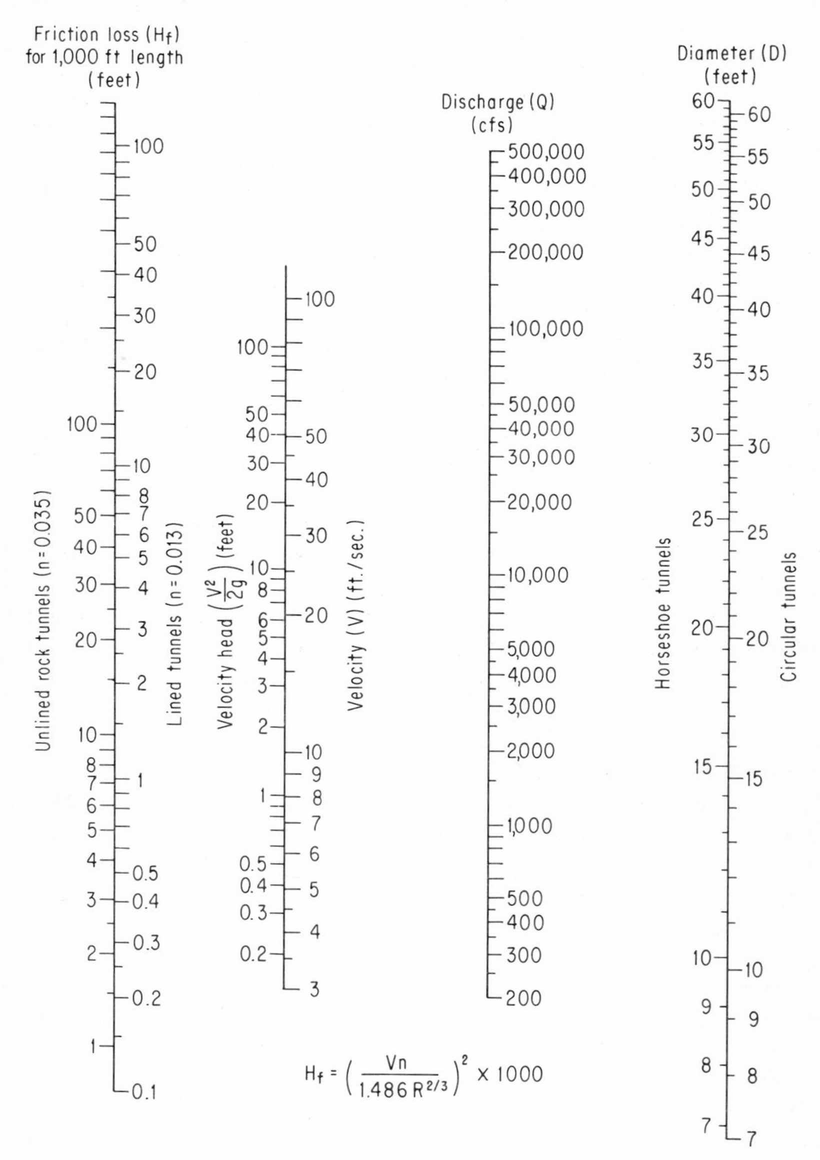

Fig. 2. Friction loss in full-flowing tunnels. (Data taken from *Handbook of Applied Hydraulics*, Calvin Victor Davis, ed., 2d ed., copyright 1952, McGraw-Hill Book Company.)

A downstream cofferdam can be of simpler construction since it will be subjected only to small tail-water heads and since its closure is generally made after the upstream cofferdam is completed. It can be of earth or rock fill, timber cribs, or steel-sheet-pile cells. Factors that influence the type of construction are the space available, whether it will be exposed to water being discharged from a diversion tunnel, and the tail-water elevation. When downstream cofferdams are exposed to the discharge from diversion tunnels, they must either be concrete or have their surfaces protected from the velocity of the discharged water, and they must be of sufficient height to provide protection against the hydraulic jump that occurs when the water is discharged from the tunnel.

When the dam has reached an elevation safe from overtopping from the maximum anticipated flood, the upstream cofferdam is removed or leveled and the diversion tunnels are closed. The dam elevation that will be safe from overtopping can be computed by evaluating the capacity of the reservoir behind the dam, the amount of water that can be handled by the permanent water outlets incorporated in the dam design, the ability to bypass flood flows through powerhouse facilities if they are incorporated in the construction, the elevation of the crest of the permanent spillway, the capacity of diversion conduits that may be placed in the dam for second-stage diversion, and, for concrete dams, the flood flows which can be passed over low blocks without damage to construction facilities or downstream installations. Whether the upstream cofferdam must be removed or leveled depends upon specification requirements.

Closure of the diversion tunnels is accomplished by using temporary portal gates or stop logs to shut off the water flow through the tunnel until permanent concrete tunnel plugs can be placed. The concrete plugs are located directly under the dam so that the plug will form part of the impervious curtain in the dam foundation. The procedure for placing this concrete plug depends on two particular job conditions: the depth of water that will be impounded in the reservoir during the closure operation and the necessity to maintain a minimum stream flow through the diversion tunnel until the water impounded by the dam reaches an elevation that permits water to pass through the permanent stream-release facilities.

If several diversion tunnels are constructed, they often have different upstream invert elevations in order to simplify the closure operation. The tunnel with the lowest invert elevation often can handle all the low-water flow, leaving the other tunnels dry. This permits the placement of the concrete plugs in these higher tunnels in the dry without the need of any temporary closure gates.

When it is not necessary to maintain any water flow through the last (or only) tunnel, temporary closure can be made with precast concrete stop logs placed in slots constructed at the portal before first-stage diversion is started. The stop logs must be designed to resist the total head of water that will be impounded by the dam during the time between temporary closure and completion of the permanent concrete plug.

Plugging of the diversion tunnels becomes more complicated when it is necessary to maintain a minimum stream flow during the time required for the water in the reservoir to reach an elevation that will allow the minimum flow to pass through permanent water-release facilities in the dam. One of the simplest methods of accomplishing this was used by Kaiser Engineers at Detroit Dam, Oregon. To maintain the required stream flow, a small tunnel was excavated which branched off from the main diversion tunnel a short distance from the upstream portal, and which daylighted adjacent to the upstream portal of the main diversion tunnel. The portal for this small tunnel was equipped with a remotely operated high head gate that was capable of resisting the water pressure. Because of the small size of the tunnel opening, this gate was not too expensive or difficult to install. Before diversion, the main tunnel was equipped with stop log slots. When the main diversion-tunnel closure was made, concrete stop logs, designed to resist the anticipated head, were dropped in the main tunnel opening. The minimum stream flow was then passed through the small branch tunnel. The concrete plug was placed in the dry by passing the minimum flow through a 36-in.-diameter pipe which extended through the area where the concrete tunnel plug was to be placed. When the upstream water reached an elevation that permitted the required stream flow to pass through the permanent river-discharge outlet, the gate was closed over the bypass tunnel, and the 36-in.-diameter pipe was filled with concrete, which completed the tunnel plug placement.

When temporary diversion conduits are located in the dam to release the required stream flow, they must be closed in the same manner as that used to close diversion conduits in cases where the entire stream is diverted through the dam. High head gates are required and upon closure the diversion openings are plugged with concrete. This concrete is often placed by prepack methods to reduce shrinkage.

When concrete dams are constructed at sites where there are great variations in flood flow, economics may make it desirable to size the diversion tunnels and build the upstream cofferdam to handle less than the maximum flood flow. The diversion facilities may only have sufficient capacity to handle floods that might occur in 5 years, but incorporated in the overall design should be provisions that will allow greater floods to be passed through the damsite without extensive damage to any facility. These provisions may consist of spillways located on the upstream and downstream cofferdams to permit overtopping, and scheduling of work so that the dam excavation and the dam concrete can be completed to streambed height during the low-water season. In lieu of cofferdam spillways, the cofferdam crest and its downstream slopes may be protected with paving or gunite so that its surface will not be eroded when the cofferdam is overtopped. Construction equipment must also be located free of the area subject to floods or be of a type that can be readily removed from the area when flooding is anticipated. The diversion facilities for the construction of Detroit Dam were designed in this manner. During construction, the upstream cofferdam was overtopped five times. This cofferdam was built of timber cribs that were not damaged during the overtopping, and when a minor portion of the downstream cofferdam was washed out, its replacement cost was negligible.

To hold diversion cost to a minimum, when earth- or rock-fill dams are constructed at sites where a tunnel is used for diversion, the tunnel may be sized to handle only the low-water flow with a low cofferdam. But, in conjunction with the storage capacity in the reservoir, it is sized to protect the dam from overtopping after the dam embankment has been raised sufficiently to allow ponding of an upstream reservoir and the creation of head on the tunnel. When this type of diversion is used for earth-fill dams, diversion is handled in four stages. First-stage diversion commences at the start of the low-water season and consists of diverting the low flow through the diversion tunnel, excavating the dam's foundation below stream level, and raising the dam embankment to streambed elevation. Second-stage diversion occurs during the high-water season when the flood waters are passed over this completed section of the embankment. Since the dam embankment was not raised above the streambed elevation, the erosion of this embankment will be slight. During this period of second-stage diversion, placement of the dam embankment is stopped but excavation for the dam's foundation on the abutments above the water level can continue. Third-stage diversion begins at the start of the second low-water season when the water is again passed through the diversion tunnel. During this low-water season, the previously placed embankment is cleaned off and the embankment is raised to an elevation that will develop enough upstream pool capacity and enough head on the diversion tunnel to store and pass flood flows and thus keep the embankment safe from overtopping. This allows the remainder of the dam to be completed irrespective of high- or low-water flow. Fourth-stage diversion is the closure of the diversion tunnel with stop logs or gates and the placement of the permanent concrete plug. Trinity Dam, California, was constructed using this method of stream diversion.

The tunnel diversion method can be modified in many ways to fit particular site conditions. This was done for the construction of Akosombo Dam, Ghana, Africa.[2] This was a rock-fill dam located at a site that was covered by a deep pool of water created by a downstream channel obstruction. The foundation rock was overlain by a layer of deposited sand. The large variation of river flow, from a minimum of 800 cfs to a maximum of 500,000 cfs, made the passage of the flood flows through the dam the most economical method of diversion. A 30-ft-diameter diversion tunnel was constructed to handle normal flows. Before starting the placement of the cofferdams, the sand was dredged from both the dam and cofferdam foundation areas. Dredging for the downstream cofferdam foundation area extended to 217 ft below water level. The cofferdams were constructed below water by dumping rock fill. The cofferdams were sealed by first placing a transition zone and then an impervious blanket on the outer side of each cofferdam. Above water level, rock was placed in the cofferdam by end dumping. After the cofferdams were closed, the enclosed area was dewatered and the dam was constructed. The cofferdams were overtopped by flood flows before they were completed, with comparatively little damage resulting from this overtopping.

Diversion-tunnel Construction

When a tunnel is used for diversion, it should be constructed with a minimum of equipment because of the small amount of required excavation. Rubber-tired excavation equipment is generally used. Since tunnel construction is a complicated procedure with many varied construction methods, the subject will not be covered in this book.[3]

DIVERSION THROUGH AN ABUTMENT

First-stage diversion through a channel or flume located on one of the dam's abutments is used at damsites that do not have wide enough streambeds to permit diversion through the dam but do have sufficient space on one abutment for a channel or flume. The topography of the abutment must permit the channel or flume to be located at a low enough elevation so that an economically sized cofferdam will furnish sufficient upstream impounded water head on the channel or flume to allow the passage of maximum flood flows. The dam must also be of a configuration that allows the use of this method of diversion without delaying construction and that will, during second-stage diversion, permit the passage of flood flows through openings in the dam, over low blocks in the dam, or over a permanent spillway.

The economical height for the first-stage cofferdam can be determined by comparing costs using this diversion method with those of other diversion methods. At Guri Dam in Venezuela, diversion was accomplished with the diversion-channel method and a large amount of material had to be excavated from one abutment to construct the diversion channel. The diversion channel was long with dikes constructed along one edge. Ten sheet-pile cells were used for erosion protection at the diversion-channel entrance. The channel location was so high on the abutment that a 145-ft-high cofferdam was required to raise the upstream water to an elevation that permitted it to flow through the channel. These diversion facilities are illustrated by Photograph 3. Even with the extreme height of the upstream cofferdam, this was the most economical type of diversion to use at this narrow damsite, where flood flows of 600,000 cfs were anticipated.

Photo 3 First-stage diversion with a channel, Guri Dam, Venezuela.

The height that concrete must be poured in the dam during first-stage diversion is controlled by the method of handling the water during second-stage diversion. If water is to be passed over low concrete blocks, these blocks only need to be poured above the downstream tailwater before starting second-stage diversion. If the water is to be handled by diversion conduits, through permanent openings in the dam, or over the dam's permanent spillway, most of the concrete in the dam must be poured before the start of second-stage diversion. When these last two methods of handling the water are used, often a second-stage cofferdam must be erected across the entrance of the diversion channel or flume.

When second-stage diversion is started, the first-stage upstream cofferdam must be removed. If this was a high cofferdam, this work may be quite extensive, as shown in Photograph 4 of the removal of the first-stage cofferdam at Guri Dam.

After removal of the first-stage cofferdam, and if required, the construction of a second-stage cofferdam across the entrance to the diversion channel or flume, the remainder of the diversion work is similar to that discussed under diversion through the dam. If the dam is all earth or rock fill, second-stage water can only be handled over the dam spillway.

Photo 4 Removal of the first-stage cofferdam, Guri Dam, Venezuela.

Depending on space available and closure velocities, the cofferdam can be of earth and rock fill, steel-sheet-pile construction, or timber-crib construction. The first-stage diversion cofferdam constructed at Guri Dam was an earth- and rock-fill cofferdam and the cofferdam was easily closed during a low-water flow of approximately 7,000 cfs.

The channel or flume method of diversion can be used to great advantage in the construction of concrete dams with earth wing walls when the diversion channel can be located through the earth-embankment section of the dam. The concrete portion of the dam and the portion of the wing walls that do not conflict with the diversion channel are constructed in first-stage diversion. Then at the start of a low-water season, the water can be diverted through the concrete portion of the dam and the earth embankment can be placed across the diversion channel. The final diversion stage is the completion of the concrete section of the dam.

In the past, such as during construction of Canyon Ferry Dam in Montana, diversion flumes were often used. Due to the rapid rise in timber and timber-framing costs, diversion channels have replaced flumes, but small wooden flumes are still used for diverting small streams past damsites.

DIVERSION THROUGH PIPE CONDUITS OR FLUMES

The pipe-conduit method is a single-stage diversion procedure widely used in the construction of earth- or rock-fill dams on small streams. A ditch is excavated through one abutment at, or slightly below, the final foundation excavation elevation. A concrete conduit is then constructed in this ditch. At the upstream end of the conduit either a permanent-gate structure or an intake structure is erected, so that the diversion conduit will also serve as a permanent river outlet. If a permanent gate is not installed on the upstream end of the conduit, a permanent valve or provision for such a valve is constructed at the midpoint of the dam. After completion of the conduit, the water is diverted through it by cofferdams. When dam construction is completed, the diversion conduit is converted to a permanent river outlet and water storage is started in the reservoir.

The pipe-conduit method has very limited application in concrete construction, but when it can be used, it is one of the simplest diversion procedures. It is applicable when small concrete dams are constructed on small streams or on streams where the flow can be rigidly controlled by upstream facilities. The stream flow must not exceed the amount that can be carried through the damsite by the conduit or a small flume. This diversion method can also be used to handle the low-water flow when the dam foundation can be excavated and concrete placed to streambed elevation in one season. Then the flood flows can be passed over low blocks in the dam.

In concrete-dam construction when the pipe or flume is carried through the dam at streambed elevation, the section that spans the dam must be supported so that the excavation for the dam's foundation can be completed alongside and beneath it. Small upstream and downstream cofferdams, usually of earth-fill construction, are required to divert the water into the pipe or flume. After concrete has been placed in the dam to the bottom of the flume or conduit, an opening is left so that the conduit or flume can remain in place until the dam is completed. Upon completion of the dam, the conduit is removed and the opening left for its passageway through the dam is filled with concrete. In some cases where pipe conduits are used, instead of blocking out for the conduit, the pipe is embedded in the concrete and used as part of the permanent stream-regulating facilities.

Photograph 5 shows low-water flow being handled by use of a flume during the construction of the afterbay dam on the American River Project. At this dam, one-half the foundation excavation and concrete placement to streambed level was accomplished during the first low-water season, and the other half was done in the following

Photo 5 Diversion with a flume, American River Project, California.

low-water season. During the first low-water season, diversion was accomplished with a large steel pipe. During the second low-water season, the water was carried by the flume (shown in the photograph), which was supported on a concrete block that had been poured during the previous low-water season.

COFFERDAMS

The number and size of cofferdams required to accomplish diversion vary with the method of diversion, the size of the river, the type of dam, and the site conditions. The available space to construct the cofferdam, the height of the impounded water, and the maximum water velocities that will be encountered during diversion and during cofferdam closure will control the type of cofferdam.

Earth- or Rock-fill Cofferdams

Earth- or rock-fill cofferdams are the most economical types to use if space is available for their construction, if the cofferdam foundation is impervious, if suitable fill material is available, and if water velocities are not too great. They can be placed on pervious foundations if a water-cutoff wall has been constructed through the pervious material. They can be constructed to any height if they are properly designed. For example, the first-stage earth- and rock-fill cofferdam at the Guri damsite in Venezuela was approximately 145 ft high. The greater the height, the more the attention that must be given to cofferdam design. The passage of water through these cofferdams is prevented either by an impervious core or by use of an upstream blanket. These water barriers must be well-designed to prevent leakage and the cofferdam must have sufficient weight downstream of the impervious core to resist the static head of the impounded water. Velocities that will be encountered in final closure must be computed; if they become critical, final closure of the cofferdam must be made with large rocks, rock necklaces, gabions; tetrahedrons, tetrapods, or other precast concrete shapes. Rock necklaces are several large rocks that have been drilled and strung on a cable. Tetrahedrons and tetrapods are names for concrete castings that have projecting arms that interlock with each other. Earth- and rock-fill cofferdam design is similar to earth- or rock-fill dam design, and design data can be found in various books on this subject.[4] Rock particle sizes suitable for final closures at different water velocities are given in other publications.[5]

Shot-in-place Cofferdams

Cofferdams have been constructed in narrow gorges by blasting the canyon walls to form a river barrier when difficult access, high stream velocities, and year-round flows prohibited other methods of cofferdam construction. This construction procedure necessitates the removal of all rock in one large blast. The upstream face can then be sealed with a blanket of impervious material. This type of cofferdam was used at Mayfield Dam on the Cowlitz River in Washington, but after a long delay in attempting to seal the upstream face, the contractor was forced to resort to an expensive program of chemical grouting before the structure would serve its purpose. However, reports on the use of this method in Russia indicate that it was quite successful.

Timber-crib Cofferdams

In the past this type of cofferdam was used when there was limited space for cofferdam construction, when the cofferdam sides were exposed to eroding water velocities, or when the cofferdam closure was made against high-water velocities. Because of the large increases in the cost of timber and timber-framing labor, steel-sheet-pile cell cofferdams are replacing timber-crib cofferdams. The latter are now used only for special purposes such as when cofferdam closure requires that cribs be floated into place and then sunk, or to prevent damage to the concrete when cofferdams are erected on concrete aprons.

A timber-crib cofferdam is framed from large timbers, usually 12-in. square, in a crib pattern. The timbers are drift-bolted or bolted together and the enclosed areas are filled with rock. Water tightness is provided either with a double layer of vertical planking across the face of the cofferdam or by placing impervious fill in some of the cofferdam pockets. If it is to be floated into position and sunk, enough pockets must be floored so that sinking can be accomplished by filling the pockets with rock.

A timber-crib cofferdam is stable when the resultant of the water pressure against the cofferdam and the weight of the cofferdam passes through the middle third of the cofferdam. To accomplish this, the cofferdam's width must equal its height. The cofferdam can be stable with a width less than its height if inside the cofferdammed area an earth fill is placed against the cofferdam. A timber-crib cofferdam can be overtopped without damage if the top is paved with large rocks, sheathed with lumber, or paved with concrete or asphalt. When timber cribs must be removed on job completion, they often are constructed with holes drilled in the crib corners to simplify the future placement of explosives.

When timber-crib cofferdams are to be floated into position and sunk, the streambed on which the cofferdam will rest should be sounded and the foundation profile determined. The bottom of the cribs can then be tailored to match the streambed. Based on the Bonneville Dam cofferdams, the amount of timber used in the construction of a timber crib is approximately 11 percent of the total crib volume. Both constructed-in-place and floated-into-position types of timber-crib cofferdams were used for the construction of Bonneville Dam as described in the reference article.[6]

Steel-sheet-pile Cofferdams

Steel-sheet-pile cofferdams are now used instead of timber-crib cofferdams when space is limited, where high water velocities occur along the cofferdam sides, and where cofferdam closure must be made against a high differential head causing high velocity during cofferdam closure. Besides having a cost advantage over timber cribs, they can give a more positive water cutoff since the piles can be driven through overburden material to rock. Steel-sheet piles are used in circular cell cofferdams, double-diaphragm-wall cofferdams, and as a single-diaphragm wall reinforced with a rock fill.

Cellular cofferdams are used when large, deep cofferdams are required since these units are individually stable and do not require bracing. The use of individual cells has an advantage in that each cell can be completely filled after erection to allow progressive construction. Such cells must be designed to prevent overturning, sliding, and rupturing of the piles in the interlocks. Design formulas for cell construction can be secured from manufacturers' literature.[7] A rule of thumb for sheet-pile cell design is that the diameter must equal the height. Cell diameter can be reduced by placing an earth berm inside the cofferdammed area against the cells. When large-diameter cells are required, overstress in the pile interlocks can be avoided by the use of high-strength piles or a cloverleaf cell. Cloverleaf cells are divided into four smaller cells and provide resistance to overturning and sliding. Pile interlock stress is reduced since it is a function of the maximum radius of the component cells. The number of piles in each of the various-sized cells and the piles in the connecting arcs between cells, the dimensions of cells, and the volume of fill required per foot of cell can be secured from sheet-pile catalogs.[8] When cells are placed on bare rock, the bottom of the cells should be sealed with sacked concrete or other impervious material to cut off water inflow, to prevent the water flowing past the cell from removing the cell fill, and to prevent cofferdam leakage when the cofferdammed area is unwatered. Sealing of the cell bottom is done after the cells are driven by cleaning out the cell and then having divers place sacked or tremie concrete against the piles on the water side of the cell. To reduce damage from overtopping, the top of the cell can be paved with large rocks, concrete, or asphalt. If overtopping is anticipated, damage can be reduced by installing gates or weirs in the cofferdam to allow flooding of the cofferdammed area before overtopping occurs. Cellular cofferdams on the Columbia River have been completely submerged during flood flows with only minor damage resulting.

The major construction problems encountered in sheet-pile cell construction are threading of the piles in the cell and maintaining the piles' position until they are driven. This problem is very critical when the cells are located in water on bare rock. Piles can be accurately threaded around the cell and held in the proper position by using a cell template. After all piles are placed, they can be progressively driven two piles at a time a distance of approximately 4 ft. This procedure is repeated until all piles are driven the required distance. To speed up pile placement, piles can be threaded in pairs on shore and then placed in the cell. Cell fill can be placed by end-dumping trucks from the last cell, or by using clamshells that place fill from piles dumped by trucks on previously completed cells, or secured from barges. The fill can also be placed with dredges. Water is used for compaction purposes during the filling operations. Drainage holes should be placed on the interior side of the cells so that hydrostatic pressure will not be developed in the cell, which would increase the stress on the pile interlocks. Also cells should be free draining so that they can better withstand overturning and sliding. This type of cofferdam has been used for diversion purposes during the construction of a large number of dams.

Double-diaphragm cells can be used instead of circular cells when water velocity during driving is not a problem. This type of cofferdam consists of parallel multiple arcs connected by cross walls at each arc intersection. The advantage of double-diaphragm cells over individual cells is that fewer piles are required. The disadvantage of this type of cofferdam is that when the cells are filled, the fill must be placed along the whole cofferdam with the fill height in each cell approximately the same as the others in order to prevent distortion of the cross walls. Because it is limited in use to locations where low water velocities occur and since the filling method is more difficult, double-diaphragm cells are seldom used for diversion purposes.

When a cofferdam is to withstand only low heads, a single-diaphragm wall of sheet piles supported on both sides by rock fill can be used for cofferdamming purposes. Since the piles transmit all loads to the rock, there will be no tension on the interlock, resulting in more leakage than with circular cells. This type of sheet-pile cofferdam was used at Rocky Reach Dam in Washington.

Concrete Training Walls

The use of concrete training walls may be economical in training the water that is discharged from downstream tunnel portals so that the downstream cofferdam will not be eroded. (This was done at Detroit Dam.) They also may be used in the dual capacity of training the water and cofferdamming. A third use is to control the discharge from diversion conduits so it will be carried past construction areas, such as those required for powerhouse construction. Concrete training walls can also be used as the dividing walls of the first-stage cofferdam at narrow damsites when diversion is through the dam and space is at a premium.

When concrete cofferdams or training walls are used, they should be constructed with holes for explosives cast in the concrete to facilitate their removal.

SUBSURFACE WATER CONTROL

If the cofferdam is erected on pervious overburden materials, large amounts of water may flow through this material into the cofferdammed area. Because such inflows have created problems at some damsites, the permeability of the local materials should be studied to determine whether the excavated area will be subject to water inflows. Sometimes river gravels are so well graded and compacted that inflows under cofferdams will be of only minor consequence. This was the case at Detroit Dam, Oregon, where excavation was carried downward through 80 ft of gravel to expose suitable foundation rock. This excavation was adjacent to the upstream cofferdam, which was placed on top of the gravel; but since the gravel was tight, there were no major water inflows through the gravel beds. At other damsites where there were shallow layers of gravel underneath the cofferdam, large water inflows into the excavated area created pumping problems and also resulted in erosion of the sides of the excavation. If it is determined that inflow either under the cofferdam or through the abutments will be encountered, provision for handling these inflows should be made prior to the start of excavation. Water inflows can be reduced by predraining the area, by increasing the path of percolation of the water, or by constructing cutoff walls.

Photo 6 Well points installed at Trenton Dam, Nebraska. (John W. Stang Corporation.)

Predraining

Most often, ground-water inflows are prevented by the installation of well points or wells to predrain the excavated area. If excavation is to be done through cohesionless material ranging from silt to gravel, the well-point system will not only predrain the material but will also stabilize the slopes of the embankment by keeping them dry. Wells may be constructed by predrilling or jetting, depending on soil conditions. Well points are from 10 to 15 ft long and are driven directly into place or are installed in sand-filled wells spaced in regular intervals around the perimeter of the excavation. Each well point is connected to a header and water is removed from the header by one or more pumps, depending on the size of the excavated area. Additional groups of well points, headers, and pumps, known as well-point stages, are installed as the excavation is deepened. In highly permeable soil, the use of perforated cased wells packed with gravel and drained with vertical pumps, or a combination of these wells and a well-point system, may be more economical than well points alone. Layout of a predraining system employing either well points or wells is a highly specialized task usually requiring the services of a specialist in the field of dewatering. A typical well-point installation is shown in Photograph 6.

Increasing the Path of Water Percolation

The volume of water inflowing through the sides of the excavation can be reduced by increasing the path of its percolation. This can be done by placing a blanket of impervious material on the bottom and sides of the river channel, starting at the upstream edge of the cofferdam and continuing upstream as far as necessary. Blankets are often used to increase the path of water percolation under and around concrete dams but are seldom used for diversion purposes since they do not provide a positive solution. Furthermore, the reduction of water inflow into a cofferdammed area may require the placement of a large quantity of material. Therefore, this method of reducing the water inflow is generally not resorted to unless unanticipated inflows are encountered after a cofferdammed area has been unwatered.

Cutoff Walls

The third method of controlling ground-water inflow is by the construction of cutoff walls. Cutoff walls can be constructed of impervious material by the *Cronese* or bentonite slurry-trench method or by the *Vibroflotation* method, or the existing overburden material can be formed into an impervious wall by grout injections or by freezing. Another method of constructing a cutoff wall is by driving sheet piles through the materials until they contact the rock surface.

The bentonite slurry-trench method of producing an impervious cutoff wall requires that a trench be excavated to bedrock. The material on each side of the trench is retained in position by stabilizing the excavation with a heavy slurry of bentonite, which is placed in the trench as it is excavated. When the trench excavation has progressed sufficiently so that backfilling operations will not interfere with excavation, the bentonite slurry is progressively displaced with impervious fill material, forming an impervious cutoff. This type of cutoff wall was used to enclose the excavation area during the construction of Wanapum Dam, Washington. In this case, portions of the slurry trenches were 90 ft deep.

Loose sand and gravel material can be compacted into a cutoff wall using a patented process called Vibroflotation. This process is based on the use of a long, slender tube called a Vibrofloat, containing an electrically driven eccentric weight to furnish the vibrating action and equipped with a water jet. The Vibrofloat is swung from a crane and inserted into the loose material at regular intervals. The water jet aids in the sinking action, and some preconsolidation takes place as the Vibrofloat descends. When it has reached the desired depth, the water jet is turned off. As the Vibrofloat is slowly retracted, the soil is compacted to approximately 70 percent density and additional sand must be added. Up to the present time this method has not been used for cofferdam cutoff walls, but it has been used to compact material under permanent earth-fill embankments. For example, it was used in the construction of the cutoff wall under the earth-fill abutments at Priest Rapids Dam in Washington. Also, the barrages (diversion dams) constructed for the Pakistan irrigation system made extensive use of this method of water cutoff.

When the rock at the damsite is of such porous nature that water can readily flow through it, water inflow into excavated areas can be reduced by establishing a grout curtain in the rock. The preferred grouting procedure is to complete the grouting before excavation is started. If grouting is done after the excavation has been started, the head differential between the water surface and the bottom of the excavation may create sufficient water velocity through the rock to wash away the grout before it can solidify. However, if grouting after excavation must be performed, admixtures can be used to accelerate the setting of the grout or the excavated area can be flooded prior to grouting to eliminate the head differential. It is difficult to form a cutoff wall with cement grout in most overburden material because it is hard to control the flow of cement grout. Recent improvements in chemical grouting have made its use feasible in overburden materials that cannot be grouted with cement. Because of these improvements, chemical grouting will probably be more widely used.

A frozen impervious wall of overburden materials can be formed by excavating wells in the material on a close pattern and then freezing the overburden materials by circulating cold brine through piping installed in these wells. In order to form an efficient frozen barrier without too great an expenditure for wells and cold-brine pumping, the flow of the ground water past these wells should not exceed 2 ft per hr. If the ground-water flow exceeds this velocity, sufficient closely spaced wells must be excavated to form a wide zone of freezing. The history of the problems encountered with this freezing method during the construction of Gorge High Dam in the state of Washington provides useful information on this type of cutoff.[9]

Steel-sheet-pile cutoff walls are seldom used by contractors because of the expense involved in their construction and because this method does not put enough tension on the pile interlocks to cut off all the water inflow.

DEWATERING

Cofferdam dewatering requires two types of operations. Initial dewatering is required when the cofferdammed area contains impounded water after cofferdam closure, when the cofferdammed area is flooded, or when cofferdams are overtopped. After the initial dewatering is completed, continuous maintenance dewatering is required to keep the area free from water.

The capacity of pumps required for the initial dewatering is dependent on the volume of water impounded by the cofferdam, the amount of cofferdam leakage, the water inflow through overburden or rock, and the time allowed for dewatering. Since time is always of critical importance to a contractor, the plant is usually sized to accomplish dewatering in the shortest practical time, often in less than one week. This initial dewatering requires more pumping capacity than is required for maintenance dewatering. Since initial dewatering pumps are only required for a short period of time, large-capacity vertical-head pumps are usually rented for this purpose. Installation of pumps of this type is shown in Photograph 7. Most such pumps are installed adjacent to the downstream cofferdam with discharge pipes running through or over the top of the cofferdam, which is the location that will result in the lowest pumping head. When the pipes are placed on top of the cofferdam and when it is not necessary to measure the water discharge, the pumping head can be reduced by extending the pipes so that they discharge below the downstream water surface.

The amount of work required for maintenance pumping depends on the type of system installed. A well-point system requires no additional maintenance pumping. When initial or horizontal pumps are used for dewatering, they are installed over sumps excavated in the cofferdammed area. These sumps must be deepened when the excavation reaches the same depth as the sumps; and as the sumps are deepened, the pumps will also have to be lowered to reduce the

Photo 7 Unwatering pumps, Oviache Dam Pumping Project. (John W. Stang Corporation.)

suction head. Also, ditches must be excavated and maintained in the excavated area to lead water to these sumps. In addition, many small portable pumps are required to pump out small excavated areas located below the general rock surface. The main dewatering pumps are usually vertical pumps since this type of pump can be easily raised or lowered. A short section of hose connecting the pump to the discharge line will provide flexibility when the pumps are lowered. The size and number of pumps required is dependent on cofferdam leakage and the amount of water that will pass under the cofferdams and through the banks of the excavation. If a cofferdam encloses a large area or if extensive areas of adjacent hillsides drain into the cofferdammed area, rainstorms or melting snow may at times contribute very large volumes of water. Estimating the quantity of water that must be handled is one of the most difficult problems to solve in dam-construction planning because it is hard to anticipate the conditions that will be encountered. For example, at Wanapum Dam very little pumping was anticipated since a slurry cutoff trench was constructed around the excavated area to reduce water inflows. Nonetheless, it was necessary to install an extremely large maintenance pumping plant to remove artesian water that entered the cofferdammed area through the holes drilled in the rock during the excavation of the dam's foundation.

When high upstream cofferdams are located adjacent to the upstream edge of the dam excavation, the combination of cofferdam height plus excavation depth may result in such a high static head that large volumes of water may enter the excavated area through the material that forms the cofferdam's foundation. This occurred at Guri Dam, Venezuela, where the water elevation on the upstream side of the cofferdam was 175 ft above the bottom of the excavation. To handle this inflow of water, a low secondary cofferdam was erected between the future concrete-dam location and the main cofferdam. The water that was collected behind this secondary cofferdam was conveyed through the excavated area by a large-diameter pipe that was left in place when the dam concrete was poured. This pipe drained the area upstream of the dam until just before the upstream cofferdam was removed. The flow in the pipe was then blocked off and the pipe was filled with concrete.

Maintenance pumping should always be closely reviewed, as the history of most dam construction is that maintenance requires a much larger and more complex water-handling system than was anticipated when construction was originally planned. A careful analysis should always be made of every possible condition that could increase pumping-capacity requirements or add to the duration of the pumping period.

REFERENCES

1. Calvin Victor Davis, ed., *Handbook of Applied Hydraulics,* 2nd ed., copyright 1952, McGraw-Hill Book Company.
2. Highest Underwater Cofferdams Keep Akosombo Dry, reprinted from *Engineering News-Record,* June 20, 1963.
3. Albert D. Parker, *Planning and Estimating Underground Construction,* copyright 1970, McGraw-Hill, Inc.
4. Subcommittee of Small Water Storage Projects of the Water Resources Committee of the National Resources Committee, *Low Dams, a Manual of Design for Small Water Projects.*
5. Frank W. Stubbs, Jr., ed., *Handbook of Heavy Construction,* copyright 1959, McGraw-Hill Book Co.
6. C. I. Grimm, Cofferdams in Swift Water for Bonneville Dam, *Engineering News-Record,* September 5, 1935, pages 315-318.
7. *Steel Sheet Piling,* United States Steel Company, design extracts from other publications.
8. *Steel Sheet Piling Catalog 433,* Bethlehem Steel Co.
9. Forming of an Ice Barrier at George High Dam, *Engineering News-Record,* March 7, 1957, page 28; January 22, 1959, page 28.

CHAPTER THREE

Foundation Excavation and Preparation

This chapter explains how the excavation and preparation of a dam's foundation are planned, scheduled, and accomplished. It describes, gives the capacities, and discusses the selection of the construction equipment that is used to perform this work.

Foundation excavation and preparation are larger factors in successful dam construction than is indicated by their share of the total cost. If they are not well-planned and scheduled, they may delay the construction of the project, causing increases in job overhead and in labor and material escalation, and a decrease in equipment salvage proportional to the additional time that the equipment will be required on the job. This work requires special consideration because the differences in specification requirements and in type of inspection will cause construction cost to vary greatly between jobs. A change of interpretation involving one short sentence of the specifications may result in thousands of dollars of cost. Since foundation excavation and preparation must be completed before the placement of the structures can start, it is important that foundation excavation and preparation be done by the most efficient system possible, and that they be scheduled to reduce interference with other construction activities.

SPECIFICATION REQUIREMENTS

Depending on the specification requirements, type of structure, and physical characteristics of the foundation materials at the site, the work required for foundation excavation and preparation may be quite extensive. The complexity of this work is suggested by Figure 3, a flow diagram of the work that may be required to excavate and prepare the foundation for a dam. The number of these operations that are required at each damsite varies with the specification requirements and the type of dam. The work necessary for a dam project can only be determined by a thorough study of the plans and specifications. The work-flow diagram can, however, be used as a checklist for such a study.

The foundations for different types of dams have to be excavated and prepared in different ways. For an earth-fill dam the excavation under the impervious core and filter blankets must be carried to sound rock. The foundation preparation of this area may require the performance of many of the items of work shown on the work-flow diagram. The remainder of the dam's foundation area need only be cleared and stripped, with the exception that any soft, unsuitable common material must be removed.

For a rock-fill dam with an impervious core, the excavation under the impervious core and filter zones must be carried to sound rock. The common material must be removed from the remainder of the foundation area, but foundation preparation under this area is seldom required.

For a rock-fill dam with an impervious facing, the common excavation material must be removed under the entire dam foundation, but foundation preparation of this area is seldom required. A narrow trench must be excavated in sound rock underneath the impervious facing to form a water cutoff.

For a concrete dam, the entire foundation area must be excavated to sound rock. The foundation preparation of the entire dam foundation is extensive in scope and often includes many of the items shown on the work-flow diagram. Powerhouse foundation excavation and preparation are similar to those required under a concrete dam. Often the powerhouse location is such that a large amount of presplitting is required.

Only part of the work required by the plans and specifications for foundation excavation and preparation is listed as unit-price bid

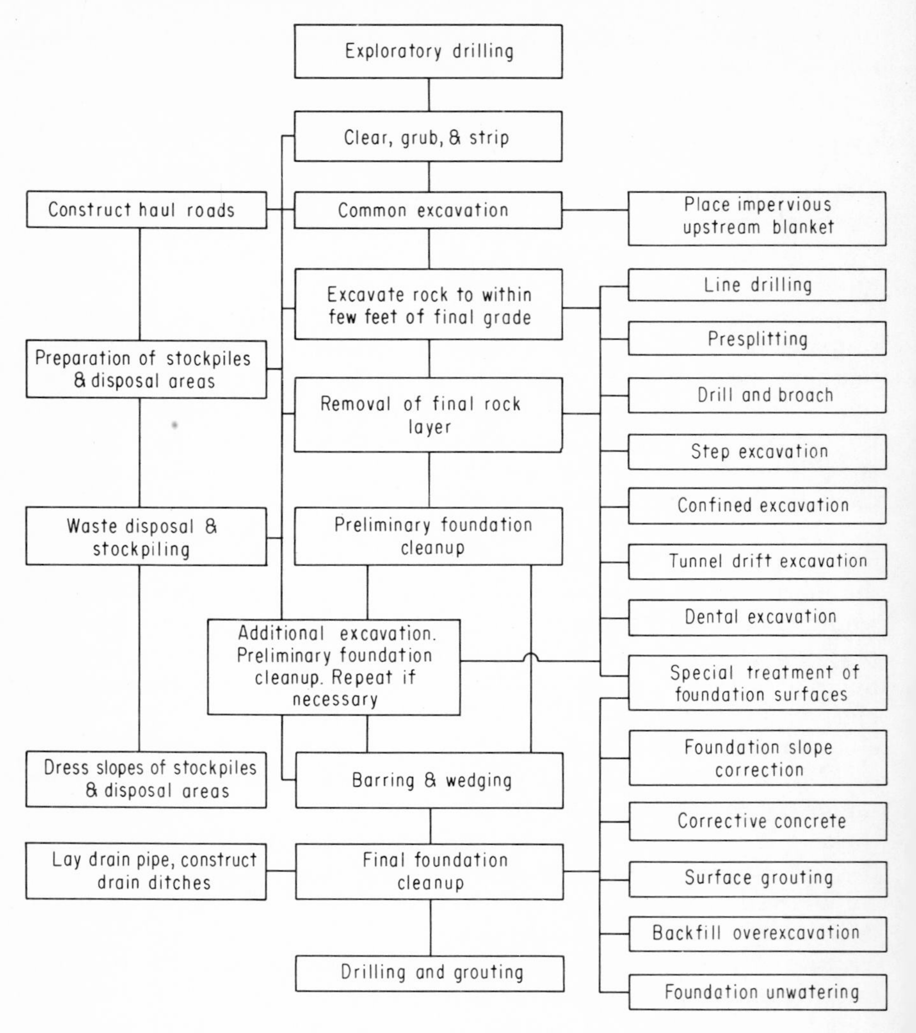

Fig. 3 Work-flow diagram for the excavation and preparation of a dam's foundation.

items in the bidding schedule. When the contractor prepares his bid, the cost of performing any work listed in the bidding schedule can be placed directly against this item. The cost of performing work that is not listed in the bidding schedule must be combined with the cost of performing listed work in order to arrive at a combined cost, which is placed against the listed bid quantities. When work covered by bid items overruns in quantity, the contractor will receive adequate compensation because payment will be received for actual quantities of work performed. When work that is not covered by bid items overruns, the contractor will not receive compensation because the bidding schedule will not reflect the change in work quantities. This procedure of combining two or more items of work into one bid item often works to the disadvantage of the contractor.

The type of foundation excavation and preparation work required by the specifications but not separately listed in the bid schedule is expensive to perform. Moreover, the extent of this work may be difficult to determine from the specifications, since it is often defined as work necessary to meet the owner's engineer's requirements. This makes it difficult for the contractor to determine what the final cost will be for these unlisted items when he submits his bid. If the owner's engineer considers essential the performance of unnecessary work, it is an unfortunate situation for both the contractor and the taxpayers or owner. In order to recover the cost of this work, the contractor must go through the trouble and expense of requesting a change order. When the change order is paid, the cost of this unnecessary or overrestrictive work is then paid for by the taxpayers or owner.

Restrictive work procedures that cannot be anticipated by the contractor may be enforced by the owner's engineer when insufficient foundation exploration has been done prior to bidding. In order to compensate for this inadequate foundation exploration, the owner's engineer may make the contractor explore the rock at the damsite while performing excavation, by ordering the contractor to excavate the rock in shallow layers and to perform preliminary rock cleanup on each successive rock surface exposed. By examination of these cleaned-up surfaces, the engineer can determine when the excavation has reached a depth that will provide a suitable rock surface for the dam's foundation. This procedure will hold pay quantities to a minimum; however, if excavation by this multiple shallow-layer removal procedure is not called for in the specifications as a job requirement, but it is later required by the engineer, it will increase the contractor's excavation cost without furnishing him adequate compensation.

Excavation in multiple and shallow layers increases the contractor's anticipated excavation cost in several ways: the drill holes are smaller in diameter and more closely spaced; there are more restrictions on the use of explosives; loading and hauling efficiency declines when shallow layers of rock are removed and excavation is repeated many times across the same area. If the square feet of preliminary rock cleanup is paid for as a separate item, the contractor will be reimbursed each time he performs this work. If it is not a separate item and since the contractor usually has not included the cost of this repetitive preliminary rock cleanup under other items of work, he will incur additional cost without any compensating revenue.

The foregoing discussion may have presented a pessimistic approach to dam foundation excavation and preparation, but each job is different and whether the contractor will be adequately compensated for this work depends on the job specifications, the number of bid items, the competency of the inspectors, the particular job conditions, and the competency of the contractor's estimator. Many jobs have been constructed without any conflicts between the contractor and the owner's engineers during the performance of foundation excavation and cleanup. However, on jobs where change orders have been requested, a great number of these change orders have been for additional work required to excavate and prepare the dam's foundation.

WORK DESCRIPTIONS AND EQUIPMENT APPLICATION

Descriptions of the work that may be required to excavate and prepare the foundation for a dam or powerhouse are given on the following pages. These descriptions appear in approximately the same order as on the work-flow diagram (Figure 3). Although these procedures are described as separate operations, in practice many of them can be done simultaneously. The construction equipment required to perform each phase of the work is described in detail, to provide assistance in the selection of proper equipment. Finally, there is an explanation of how the required number of each type of equipment unit is established.

Exploratory Drilling

When insufficient exploratory work has been done at the damsite to determine the characteristics of the dam's foundation material, the specifications may require the contractor to do exploratory drilling so that the owner's engineer will have more information on foundation conditions. When this is required, bid items are usually included in the bidding schedule to reimburse the contractor for this work. Since this work is of a special nature, it is usually subcontracted to specialists. If the dam contractor wishes to do the work using his own resources, a diamond-drilling setup is required.

Clearing, Grubbing, and Stripping

This step involves the removal of trees or other types of vegetation, the removal of all stumps or roots, and the removal of all topsoil containing organic material. Whether the topsoil must be removed before common excavation can commence is dependent upon the extent of the vegetation and whether the common excavated material is to be used for backfill or fill in the permanent construction. Specifications may require that the contractor cut, pile, remove, and sell all merchantable timber. This costs the contractor more than he can recover on the sale of timber, but there is no other solution to the problem.

If the area to be cleared is small and sparsely timbered, this work can be done with chain saws and standard excavation equipment. If there is a large timbered area to be cleared, then special equipment is required for clearing and grubbing. This specialty item is often subcontracted to clearing contractors who own this type of equipment and can rapidly mobilize it and move on the job. If the prime contractor wishes to do this work with his own forces, economical cost can seldom be obtained without the use of tractors equipped with clearing blades, clearing cabs, and root rippers. Tractors equipped with logging winches may be required to snake logs down from steep slopes. Rubber-tired front-end loaders equipped with log tongs are needed for loading logs. On highways, logging trucks and trailers are required for hauling logs to sawmills. Bulldozers and tractors equipped with brush rakes are used for general cleanup.

Any stripping that is required can be accomplished by the same equipment used on the common excavation.

Construction of Haul Roads

Haul roads are required to provide access to the excavation areas and from the excavation areas to the waste disposal and stockpiling areas. This is a nonpay item of cost, so the contractor must spread the cost of road construction to the pay items of excavation.

Haul-road construction is a priority work item, since the various required accesses must be provided before the excavation of the dam's foundation can be started. Road construction must continue throughout the excavation process, since access roads have to be relocated as excavation progresses and haul roads to the disposal areas changed as excavation advances and disposal areas are filled.

Bulldozers and air-powered, crawler-type percussion drills are used for pioneering the roads. The roads can then be widened by drilling and shooting and side casting the excavated material over the bank. Specifications should be closely checked concerning the disposal of material from access-road construction. In some areas the excavated material cannot be side-casted but must be loaded into trucks and hauled to fills or to designated disposal areas.

Preparation of Stockpile and Disposal Areas

It may be necessary to clear, grub, and provide drainage for stockpile or disposal areas. As previously mentioned, clearing and grubbing are often subcontracted, since these are specialty work items. If disposal areas are located over natural drainage channels, considerable expense may be incurred by installing culvert pipe or by providing other methods of maintaining the natural drainage channels. Selected rock pavement may be required under stockpile areas to prevent the stockpiled material from being contaminated by the local material. This work can be performed by the same equipment selected for access-road construction or for performing the foundation excavation for the dam and powerhouse.

Common Excavation

Common and rock excavation are the basic work items for foundation excavation. Common excavation is the removal of all loose, small-size, unsuitable material which overlies the dam foundation. Specifications define the difference between common and rock excavation, and the definitions vary with the contracting agency. In some cases, common and rock excavation are lumped together in the bid schedule and can be removed together and combined as *unclassified excavation.*

The construction methods and equipment used for performing common excavation vary with site conditions and the thickness and extent of the deposit. Thin layers of common material can be removed from steep slopes by using large, open-bottomed, crescent-shaped drag scrapers. A typical crescent scraper setup consists of an anchor on top of the abutment and a tractor, equipped with a logging winch, at the bottom. Cables are reeved from the logging winch to the anchor and scraper so that the scraper can be pulled up and down the slope. The anchor on top of the abutment may be a fixed anchor, a tractor, or an adjustable bridle attached to two fixed anchors. Another method often used to remove thin layers of common material is high-pressure water jets discharged from hydraulic monitors. When these are used, it must be permissible to discharge the muddy water into the natural drainage channel or to clarify it in settling ponds. To provide flexibility, the monitors can be mounted on crawler tractors. A typical hydraulic monitor is shown in Photograph 37 in Chap. 7. When thin layers of common material occur on abutments with flat slopes, then bulldozers can strip and push the common excavation into piles. These piles can be loaded into trucks with front-end loaders.

Scrapers are used to remove deeper deposits of common material from damsites where abutment slopes are not excessive and haul distances are not great. The most economical method of loading common excavation is by scraper, since the scraper participates in the loading operation. The most economical method of hauling common material is by bottom-dump truck, as this type of haulage vehicle has the best truck weight to payload ratio. Scrapers are economical to use on short hauls when the savings on loading cost are greater than the increased haul cost. Bottom-dump haulage is most economical on long hauls when the savings in haul cost are greater than the increased loading cost. When bottom-dumps are used on long hauls, they can be top-loaded with portable-belt loaders, front-end loaders, or shovels. Haul trucks are loaded by portable belts when thick deposits of common material are located on side-hill slopes. With this type of deposit, maximum production can be secured by using bulldozers to shove material onto the belt loader. Heavy-duty belt loaders can be secured in 48-, 54-, and 60-in. widths. Production depends on the number of feeding bulldozers, length of push, type of material, width of belt, and capacity of hauling units. Photograph 8 shows bottom-dump trucks being loaded by a belt loader.

Photo 8 Loading bottom-dump trucks with a portable belt loader. (C. S. Johnson Division of Koehring Company.)

Front-end loaders are used when conditions are not perfectly suited for portable-belt loaders. Loading trucks with shovels is a standard excavation method when small amounts of materials are involved and when shovels that have been purchased for rock excavation are available. Shovels are also used when hard-digging material is removed from confined areas.

Draglines are used on common excavation when the excavated material is removed from below the level of the loading equipment or when material is excavated below a water surface. Small hydraulic backhoes can be used to good advantage in removing pockets of common material occurring in the rock formation.

When large quantities of submerged common materials cover the dam's foundation, dredges for the removal of this material may be economical. Dredges were used on Akosombo Dam in Ghana, Africa, to remove thick beds of submerged sand from the dam's foundation in areas that were slightly more than 200 ft below the water surface.

Rock Excavation to within a Few Feet of Final Grade

Rock excavation is the major work item for foundation excavation. It consists of the removal of all weathered and unsuitable rock material to within a few feet of final grade. Sound rock may have to be removed to provide a stepped surface on the dam's abutments or to eliminate abrupt irregularities in the rock surface. When nearly vertical sidewalls must be maintained, line drilling, presplitting, or drilling and broaching may be required. When a narrow zone of unsuitable rock passes through the dam's foundation, the rock must be removed with equipment of limited maneuverability. These circumstances define *confined excavation.*

At most damsites, rock excavation is a drill, shoot, load, and haul operation, but at some locations, ripping of the rock may be practical. If the rock appears rippable, it should be checked with a refractory seismograph, which checks the time it takes for seismic waves to travel through the material. Seismic velocity tests are performed by some equipment suppliers or by consultants who specialize in this field. Almost all rocks are rippable when the seismic velocity in feet per second is 6,500 or less, and some sedimentary rocks with seismic velocities of 8,500 ft per sec are rippable. When hauls are short, the ripped rock can be loaded and hauled with rock scrapers pulled by crawler tractors. For longer hauls, rock scrapers pulled by rubber-tired tractors are employed. When this type of unit is used, the requirement that the scrapers be push-loaded with crawler tractors must be rigidly enforced. Any wheel spinning of the scraper power unit will cause extreme tire wear and skyrocket the costs. Portable-belt loaders are also used for loading this material into trucks. Another method is to use bulldozers to push the ripped material into piles and then front-end loaders or shovels to truck-load this material. Further information on rock ripping is available in manufacturers' publications.

If drilling and shooting are required, the rock is seldom deep enough to warrant using drills larger than drifter-type percussion drills mounted on booms erected on air-powered crawler tracks (see Photograph 9). This type of drill is used for holes from 2 $^3/_4$ to 6 $^1/_2$ in. in diameter and for depths up to 50 ft. One of the latest improvements in these drills is the addition of masts that adjust 5 ft outward from their normal position. This increases the number of holes that can be drilled from one carriage location and also permits hole drilling in previously inaccessible locations.

Drilling for rock-foundation excavation requires small diameter holes at a close spacing, so that good fragmentation and bottom breakage will occur. The harder and more massive the rock, the closer the drill hole spacing must be and the greater the amount of

Photo 9 Medium-sized drifter drills mounted on air-powered tracks. (Joy Manufacturing Company.)

explosives needed to fragment the rock. In rock of normal breaking characteristics, when average fragmentation is desired, the spacing and burden of small-diameter blast holes are related to the depth of rock as long as the proper powder factor can be maintained. It is desirable to do subgrade drilling of from 1 to 3 ft, depending on the hole spacing; however, this may be controlled by the specifications. Following are the recommended hole spacings for holes 3 1/2 in. and smaller in diameter for various depths of rock:[1]

Hole Spacing for Various Depths of Rock

Rock depth, ft	*Hole spacing, ft*
2 1/2 to 3	2 1/2 × 2 1/2
3 1/2 to 5	3 × 3
5 to 8	3 1/2 × 3 1/2 to 4 × 4
8 to 12	4 × 4 to 4 1/2 × 4 1/2
12 to 18	5 × 5
18 to 24	5 1/2 × 5 1/2
24 to 30	6 × 6 to 8 × 8

The powder factor that can be obtained for different diameter holes on different hole spacings can be determined using the following two tables:[2]

Hole Burden

Hole spacing, ft	*Cu yd rock per lin ft of hole*
2 1/2 × 2 1/2	0.23
3 × 3	0.33
3 1/2 × 3 1/2	0.45
4 × 4	0.59
4 1/2 × 4 1/2	0.75
5 × 5	0.93
5 1/2 × 5 1/2	1.12
6 × 6	1.33
8 × 8	2.37

Pounds of Explosive per Lin Ft of Hole

Blasting agent	ANFO	ANFO	Slurry
Placement method .	Gravity	Pneumatic	Pumping
Density	0.85	0.90	1.15
Hole diameter, in.			
2 1/2	1.81	1.92	2.45
3	2.60	2.76	3.52
3 1/2	3.55	3.75	4.80
4	4.63	4.90	6.26
4 1/2	5.86	6.21	7.93

The most economical type of explosive for blasting foundation rock is the blasting agent ANFO (ammonium nitrate fuel oil mixture). This can be loaded in loose form into dry drill holes by gravity or pneumatic methods. It must be contained in plastic bags when it is loaded into wet holes. It can either be purchased in bags or loaded into the plastic bags at the jobsite. After the holes are loaded, the tops of the holes are stemmed with drill cuttings, sand, or other inert material. Detonation can either be by electric blasting caps inserted in dynamite and placed near the bottom of the hole or by dynamite that is exploded by the use of primacord. If access to the drilled area is provided to an explosives truck equipped for bulk loading, fast loading of ANFO for dry holes or slurry for wet holes may be warranted, depending upon the extent of the work.

After the rock has been drilled and shot, it is loaded by shovels or front-end loaders into rear-dump haulage units. Because of their lack of maneuverability, the large-sized rear-dump units are seldom used on dam foundation rock excavation; instead, trucks of the 35- or 50-ton class or smaller are used. If the rock can be loaded with a front-end loader, this method is preferred to shovel loading because of savings in capital and operating cost. Only one operator is needed on a front-end loader, while small shovels require an operator and an oiler and large shovels require an operator and two oilers. Another advantage in the use of the front-end loader is that its purchase price is much less than that of a shovel of similar production capacity. Truck capacity and the size loader or shovel should be keyed together so that two or three loaded buckets will fill the truck to capacity.

Compressed air for the drills can be piped from a stationary compressor plant or secured from portable compressors. Compressed air from electrically driven, stationary compressors is cheaper than air from portable compressors, since electric motors cost less to purchase and operate than do diesel motors. Portable compressors are used prior to the erection of the stationary compressed-air plant and when their maneuverability warrants their usage.

Rock excavation should be planned so that the face of the excavation has sufficient height for efficient loading. This also results in more efficient utilization of drill holes and explosives. Shovel loading is maintained at maximum efficiency when the bank of shot rock is the same height as the vertical distance from dipper shaft (dipper stick pivot shaft) to ground level. This distance varies with size and make of shovel.

If thick layers of rock must be removed, the rock can be drilled with large drills, explosives can be bulk-loaded, and large-sized shovels and trucks can be used in the excavation and hauling operations. Some rock-fill dams are designed so that the rock fill is secured from large rock cuts required for the development of either a spillway or a power-intake channel. The procedure used to remove rock from massive cuts is similar to the procedure used in a rock quarry, as explained in Chap. 7.

Waste Disposal and Stockpiling

This forms a part of all excavation work items since all excavated material must be directly placed in embankments, utilized as backfill, used as a source of aggregate material, stockpiled for future use, or placed in waste-disposal areas. The disposition depends on the characteristics of the excavated materials, the type of dam, and the specification requirements. Equipment for this type of work is the haulage equipment used in the excavation process and graders, water trucks, or similar road maintenance equipment. At the disposal or stockpile locations, bulldozers are required to spread the truck-dumped material.

Line Drilling

This is the drilling of regularly spaced, small-diameter holes along the desired vertical, or nearly vertical, rock surface of the area to be excavated. Holes are usually 2 to 3 in. in diameter, spaced apart from two to four times the hole diameter. The drilling operation is performed with air-powered, crawler-type drills. After the holes are drilled, the enclosed rock is drilled and shot. By proper location of the drill holes and by proper use of explosives, the rock is broken to the plane of weakness formed by the line-drilled holes, leaving part of the circumference of the holes visible. Because of the large amount of drilling resulting from the close hole spacing, line drilling has declined since the development of presplitting.

Line drilling may be listed as a separate item in the bidding schedule or it may be designated as part of the cost of performing rock excavation.

Presplitting

Presplitting is a technique that extends line drilling a step farther. The line-drilled holes are widely spaced and light explosive charges are loaded in the holes and exploded with primacord. When these are set off, the explosion cracks the rock along the line of the drill holes. When the remainder of the rock is later drilled, shot, and excavated, this presplit rock surface forms the side walls of the excavated areas. Recommended hole spacing and explosive charges for different-sized holes are given in the following table:

Presplitting Hole Spacing and Loading

Hole diameter, in.	Hole spacing, ft	Explosive charges, lb/lin ft of hole
$1^1/_2$ to $1^3/_4$	1 to $1^1/_2$	0.08 to 0.25
2 to $2^1/_2$	$1^1/_2$ to 2	0.08 to 0.25
3 to $3^1/_2$	$1^1/_2$ to 3	0.13 to 0.50
4	2 to 4	0.25 to 0.75

If this is not a separate pay item in the bidding schedule, the contractor must include the cost of this work with the rock excavation.

Drilling and Broaching

This is a technique that will also produce a smooth, excavated rock surface. Final trimming of the sides of the excavation is accomplished by drilling holes along the desired excavation surface and breaking the rock along the plane of the holes, using expansion by mechanical means. Again, since the advent of presplitting, drilling and broaching are seldom used.

Confined Excavation

At many damsites, a zone of unsound or fractured rock may cross the damsite at any angle. This zone is often called a *gut section.* The zone of unsound material may be narrow, with vertical side walls forming restricted areas for equipment operation. When equipment operating space is restricted, excavation cost increases and is often set up in the bidding schedule as a separate bid item called *confined excavation.*

If there is any reason to anticipate that work of this nature will be required, it should be listed as an item in the bidding schedule. Otherwise, when this work is necessary, the contractor must submit a change order to receive proper compensation.

Step Excavation

Step excavation may sometimes be required for the foundations of concrete dams. This necessitates that the sloping rock surfaces of the dam abutments be excavated in horizontal and vertical stepped surfaces. This is always a nonpay item and is considered a part of rock excavation.

Special Treatment of Foundation Surface

Special excavation procedures and surface treatment may be needed when unstable types of rock are used for the dam's foundation. Special treatment is often requested when the dam's foundation is formed by shales that will air-slake upon exposure. Specifications often require that the foundation be protected by leaving the last foot of material in place until 24 hr before concrete placement. Or they may require that the foundation surface be covered with gunite or sprayed with bitumen as soon as it is exposed.

When the rock in the dam's foundation consists of stratified rock layers, specifications may require that these rock layers be tied together by closely spaced vertical anchor bolts. Specifications may prohibit blasting when excavation is being performed for other types of foundations. In this case, rock excavation must be done by using mechanical methods.

Removal of Final Layer of Rock Excavation

Specifications often require that the final few feet of rock be removed by controlled drilling and blasting, using small holes drilled with close spacing. Subdrilling is not permitted. To secure adequate breakage, the hole spacing should not exceed the depth of rock to be excavated. A large amount of this drilling is done with jackhammers. Barring and wedging is required to level the foundation and to remove any loose rock. Small backhoes are useful in this work to eliminate some of the hand labor. Hydraulic-piston control of backhoes has improved their operation to such an extent that they are very economical to use in subsurface excavation. This thin, excavated rock layer is pushed into piles by a bulldozer and loaded in rear-dump trucks by either a front-end loader or a shovel. Final rock removal is shoveled or placed by hand into skips which are picked up by cranes and dumped into trucks. Photograph 10 shows barring and wedging of the rock foundation of a concrete dam. A rock skip is shown in the background.

Dental Excavation

Dental excavation is the removal with hand-held tools of small particles of soft rock or other unsound materials that extend into the dam foundation. If the unsound rock occurs in large pockets, small backhoes can be utilized for its removal. It may be necessary to excavate shallow shafts when the unsound materials extend in depth. To avoid requests for change orders, this work should be included as a separate item in the bid schedule.

Photo 10 Barring and wedging of the rock foundation for a concrete dam.

Tunnel Drift Excavation and Backfill

When unsound rock extends into the abutments, specifications often require that this rock be removed by following it with tunnel drifts. Tunnel drifts may also be driven for use as grouting or drainage galleries when extensive grouting or drainage is necessary for stabilization of the rock in the dam abutments. Drifts can often be driven with small, air-track drills and by using a front-end loader to load and haul the excavated material out of the tunnel. This item of work is usually paid for under appropriate bid items.

Foundation-slope Correction

Foundation-slope correction may be needed in the preparation of the foundation surfaces for the impervious core of earth-fill and rock-fill dams. When it is a requirement, any rock surface that has a downstream slope in reference to the axis of the impervious core must be changed to slope upstream. This work is included under the pay item of rock excavation. It generates small pay quantities while requiring relatively great expenditures for labor since it must be done with hand-held equipment. Prior to bid submittal, it is difficult to estimate the amount of work that will be required since the extent can only be determined when the foundation surface has been exposed. When this work is more extensive than was estimated, the contractor can only recover his additional cost by requesting a change order.

Barring and Wedging

Barring and wedging involve the removal by bars, picks, and wedges of the last layer of rock on the foundation surface. The material usually removed is loose rock or drummy rock that has been loosened by other excavation procedures. Backhoes can be used in this operation, and a great amount of hand labor is required. The contractor is often paid for this work under the rock-excavation item.

Preliminary Foundation Cleanup

The owner's geologist and inspectors often request that the foundation be given a preliminary cleanup so that they can examine it and determine if it is suitable for the dam foundation. After this has been done and the engineers have examined the rock surface, they may find that it is not suitable for the dam foundation. Then another layer of rock must be removed and the engineers will again request a preliminary cleanup of the rock surface. This procedure may be repeated many times until a suitable foundation surface is established. If bid items and pay quantities for this work are not included in the bid schedule, the inspector's extensive use of this procedure will be a large item of cost to the contractor.

Preliminary foundation cleanup requires the removal of all loose material and all water from the foundation surface by air and water jets. In some areas removal of the water may necessitate construction of small cofferdams, ditches, and sumps with the installation of small pumps. Some of this material can be removed with backhoes, but the majority of the work must be accomplished by hand labor.

Final Foundation Cleanup

Final foundation cleanup is required when a suitable foundation surface has been established. The work requirements are the same as for preliminary foundation cleanup except that final cleanup must be completed more thoroughly. This should be a separate pay item in the bidding schedule; otherwise the contractor must include this cost under rock excavation.

Surface Grouting

When the foundation surface under earth-fill dams contains cracks or small fissures, it may be required that these be filled with hand-placed cement grout. The specifications should be checked to determine whether this is a requirement and whether it is a separate pay item.

Corrective Concrete

Corrective concrete may be required to treat the foundation for the impervious core of earth- or rock-fill dams. The impervious core cannot be effectively placed and compacted underneath overhanging rock ledges, which form when unsound material is removed from the foundation surface. To provide a suitable foundation surface, specifications may require that these spaces be filled with concrete. If this work is scheduled before the mixing plant is ready for operation, it may be necessary to provide a small, portable mixing plant to furnish the concrete. This is usually a separate bid item.

Backfill Overexcavation

If the contractor overexcavates, this overexcavation may have to be backfilled to grade. Depending on the type of dam structure and specification requirements, selected backfill or lean concrete may be used. Often in small areas refilled with backfill, compaction can only be accomplished with hand compactors. The cost of this work is the responsibility of the contractor.

Unwatering of the Foundation Surface

The dam's foundation surface must be free of water before placement of the dam material can commence. This work may be quite extensive in the bottom of river channels and may require the construction of special cutoff cofferdams, drainage ditches, small sumps, and localized pumping. Job supervision often describes the work as "being performed with tablespoons and mops."

Drainage Ditches and Drain Pipe

This job may require the construction of drainage ditches backfilled with gravel or ditches containing drainage pipe and backfilled with gravel or the installation of half-round tile drains. When required, they are usually paid for under separate bid items. Ditch excavation is done with hand-held jackhammers and small backhoes; a considerable amount of hand labor is also required.

Placement of Impervious Blanket on Sides of Channel Upstream from the Dam

It may be necessary to place an impervious blanket on the sides of the channel upstream from the dam to cut off any water flow through the abutments when dams are constructed on abutments containing pervious material. If required, bid items and estimated quantities are included in the bid schedule.

Dressing the Slopes of the Stockpiles or Disposal Areas

Dressing the slopes of the stockpiles or disposal areas is often required and can be performed with bulldozers. Sometimes specifications call for mulching and seeding of dressed waste disposal surfaces. In this case, equipment for seed broadcasting is needed. Cost of this work often must be included in the bid items for common and rock excavation.

Drilling and Grouting

If a concrete dam is to be constructed, only a shallow grout curtain is placed prior to covering the rock surface with concrete. This curtain consists of holes approximately 25 ft in depth filled with low-pressure grout. The main grout curtain is formed after the foundation is covered with concrete. (This procedure is explained in detail in Chap. 5.) For earth- or rock-fill embankment foundations, the complete grout curtain is installed before any fill is placed. Lengths of holes and the required grouting pressure vary with local rock conditions.

Drilling and grouting are normally covered by separate bid items in the bidding schedule. Since the technique is a specialty work item, it is often subcontracted. If the contractor wishes to do the work with his own work force, he must provide diamond drills and a grouting setup. The equipment for drill-hole grouting is illustrated by Photograph 11.

THE EXCAVATION-CONSTRUCTION SYSTEM

The planning and scheduling of the excavation and preparation of a dam's foundation control the selection of the number of operating construction equipment units necessary. Planning is started by making a takeoff of the work quantities for each section of the dam's foundation that is scheduled as a separate operation. Access roads and haul roads should be laid out, and haul distances and grades established. The excavation sequence should be established by scheduling the work so that a minimum amount will form part of the critical path for the dam's construction and so that excavation equipment requirements can be minimized without delaying the project's completion. These data are then used to select the types and number of equipment units.

QUANTITY TAKEOFFS

Quantity takeoffs are made to check the bid quantities, to list all the work essential for foundation excavation and preparation, to compute

Photo 11 Grouting equipment. (Gardner-Denver Company.)

the quantities of work that must be done during each phase of the excavation, and to check how reimbursement will be received for each item of work. Dam excavation may be scheduled so that portions of the dam's foundation are completed and dam placement is started in these portions before the remaining excavation is completed, or all foundation excavation may be completed before any dam placement is started. The sequence of work performance should be established before starting the quantity takeoffs for foundation excavation so that quantities can be tabulated in accordance with the planned performance of the work. If a powerhouse is included with the dam contract, these excavation quantities should be listed separately.

Bid item descriptions and bid quantities shown on the bidding schedule cannot be used as the scope of the work. Work scope must be determined from the specifications since many items of work may be called for by the specifications but will not be listed as bid items. As an aid in determining the extent of the foundation work for any project, the specifications can be reviewed and compared to the work-flow diagram presented at the beginning of this chapter.

Quantity takeoffs should properly designate whether the quantities are bank or loose yardages. Bank yardages are yardages of material in its natural state. Loose yardages are the yardages which the material will occupy after it has been excavated. Swell is the increase in yardage from in-place to loose yardages. The loading equipment loads bank yardages. The transportation equipment hauls loose yardages, and the capacity of this equipment must be computed using loose yardages and weights of the loose material. Payment will be in bank yardages. For estimating purposes, all yardages should be converted to bank yardages, which will prevent confusion and simplify computations.

Materials vary so much that the only method of accurately determining the relation of bank yardage and weight to loose yardage and weight is by actual measurement. An approximation of these data for certain materials is listed in the following table:[3]

Comparison of Bank Yardages to Loose Yardages

	Bank cu yd, lb/cu yd	Swell, %	Swell factor	Loose cu yd, lb/cu yd
Clay, dry	2,300	25	0.80	1,840
Clay, wet	3,000	33	0.75	2,250
Earth, dry	2,800	25	0.80	2,240
Earth, wet	3,370	25	0.80	2,700
Earth, sand, gravel	3,100	18	0.85	2,640
Granite	4,500	50–80	0.67–0.56	3,000–2,540
Gravel, dry	3,250	12	0.89	2,900
Gravel, wet	3,600	14	0.88	3,200
Limestone	4,200	67–75	0.60–0.57	2,620–2,760
Loam	2,700	20	0.83	2,240
Sand, dry	3,250	12	0.89	2,900
Sand, wet	3,600	14	0.88	3,200
Sandstone	4,140	40–60	0.72–0.63	2,980–2,610
Shale, soft	3,000	33	0.75	2,250
Slate	4,590–4,860	30	0.77	3,530–3,740
Trap rock	5,075	50	0.67	3,400

The quantity takeoff should list the thickness of both common and rock excavation so that the most economical use of excavation procedures can be developed. The common excavation takeoff should tabulate separately the quantities of silt, sand, clay, boulders, etc., that will be encountered. The takeoff of rock excavation should show the various types of rock that will be encountered so that drilling speeds and explosive consumption can be properly estimated. The capacities of the disposal areas should be checked to prove their adequacy. Dams have been put out for bids when the specified disposal areas were inadequate. Depending on the job specifications, the cost of providing more disposal capacity could be the contractor's responsibility. If additional disposal areas are required, the cost involved in their development and the cost of any additional haul should be determinable.

LAYOUT OF CONSTRUCTION ROADS

Construction roads are required for access from the contractor's installations to the excavation area and from the excavation area to the stockpiles or to the waste-disposal areas. Depending on the particular site conditions, the cost of providing the roads may be a major cost to the contractor. It is necessary to prepare a road layout to decide the amount of work required for their construction. It is important not only to locate the original haul roads required but to determine the relocation that may be necessary for completion of the excavation.

The other reason for preparing a road layout is that it is needed to estimate the turnaround time for scrapers or trucks used for hauling excavated materials. Haul-road layout is the most important factor in determining truck-turnaround time as most of the errors in this computation can be attributed to actual haul-road grades' being different from those originally estimated. Errors due to assigning the wrong truck speeds to specified grades in comparison are usually of a minor nature.

SCHEDULING

The excavation of the dam's foundation should be scheduled to minimize the excavation time that is part of the project's critical path, to keep the amount of excavation equipment that must be purchased to a minimum, and to perform excavation in a manner best suited to site conditions.

The less the amount of excavation time that is part of the critical path time, the less the time required for the project's completion. Reducing the overall construction time reduces the overhead cost and amount of escalation, for both costs vary in direct proportion to construction time. Similarly, reduction in overall construction time diminishes plant and equipment write-off, since the shorter the period that equipment is on the project, the greater its resale value and the smaller the amount of equipment depreciation that must be charged to the job.

The critical path time for foundation excavation and preparation will be reduced if the work is done while other critical work items are being performed. Among the factors influencing the critical-path construction time are considerations of whether excavation can be done concurrently with the construction of the diversion facilities, whether all the dam's foundation must be excavated before dam placement can commence, whether one section of the dam's foundation can be excavated and dam placement done in this section while the remainder of the dam's foundation is excavated, and whether powerhouse excavation can be delayed until the dam's foundation is excavated.

Specifications may require that the dam placement start at the lowest point in the dam's foundation and that the total length of the dam be placed continuously, with a minimum difference in elevation between any two points. If so, the dam's foundation must be completely excavated before dam placement commences; therefore, the schedule cannot readily be arranged to shorten the critical path.

If the damsite is in a steep canyon and if tunnels are used for river diversion, the dam abutments can often be excavated from the top of the abutment down to stream level while the diversion tunnels are being constructed. The stream must be diverted before the remainder of the dam's foundation can be excavated. Under these conditions, only the amount of excavation that is below stream level becomes part of the critical path.

Some damsites are wide enough to permit one side of the dam to be excavated before excavation is started on the remainder. When specifications permit, dam placement can be done in this side section while the remainder of the dam is excavated. With this procedure, the excavation of the first section of the dam's foundation is the only part of the excavation that is part of the critical path, and the remainder of the excavation is a noncritical item.

When the river channel is wide and diversion can be made through the dam, the contracting agency may divide the construction work into two or more contracts, with each diversion stage let as a separate contract. Then the excavation sequence for the entire project is taken out of the control of the contractor. The excavation for each contract becomes part of the critical path of each contract, and the total time required for the dam construction will be extended.

There are too many ways to schedule excavation to reduce the critical path time to discuss them here. The point is that excavation is such an important item in job planning that considerable effort should be spent on arriving at the best possible solution.

When the excavation work is scheduled, it is necessary to correlate this work with other similar project work so that the excavating equipment can be used on as much work as possible; this will reduce the amount of excavating equipment required for the total project and the amount of equipment write-off. It may be possible to schedule the work so that the same equipment can be used for foundation excavation, in the aggregate pit or quarry, for construction of earth- or rock-fill embankments, for construction of diversion facilities, or for road construction. Proper job scheduling will increase the quantities of work done by a piece of equipment and thus decrease the unit cost of equipment write-off. An equipment use schedule should be prepared so that equipment usage can be planned for the total project. This scheduling of equipment use is called *resource leveling* by technologists.

Work should be scheduled to fit site conditions so that excavation can be done for a minimum of cost. At steep damsites, the preferred method of performing excavation is to start work at the top of each abutment and excavate downward until the bottom of the dam's foundation is reached. This method of excavation prevents material from sliding or being shoved into completed areas during execution of the work, and the excavation crew does not have to return to any completed areas.

When the abutments have an abrupt slope, when the excavated material can be shoved over the edge and there is space on the canyon floor to receive it, then shelf excavation can be started at the top of each abutment. This shelf is lowered as each excavation round is drilled, shot, and shoved over the bank. The equipment remains on this shelf until the shelf is excavated to the canyon floor. After the excavated material falls on the canyon floor, it can be loaded into trucks and hauled to disposal areas. Excavation cannot be done below streambed level until the stream has been diverted.

If the abutments are too steep for haul-road construction and if there is insufficient space on the canyon floor to receive excavated material, abutment excavation must wait until diversion has been completed. After diversion, the excavated material can fall into the cofferdammed area, where it can be loaded into trucks and hauled to disposal areas. When abutments are flatter, excavation can still be started at the top of each abutment, but roads must be constructed to furnish access to the excavation areas and to provide haul roads for the transportation of the excavated material to the disposal areas. On flat damsites, the absence of grades and the ease of haul-road construction allow foundation excavation to be performed by any method.

SELECTION OF EQUIPMENT

Before the final selection of excavation equipment is made, the use of this equipment in the concrete subsystem, embankment subsystem, etc., should be reviewed so that the type that is capable of performing as many tasks as possible can be selected. This will keep expenditures for plant and equipment to a minimum and will result in a low project write-off for the equipment. On the other hand, excavation equipment must be selected that will perform the excavation economically. The preferred equipment and methods used for each item of foundation excavation and preparation were systematically described in the first part of this chapter.

To achieve low direct cost, the equipment must be selected to fit the depth of the excavation and the type of haul roads. Shallow common excavation requires different equipment from that suitable for the excavation of deep massive flat layers. Small drills are used on shallow rock layers, but large drills can be utilized when rock excavation must be done in depth. Length and steepness of haul roads influence the selection of haul units.

When the equipment can only be utilized on foundation excavation, the yardage of common and rock excavation will control equipment selection. When there are only small quantities of common excavation but large quantities of rock to be excavated, the purchase of special equipment for the small amount of common excavation cannot be justified. One solution to this problem is to excavate common and rock as one operation, if the specifications or job planning do not require their separation. When this is possible, the same bid price should be submitted for common and rock excavation so that separate quantity surveys for each class of excavation will not be required. If common and rock excavation must be removed separately, then the rock-loading and haulage equipment can often be used to excavate both materials. Use of rock excavation equipment on the common excavation may slightly increase the direct cost of common excavation, but the total cost of common excavation will be low because the amount of equipment write-off chargeable to common excavation will be held to a minimum.

When foundation excavation is the main use of the excavation equipment and there are large quantities of common but only small quantities of rock, then expenditures cannot be justified for the purchase of rock excavation equipment such as drills, shovels, etc.; instead, it should be rented. Or rock excavation can be subcontracted to contractors who specialize in this work and therefore own this type of equipment.

At flat sites, it may be desirable to excavate one abutment, start dam placement in this abutment, and, as dam placement continues, excavate the remainder of the dam. The quantity of material to be excavated in the first abutment then controls equipment selection. The remainder of the dam foundation can be excavated with less emphasis on the time required for completion, which may result in more work being performed with the equipment, which thus reduces write-off.

After the types of excavation equipment have been selected, the number of required units of each type of equipment can be determined by establishing what the production rates will be for each type of unit and then by dividing the scheduled rate of production by these production rates to establish the number of equipment units that must be in operation. This number must be increased by one-third to arrive at the number of units to be purchased. This one-third factor is required because only 70 to 80 percent of the equipment will be available for operation at one time. The remainder will be out of service, being repaired, waiting to be repaired, or being serviced. To this equipment fleet must be added the required servicing equipment and facilities, i.e., service trucks, tire trucks, repair shops, tire shops, and rigging loft; road maintenance equipment such as water wagons and graders; and pickups and man-haul trucks for transportation of supervisory personnel and workmen. This last group must be treated as project equipment since the same service equipment and facilities may be suitable to use for all the project work and need not be duplicated in every construction subsystem.

The number of rock drills required for excavating rock from the dam's foundation is readily computed. First, the scheduled quantity of bank cubic yards of rock to be excavated per day is multiplied by the linear feet of drill hole required to break one cubic yard of rock. The number of rock drills that must be in operation is then determined by dividing this product by the linear feet of drill hole produced by one drill in one day. Of these three quantities, the one that is most difficult to determine accurately is the linear feet of hole produced by a drill in a working day. The best method of determining this is to make drilling tests on the rock; the next best method is to base production on past drilling records for similar rock. The third choice is to send rock samples to drill manufacturers so that they can run laboratory drilling tests on the rock. The last and least satisfactory solution is for the estimator to base drill production on his judgment and published data.

Drill manufacturers publish penetration rates for their drills, based on drilling a medium-hardness granite 100 percent of the time. Therefore, penetration rates have to be reduced to allow for the nondrilling time required for moving drills, changing steel, and freeing bits, and for time wasted by the drilling personnel. The amount that penetration rates should be reduced varies with the hardness of the rock. The harder the rock, the longer it will take to drill one hole and the greater the amount of available time spent on drill penetration. The linear feet of hole produced per drill per shift will vary between 70 percent of the penetration rate in hard rock and 40 percent in soft rock.

The following figures are presented to illustrate how rock type influences drill penetration rates, hourly drilling rates, and bit life. The drill penetration rate in medium-hard Sierra granites will be approximately 12 in. per min; a drill will produce 45 lin ft of hole per hr; and a bit with 10 resharpenings (the maximum) will drill 260 lin ft of hole. In extremely hard black granitic rock, the drill penetration rate will be 4.5 in. per min, drill production will be approximately 15 lin ft of hole per hr, and bit life will be approximately 120 lin ft of hole with 10 bit resharpenings. For soft argillite, the drill penetration rate will be approximately 4.5 ft per min, the drill production will be approximately 100 lin ft of hole per hr, and the bit life approximately 600 lin ft of hole with 10 bit resharpenings. These drill speeds and bit lives are based on those of percussion drills using carbide insert bits.

The amount of compressed air required for excavation can be found by multiplying the number of operating drills by the air consumption. To allow sufficient air for leaks, hole blowing, and drill inefficiency, 125 cu ft per min of compressed air should be supplied to a jackhammer, 550 to 600 cu ft per min to a crawler-mounted $4^1/_2$-in.-bore-diameter drifter drill, 800 to 900 cu ft per min to a crawler-mounted 5-in.-bore-diameter drifter drill, and 1,000 to 1,200 cu ft per min to a crawler-mounted $5^1/_2$-in.-bore-diameter drifter drill.

If a shovel is used for loading material, the preferable method of estimating its production is to evaluate performance records on similar work. Lacking this information, shovel production rates can be secured from tables in various publications or from production slide rules.[4,5] Production figures given in these publications are for perfect working conditions and for 100 percent work efficiency. These production figures must be reduced for the anticipated equipment efficiency, the types of material, the height of bank, the extent of shovel swing, and the efficiency of the operator. Following are estimated shovel production figures suitable for foundation excavation with 70 percent efficiency, 90-degree boom swing for truck loading, and, since efficient bank heights are seldom encountered in work of this type, an assumed bank height of 40 percent of optimum. If operations are such that the bank height will always be at optimum, these figures can be increased by 25 percent. Optimum bank height for any shovel for maximum loading capacity is level with the dipper stick pivot shaft.

Shovel Production in Bank Yd per hr for Dam Foundation Excavation; 70% Efficiency, Bank Height 40% of Optimum, 90° Swing

Bucket size, cu yd	Sand and gravel	Common earth	Clay		Rock	
			Hard	Wet	Well broken	Poorly broken
1/2	50	45	35	25	30	20
1	110	90	70	50	55	40
2 1/2	210	170	140	110	135	80
3 1/2	280	260	220	160	170	120
4 1/2	350	310	260	190	240	160

A rule of thumb for determining the production of articulated front-end loaders is that their loading capacity will equal that of a shovel whose bucket size is from one-half to two-thirds the bucket size of the front-end loader. The cycle time of a front-end loader is greater than a shovel cycle since the loader must accomplish with travel and turning that which is done by the swing of the shovel boom.

To illustrate the production rates that have been achieved with front-end loaders, let us consider the 10-cu-yd front-end loader that was used during the construction of a freeway south of Portland, Oregon. When it loaded blasted rock into rear-dump trucks, it reached a production high of 465 bank cu yd per hr. When it loaded fines and small rocks, production increased to 560 bank cu yd per hr.[6] Photograph 12 shows a 10-cu-yd loader of this type.

The production achieved with portable belt loaders varies with the ease with which the material can be bulldozed onto the loader, the pushing distance, the grade on which the bulldozers operate, the number of bulldozers, and the availability of trucks. At Portage Mountain Dam 60-in.-belt loaders could load continuously because they discharged onto conveyor belts. Loader production in the sand and gravel moraine was 3,000 tons per hr.[7] Converted to bank yardage this was 1,850 cu yd per hr. With easily pushed material, 115-ton bottom-dump trucks have been loaded at a rate of 1,800 bank cu yd per hr by a portable 60-in.-belt loader fed by four D-9 bulldozers.[8] Any change from these optimum conditions will reduce this loading capacity. Loading capacity can always be computed with bulldozer production, which is the controlling factor. Bulldozer capacity is given in manufacturers' literature.[4] Again, this production must be adjusted for job efficiency and other operating conditions.

Photo 12 Ten-cu-yd front-end loader. (Caterpillar Tractor Company.)

After production computations are completed, the result should be adjusted as judgment and experience dictate.

The number of operating trucks or scrapers required to service loading equipment is determined in the following manner. The cycle time of this equipment is computed and then divided into a 50-min hr to establish the number of cycles per hr per unit. The unit's cycles per hour are then multiplied by the unit's capacity to establish the hourly production per equipment unit. This can then be divided into the scheduled hourly production rate to establish the number of required equipment units. A check of these figures can be secured from production charts published by equipment manufacturers.[4] Again, these charts must be adjusted for job efficiency and other factors.

In order to compute a turnaround cycle, it is necessary to know the swell and the weight of the loose material so that the allowable payload can be determined. The turnaround time is a combination of waiting time at the load out point, loading time, haul time, waiting time at the dump, dumping time, and return time. Truck loading time is controlled by the loading capacity of the loading equipment. Scraper loading time varies with the type of pusher and type of material to be loaded. The haul and return time is computed from haul distances, grades, type of roads, weight of the empty and loaded vehicle, and the vehicle's operating characteristics. Particulars and examples of turnaround time are very well described in equipment manufacturers' publications.[9,10,11]

To illustrate how turnaround cycles are computed, the following cycle is for a 35-ton rear-dump truck hauling broken granite. Haul roads are assumed to be well maintained and covered with gravel surfacing. The first 3,000 ft of haul will be down a 10 percent grade, and the remainder will be level. Truck specifications pertinent to this haul are payload capacity of 35 tons and heaped cu yd capacity, on 3 to 1 slope, of 27 cu yds. The granite weighs 4,500 lb in the bank; therefore, if a 60 percent swell factor is used, 1 bank cu yd will be the equivalent of 1.6 broken cu yd and each broken cu yd will weigh 2,700 lb, or 1.35 tons. Assuming that the average truck load will be 25-cu-yd loose measurement, then this broken yardage will weigh 34 tons, which does not exceed the truck's payload weight capacity. This establishes the truck's payload at 15.6 bank cu yd. Trucks will be outfitted with dynamic braking so that a loaded truck can maintain a speed of 10 mph down the 10 percent grade; for the level hauls the truck speed will be 30 mph. The empty truck will be able to maintain a speed of 30 mph on the level and when it climbs the 10 percent grade, its speed will drop to 15 mph.

Truck cycle	*Min*	
Load:		
Wait	0.5	
Load	4.0	
Pull out	0.2	
Subtotal		4.7
Haul:		
3,000 ft down a 10% grade at 10 mph	3.4	
2,000 ft level at 25 mph	0.9	
Subtotal		4.3
Dump:		
Wait	0.5	
Dump	1.0	
Pull out	0.2	
Subtotal		1.7
Return:		
2,000 ft level at 25 mph	0.9	
3,000 ft up a 10% grade at 15 mph	2.3	
Subtotal		3.2
Total		13.9 min

Number of cycles in a 50-min hr	3.6
Bank yd hauled per hr at 15.6 cu yd per truckload	59 cu yd

When all the foregoing tabulations and computations have been completed, the list of plant and facilities required for this operation can be compiled.

REFERENCES

1. *Blasters Handbook*, copyright 1955, reprinted 1959, Canadian Industries Limited, Montreal, Quebec.
2. *Blasters Handbook*, 15 ed., copyright 1966, E. I. du Pont de Nemours & Co. (Inc.), Wilmington, Delaware, page 492.
3. *Terex Equipment Handbook*, published by the Earthmoving Equipment Division of General Motors Corporation, Sales Data Section III.
4. *Caterpaillar Earthmoving Performance Handbook.*
5. P & H *Excavation Production Calculator*, copyright 1956, Harnischfeger Corporation, Milwaukee, Wisconsin.
6. 10-Yard Wheel Loader/Truck Team Speeds, Oregon Rock Cut, *Roads & Streets*, September, 1969.
7. Irvine Low, M. ASCE, Portage Mountain Dam Conveyor System, *Journal of Construction Division*, Proceedings of the American Society of Civil Engineers.
8. Modified Equipment Moves 45000 Yards Per Shift, *Roads & Streets*, April, 1969, page 48.
9. *Estimating Production Cost of Material Movements with Euclids*, copyright 1953, by the Euclid Road Machinery Co., reprinted 1955, Euclid Division, General Motors Corporation.
10. *Earthmoving Selection*, Caterpillar Tractor Co.
11. *Basic Estimating*, Sales Development Construction Equipment Division, International Construction Equipment.

CHAPTER FOUR

Concrete-placing Equipment

This chapter describes the selection, the application, and the capacities of concrete-placing equipment. The remaining components of the concrete-construction system are discussed in Chap. 5. The use of concrete subsystems to pour powerhouse concrete is covered in Chap. 8.

Concrete-placing equipment is discussed prior to the other components of the concrete-construction system because its selection controls the location and selection of the remainder of the system. The type of placing equipment establishes the required capacity of all the other system components, since their main purpose is to furnish sufficient mixed concrete to the placing equipment and to supply it with enough formed areas and placement crews to permit it to operate at maximum capacity.

Equipment for placing mass concrete consists of cableways, cranes supported by trestles, and cranes operating from the ground. There are many different types of equipment for placing miscellaneous concrete.

When large concrete projects are constructed, the mass-concrete-placing equipment controls the concrete-construction system, the construction time, and the concrete cost. It influences and is influenced by the diversion plan, the excavation plan, and the concrete-pouring sequence. The types of equipment used for mass-concrete placement have not changed since Grand Coulee and Hoover Dams were constructed in the 1930s. Improvement in mass-concrete-placing efficiency since then results from increases in speed, in the area of coverage, and in the lifting capacities of the same types of placement equipment.

In contrast, there have been many new developments in miscellaneous-concrete-placing equipment. These new developments, however, have had a minor effect on the concrete-construction system since miscellaneous-concrete placement forms only a small portion of the total cost of a large concrete dam. Therefore, the selection of miscellaneous-concrete-placing equipment is governed by the criterion that it must function with the mass-concrete-placing equipment.

Since this book is written for contractors' engineers who usually do not design this equipment but rather rely on equipment manufacturers to design and furnish concrete-placing equipment to their specifications, the descriptions of concrete-placing equipment are given to furnish only the necessary background for dam-construction planning but not in sufficient detail for equipment design. If design data on this equipment are desired, the reader is referred to equipment manufacturers' literature or other publications.[1,2]

In these discussions, the terms *mass concrete* and *miscellaneous concrete* are used. Mass concrete occurs in structures of such large cross-sectional dimensions that large volumes of concrete are placed in every pour. It consists of concrete in large concrete dams, in combination dams that contain both concrete and earth- or rock-fill sections, and in earth- or rock-fill dams that contain power intake sections, massive bottom cutoff walls, or other large concrete structures. In this book all concrete that is not mass or powerhouse concrete is called miscellaneous concrete. This is concrete in structures of such small cross-sectional dimensions that only a few cubic yards of concrete can be placed in any pour. It occurs in small structures, in small dams, or in the facing on rock-fill dams.

CABLEWAYS

Description

A cableway used for placing the mass concrete in a dam consists of a large-diameter track cable which is suspended from two towers and spans the length of the dam. A traveling carriage rides on the track cable and from this carriage is suspended a concrete bucket that can be raised and lowered as desired. Both of the towers supporting the track cable may be movable, or one tower may be fixed and the other movable. The mobility of a cableway with its traveling towers, the travel of the carriage, and the raising and lowering of the concrete bucket assure that concrete can be placed at any location in a dam.

Cableways are designed to have the capacity to handle various sizes of concrete buckets. Small cableways can handle 4-cu-yd buckets and are known as 12½-ton cableways. Except for one installation, the capacity of the largest cableways is 8 cu yd, or 25 tons. The exception was at Glen Canyon Dam where two cableways were used that had a capacity sufficient to handle 12-cu-yd buckets. Cableways with the capacity to handle 20-cu-yd buckets have been designed but have never been placed in service. Job specifications may limit to 4 cu yd the amount of concrete that can be dumped at one time. If a dam is constructed under this type of specification, larger-sized buckets can be used, but these buckets must be divided into 4-cu-yd compartments, and the concrete in each compartment must then be dumped separately.

In tower layout and tower travel, with the exception of a few installations, two general types of cableways are used for dam construction. The first type, with one fixed and one movable tower, is called a *radial cableway*. The runway for the movable tower is constructed as a circular arc of constant radius from the fixed tower. This allows travel of the movable tower at a constant track-cable distance from the fixed tower. A typical radial cableway is shown in Drawings 1A and 1B.

When a radial cableway is planned at damsites with the fixed tower located on an abutment that has a steep rock face, the fixed tower can often be replaced with an anchorage excavated into the rock at the proper elevation. Examples of this type of anchorage for the fixed end of a cableway were the cableways used for the construction of Donnells Dam, California, and Yellowtail Dam in Montana.

The other type of cableway has two movable towers operating on parallel straight runways. When these towers travel, they both must move in the same direction and at the same speed so that the track cable is always at right angles to the runway, with a constant track-cable length. As an illustration, two head towers for parallel cableways are shown in Photograph 13.

It is theoretically possible to construct a parallel cableway with the towers operating on parallel runways curved to the same radius or on concentric curved tracks, but as nearly as can be determined, installations of this type have not been made.

As previously stated, there are a few exceptions when dam

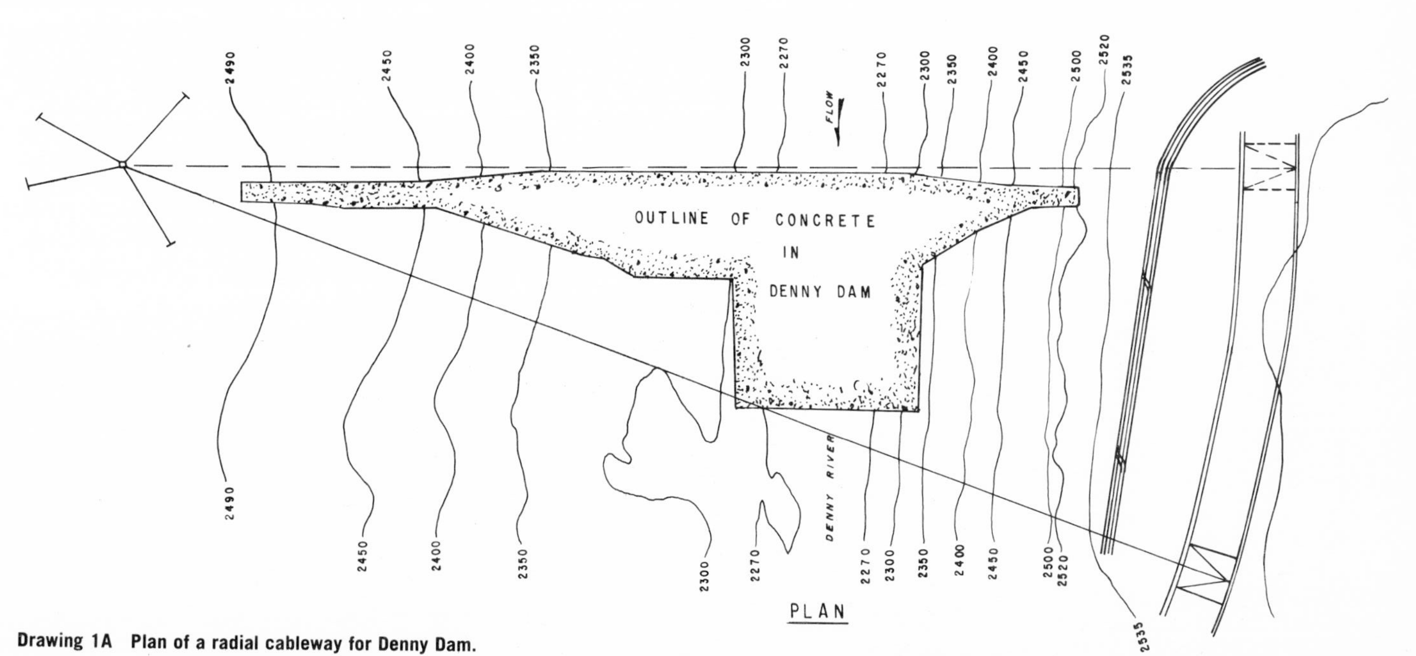

Drawing 1A Plan of a radial cableway for Denny Dam.

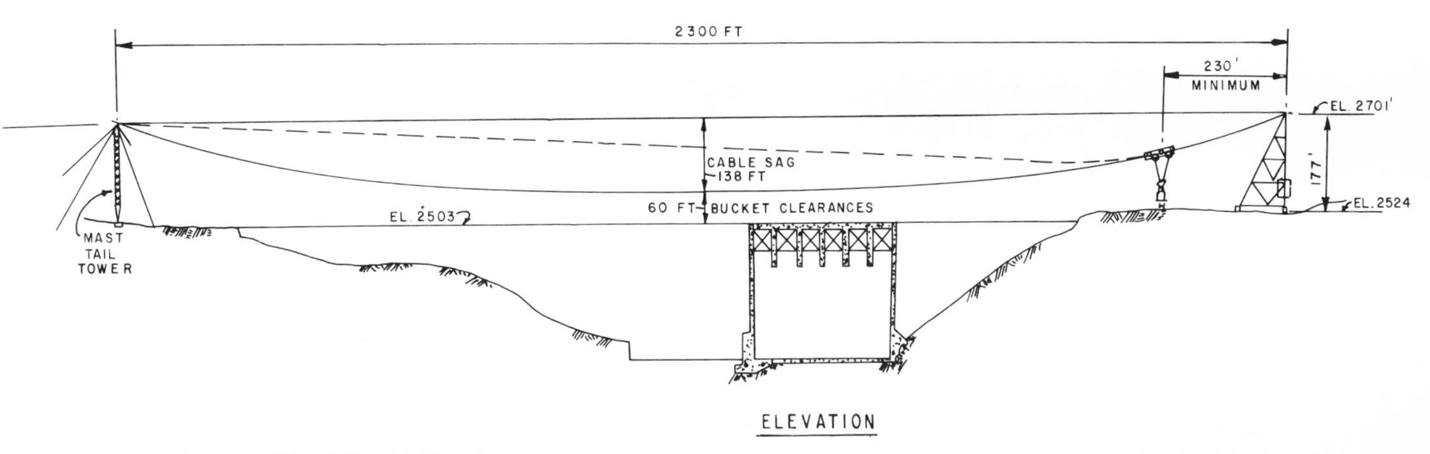

Drawing 1B Elevation of a radial cableway for Denny Dam.

Photo 13 Two parallel traveling tail towers used for the construction of Detroit Dam, Oregon.

concrete is placed with cableways that have fixed towers. This may be the case at Dworshak Dam, Idaho. The major portion of the concrete was placed with three parallel cableways with movable towers. To keep their span under 3,000 lin ft, they were located at an elevation that was too low to top out the dam. The dam may be topped out with a luffing-type cableway with a span of over 3,400 lin ft and stationary mast towers approximately 200 ft high, with a luffing coverage of approximately 50 ft. Luffing cableways are commonly used for bridge construction. Hook coverage is achieved by movement of the top of the stationary towers as the side guys are shortened and lengthened.

Cableway towers are designated *head* and *tail* towers. The tower at the end of the cableway where the hoist machinery is located, irrespective of whether it is a fixed or movable tower, is called the head tower. The other tower is the tail tower and, depending on the cableway layout, can also be either movable or fixed. All head towers contain sheaves for the hoist and travel lines and facilities for anchoring the track cable. If the head tower is a movable one, then it also contains the drums and motors that control the load and hoist lines, a small air compressor for control of the drum brakes, the motors for moving the tower and supports, and a counterweight to resist the overturning moment caused by the pull of the track cable. The tail tower contains the track-cable tightening facilities, sheaves for the travel line, anchorage for the load line, and,

the tail tower is movable, motors for tower travel and a coun-
erweight. If one tower is fixed, it may be a fabricated tower, an A
ame, or a mast, or the sheaves may be fastened to structural
embers anchored into the canyon walls. Fixed towers must be
quipped with fairlead sheaves so that the sheave is always centered
the direction of the movable tower. Fixed towers may support
ables from any number of movable towers. At Shasta Dam, a 460-ft
xed tower was used which supported track cables from seven mov-
ble towers. All cableways were of 8-cu-yd capacity with spans vary-
ng from 720 to 1,670 lin ft.

If the head tower travels, the hoist machinery is located in the
ower. If the head tower is stationary, the hoist machinery may be in
he tower or in a separate hoist house. The cableway controls may
e located in the head tower or at a control observation point that
ermits the operator to view both the pickup point and the pour area.
t Bullards Bar Dam, California, radio controls were used; the con-
rol box was small and portable and, when in use, was set on a pillow
o cushion it from vibrations. Also at Bullards Bar, closed-circuit
elevision was used to monitor the carriage and bucket travel.[3] On
nost projects, final spotting of the bucket over the pour area is done
y the bellboy, who gives both voice and signal directions to the op-
rator by separate communication systems. Voice signals are used to
ontrol the bucket's horizontal movement, and an electronic signal
s used to control its vertical position. Photograph 14 shows the
ontrol house for the Guri Dam cableway in Venezuela, located so
hat the operator had a view of the concrete-bucket dock and also of
he pour area.

The movable towers for a cableway are supported on rail trucks
that travel on two parallel sets of tracks, constructed at a constant
elevation. The trucks under the inclined tower leg must be sturdier
than those under the vertical leg, because of the greater force transmitted to the track. Rail sizes used for the runway tracks are approximately 132 lb with 3-in.-wide heads. The trucks on each track are
propelled by geared motors; their sizes depend upon climatic conditions and tower height, but generally they are 75 or 100 hp each.
The track under the inclined tower leg is constructed perpendicular
to the leg to permit resistance of the horizontal component of track-
cable stress in the tower.

The track cable, often called the *main gut*, is stretched between
the head and tail towers and provides a track and support for the carriage as well as for the slackline carriers. The cableway's hook is
suspended from the carriage and the slackline carriers support the
operating cables. Track cables vary in diameter from $1\frac{3}{4}$ to 3 in. for

Photo 14 Carriage, bucket dock, transfer track, transfer car, and the control house for the Guri Dam cableway. (Washington Iron Works.)

small cableways and from 3 to 4 in. for 8-cu-yd and larger cableways. The cable's weight varies correspondingly from 22.2 lb per lin ft for a 3-in.-diameter track cable to 38.6 lb per lin ft for a 4-in.-diameter track cable.[1] The cable is made with a track lay, which means that the surface strands are flat and keyed together to resist the wear of the carriage wheels. The size of track cable used for any application depends on the stress in the cable. Stress in the cable is a function of cable sag, cableway capacity, and the cableway span, since the weights of the track, travel, and hoisting cables, carriage, hook, and load enter into the computations. For those interested in computing a track-cable size, formulas are given in cable manufacturers' publications.[1]

There are many different methods of rigging the operating cables of a cableway. For those interested in this subject, manufacturers' literature should be obtained and other books reviewed.[2] The following general description of a cableway's rigging is one often used that has produced satisfactory results.

The operating cable that controls the movement of the carriage is called the *travel line, endless line,* or *conveying line.* (It is not endless, but since both ends are attached to the carriage, it is operated as an endless line.) Both spans of the cable, from the carriage to each tower, are supported by *slackline carriers* which ride on and are supported by the track cable. At the tail tower, the travel line passes through a sheave and then is returned to the head tower; at the head tower both parts of the travel line (the one from the carriage and the one from the tail tower) are passed over sheaves and carried down to the hoist house. There the loop of the cable is wrapped for several turns around the gypsy drum that controls the movement of the cable. Typical cable diameter for an 8-cu-yd-capacity cableway is between $1^1/_8$ to $1^1/_4$ in.

The operating cable that suspends the hook from the carriage and controls the vertical movement of the hook is called the *hoist line*. One end of the cable is anchored to the tail tower. Its span from tail tower to the carriage is supported by the slackline carriers. At the carriage it is reeved through a sheave down to and through the bucket hook's sheaves, thence back to the carriage and through another sheave. Its span from the carriage to the head tower is again supported by slackline carriers. At the top of the head tower, it is reeved through a sheave and down to the hoist house where the remainder of its length is wrapped on a large-diameter, large-capacity drum. For an 8-cu-yd-capacity cableway this cable diameter will vary from 1 to $1^1/_4$ in.

At the tail tower, a cable for tightening the track cable is reeved through blocks to get a multipart line. Typical diameter for this line is $1^5/_8$ in.

If a mast is used as the fixed tower for a radial cableway, then the mast is guyed with two backstays of approximately $3^1/_2$-in.-diameter cable and with two forestays of approximately $2^1/_2$-in.-diameter cable. The backstays must have the capacity to resist a stress equal to or larger than that in the track cable. For many applications, the stays are made from used track cable.

To prevent excessive wear on all operating cables, cableways are equipped with large-diameter sheaves and drums. The diameters of these sheaves and drums are determined by the cableway manufacturers. As an indication of their sizes, cable manufacturers recommend that when heavily loaded, fast-moving ropes are used on fixed equipment, sheaves and drum diameters should be 800 times the diameter of the wires in the outer course of the cable.[1]

Cable life varies greatly between cableways. The track-cable life shortens as the cableway span increases and the distance from the head tower to load pickup point decreases. Other factors influencing cable life are the diameter of sheaves and drums, method of cable lubrication, load on the cable, frequency of use, type of cable, and the particular cable manufacturer. W.M. Bateman, in the *Handbook for Heavy Construction*, gives the following life for cables: life of a track cable is between $1^1/_2$ and 2 million tons of payload; travel ropes will place 200,000 cu yd of concrete; and hoisting ropes need replacement after they have placed from 40,000 to 60,000 cu yd of concrete.[2] These recommendations are for standard conditions. In some instances, cables have remained in use much longer, and for long-span cableways, the track cable has required more frequent replacement.

To provide more flexibility in speed control and to give better power-consumption characteristics, direct-current motors with Ward-Leonard controls are preferred over alternating-current motors for driving the cable drums. Motor generators are used for conversion of alternating current to direct current. To permit the carriage to travel horizontally at the same time that the concrete bucket is hoisted or lowered, separate motors supply power to the travel-line drum and the hoist-line drum. Motor sizes vary with speeds of operation. As an example of motor sizes, for an 8-cu-yd cableway with an average hoisting speed of 640 ft per min, an 800-hp motor is required to drive the hoisting drum. If this hoisting speed for a loaded bucket is increased to 950 ft per min, two 625-hp motors are used. Similarly, with a carriage travel of 1,600 ft per min, a 500-hp motor is used to control the travel line; if the carriage speed is increased to 2,100 ft per min, a 625-hp motor is used.

The travel line, which controls and moves the carriage, is operated either by a gypsy of approximately 72 in. in diameter or, preferably, by a multiple-sheave unit. These units are equipped with a weight-set, air-release brake. A gypsy-type drum for driving the travel line requires that the several wraps of wire rope slide on the drum face as the line inhauls or outhauls. This results in a wedging of the incoming line between the flange and the wraps already on the drum, which causes rapid line wear. A *multiple-sheave* drive for the travel line eliminates the sliding and wedging that are so destructive to wire rope. This drive consists of a conventional reversible motor with a pinion bull gear and brake arrangement, which applies power to a drum on which there are typically six separate grooves sized to fit the travel line. A six-groove idler drum is mounted on the same frame, with the axis of this drum tilted slightly from horizontal. The travel line is reeved in six wraps around the two drums, without taking a full turn around either drum. The tilted axis of the idler drum ensures that fleet angle within the drive unit is eliminated. This refinement reduces wire-rope wear.

The hoist line that controls the bucket hook is spooled on a large-diameter, large-capacity drum also with a weight-set, air-release brake. Typical dimensions of the drum are 84 in. in diameter and 84 in. long, depending on cableway size and cableway span. The drum is grooved to fit the diameter of the hoist line, and the large cable capacity on the drum allows spooling of all the take-in of the hoisting rope on the first wrap around the drum, thus reducing cable wear.

The travel line and the hoisting line are maintained in elevation across the cableway span by slackline carriers that suspend them from the track cable. Two types of slackline carriers are used. The two carriers at each end of the cableway, which are located outside the area of normal carriage travel, are only moved when the carriage is brought into the tower for repair or servicing. Because of this, these two carriers are of simple design and are often called *old-man* slackline carriers. The remaining carriers are known as *proportional* slackline carriers because of their spacing along the cable and their more complicated design due to their operation in conjunction with the carriage travel. Present day cableway-carriage speeds would not be feasible without the use of proportional slackline carriers. Older machines used slackline carriers moved by the carriage and spaced by a *button line*. In order to avoid rapid destruction of the slackline carriers, it was necessary to slow the carriage as each slackline carrier was picked up and another dropped off the carriage. In contrast to this, the proportional slackline carriers ride freely on the track cable, propelled by gearing which is actuated by the carriage travel line. When the cableway is pouring concrete, they do not come into contact with the carriage.

Proportional slackline carriers are purchased in sets. A typical set might consist of ten slackline carriers: five for the inhaul side of the span and five for the outhaul side. The carrier nearest the head tower and the carrier nearest the tail tower, in this case, would each be geared to move along the track cable at one-sixth of the carriage speed, the next pair at two-sixths of the carriage speed, and so forth, so that the two carriers nearest the carriage would be geared to move along the track cable at five-sixths of the carriage speed. Thus, the carriers are spaced at constantly varying distances as the carriage travels along the cableway span. None are ever in contact with the carriage except when the carriage is moved all the way to the head tower or all the way to the tail tower. When this occurs, contact between carriers disengages spring-loaded clutches. As the carriage moves away from the towers, the clutches reengage, and the carriers again space themselves along the track cable.

The carriage riding on the track cable has a large number of wheels, from 4 to 16, depending on the size of the cableway. Multiple wheels allow smooth travel of the carriage and reduce track-cable wear by distributing the weight and preventing sharp deflections. The carriage contains provisions for picking up and releasing the slackline carriers and contains two large-diameter sheaves over which the hoisting line passes for control of the concrete bucket. The carriage used on the cableway for construction of Guri Dam is shown in Photograph 14 in this chapter.

There are many makes of concrete buckets available and suitable for cableway use. The main difference in buckets is in the method used to discharge the concrete. This can be done by tripping the release gate by hand, or by using an air-activated gate, or by using a gate activated by hydraulic pressure.

The number of buckets in use at any one time depends on specification requirements. Some specifications do not allow the transfer of concrete from one container to another. With this type of specification, concrete must be transported, in the same bucket, between the mix plant and the cableway and then to the point of pour. The loaded concrete bucket is transported from the mix plant to the point of transfer under the cableway on special self-propelled bucket cars or on trailer-type bucket cars pulled by diesel locomotives. The bucket cars have space for two to four buckets. Along three sides of the bucket spaces are runways at the proper elevation for the operator to hook and unhook the buckets from the cableway. One space on the car is always left empty so that the cableway will have an empty slot for placement of the empty concrete bucket, when the cableway carriage returns it to the point of transfer. The procedure for bucket pickup is: the cableway drops an empty bucket into the empty bucket space on the bucket car and then picks up a loaded bucket, transports it to the pouring area where the bucket is dumped, and then returns this empty bucket to the bucket car. This procedure is repeated until the last loaded bucket has been hooked onto the cableway. When this has been done, the bucket car returns to the mix plant, while another bucket car takes its place under the cableway. This pouring procedure makes it necessary to hook and unhook every bucket from the cableway, either manually or by an automatic hook. Photograph 15 shows manual bucket changing for the cableway used to construct Chief Joseph Dam, Washington. This photograph also shows the old-type slackline carriers used with a button line.

Other specifications allow the use of a transfer car. The transfer car is constructed with one or more compartments, each of which is

Photo 15 Changing the cableway bucket using a bucket car, Chief Joseph Dam cableway, Washington. (Washington Iron Works.)

capable of holding the same volume of concrete as the bucket used on the cableway. Each concrete compartment is equipped with fast concrete-dumping chutes or container dumpers for quick loading of buckets at the bucket dock. The transfer car can be either a self-propelled unit or a trailer unit pulled by a diesel locomotive. The transfer car is loaded with concrete at the mix plant and then travels to a position above a bucket dock located alongside the transfer track, under the cableway. Here the concrete is dumped from the transfer car into the concrete bucket which has been spotted on the bucket dock by the cableway. The bucket dock is constructed lower than the transfer track, at an elevation that will place the top of the concrete bucket at the proper height to receive the concrete from the transfer car. The dock must be of rigid construction to resist the shock caused by the abrupt stopping of the bucket on the bucket dock. Rubber tires are often used along the dock to cushion this shock. Hauling concrete from the mix plant to the cableway by the transfer-car method saves time in comparison to the bucket-car method because it is not necessary to hook and unhook every concrete bucket to the cableway. Photograph 14, in this chapter, shows the bucket dock, transfer track, and the transfer car used on the cableway at Guri Dam.

Whether a concrete transfer car is used with a bucket dock or whether bucket cars are used, a concrete transfer track is required. Rail-mounted equipment is usually used to transport concrete from the mix plant to the cableway, since it is more efficient and less expensive to operate for this service than is rubber-tired equipment. The transfer track and the bucket dock, if a transfer car can be used, must be of sufficient length that the bucket may be spotted under the cableway when the movable cableway towers are in any position. In order to permit fast switching of the transfer of bucket cars, the transfer track is double-tracked with switches at frequent intervals.

Layout

Layout and design of a cableway are, in the majority of cases, the products of a joint effort between the contractor's engineer and the cableway manufacturer. The contractor's engineer must choose between radial- and parallel-type cableways; the runways must be located in plan and the cableway span must be determined; and the sag must be computed in order to determine the tower height. Next, he must select the size of cableway, i.e., 12, 8, or 4 cu yd, and he must determine how many cableways will be required. This information is then furnished to the cableway manufacturers who will furnish quotes on the cableway or cableways. If the contractor is the low bidder and purchases the cableway, then the manufacturer will design and furnish towers, cables, carriage, hook, motors, drums, sheaves, etc., and recommend the weight and position of counterweights.

For estimate preparation, the engineer will have the cableway manufacturer's quotation for the cableway equipment. He must estimate the cost of constructing the tower runways, including any required trestles; the cost of the trackage; the cost of the concrete in the tower counterweight; if a fixed tower is used, the cost of anchors and the tower foundation; and the cost of erecting the cableway towers, cable, cableway machinery, and parts. If the contract is awarded, the portion of cableway design usually handled by the contractor's engineer is the design of any trestle required in the cableway runway, the design of transfer track, and, if used, the design of the bucket dock.

To determine the suitability of cableways, to decide between radial or parallel cableways, and to secure other estimating data, it is necessary to lay out many possible solutions in plan and elevation. The plan should show the location of the towers and runways; the elevation should show the downstream view of the dam, elevation of the top of the dam, the necessary bucket clearance, cableway span, cable sag, and required tower height. The topography of the dam abutments and the curvature of the dam influence the choice between a radial or a parallel-track cableway. It is necessary to lay out both types with different span distances and with different runway elevations to determine the best solution. The solution selected should be the one which has the shortest span and the shortest towers and which requires the least amount of expense in constructing the runways. For straight concrete dams, radial cableways are usually more economical because of the savings in using a fixed tower, compared to having movable towers. However, if large placing capacities are required, more parallel cableways can be used than radial. Traveling-tower runways may be located in a cut, on a fill, or, if the slope of the ground makes it necessary, on a trestle. An example of runways on a cut and on a trestle are the runways built for the cableways used to construct Detroit Dam, Oregon, as shown in Photograph 13 in this chapter.

The movable towers should have sufficient travel to service all the large pours in the dam. It is not necessary to provide cableway coverage for small pours such as those in the powerhouse, the downstream spillway aprons,or small blocks located at the dam abutments when the tower runway extensions necessary to service these pours require a large amount of fill or excavation or require the construction of a trestle. The expense of runway extensions is not justified and other means of concrete transport and placement should be used.

In preparing the elevation layout of the cableway, it is accurate enough to expect a sag in the track cable of 6 percent of the length. (This will probably be changed to a more accurate figure by the cableway manufacturer in his final layout.) From the bottom of the sag to the top of the concrete, approximately 60 ft should be allowed for hook and bucket clearance. If the cableway tower heights become critical, the sag can be reduced when concrete is being placed in the top of the dam by using a smaller-capacity bucket during the topping-out process. In determining movable-tower height, the minimum height of a head tower should be approximately 60 ft to accommodate the machinery, cable blocks, and counterweight. Head-tower height is also required to furnish concrete-bucket clearance at the concrete-transfer point, to provide hook clearance at yarding areas usually located next to the head tower, and to provide hook clearance when the carriage is brought into the head tower for servicing.

The minimum height required for traveling tail towers is approximately 30 ft. This is illustrated by Photograph 16, which pictures the tail tower used at Yellowtail Dam. Movable towers are preferably under 150 ft in height, since taller towers create increased operating troubles. When it has been necessary to use taller movable towers, they have been used in spite of their disadvantages. In the construction of Guri Dam in Venezuela, it was necessary to use a traveling head-tower height of 210 ft.

Drawing 1B, in this chapter, of a radial cableway for Denny Dam shows the relationship between cable sag, bucket clearance, tower height, and runway elevations. In this layout, the tops of the tail and head towers are at the same elevation. This is not a requirement and if site conditions make it necessary, cableways can have towers with their tops at different elevations so long as bucket clearance is provided for the full length of the dam. The dam shown is of such a size that only one cableway is required and the most economical type was a radial cableway.

After the cableway tower heights have been determined, the tower widths can be computed. For traveling towers, the tower width determines the distance between the two runway tracks. Towers are designed so that the resultant of the stress in the track

Photo 16 Tail tower for one of the Yellowtail Dam cableways, Montana. (Washington Iron Works.)

cable and the weight of the counterweight pass through the sloping leg of the traveling tower and through the trucks supporting this leg. For preliminary layout purposes, the width of a cableway tower can be taken as 55 or 60 percent of its height. For final estimating purposes, tower widths and tower weights should be secured from the cableway supplier. Again, for preliminary estimating purposes, the weights of towers and trucks are as follows:

Tower and Running Gear Weights for One 25-ton Cableway

Height of tower, ft	Traveling tower, weight in tons	Fixed-framed tower, weight in tons
30	39	12
40	52	16
50	65	20
60	78	24
70	91	28
80	104	32
90	117	36
100	130	40
110	143	44
120	156	48
130	169	52
140	182	56
150	195	60
160	...	64
170	...	68
180	...	72
190	...	76
200	...	80

The weight of a fixed tower common to two cableways is 1.3 times the fixed tower weight for one cableway.

The weight of a fixed tower common to three cableways is 1.9 times the fixed weights for one cableway.

The weights of A-frame fixed towers is 0.64 times the weights of framed fixed towers.

The weights of mast-type fixed towers is 0.33 times the weights of framed fixed towers.

Running-gear weights for movable towers vary with the stress in the track cable. The running gear will weigh 36 tons for a track cable stress of less than 250,000 lb, 45 tons for stresses between 250,000 and 350,000 lb, and 50 tons for stresses over 350,000 lb.

The counterweight volume and location can be secured from the cableway supplier or can be computed by balancing the moment resulting from the cable stress and its moment arm around the center line of the inside tower track, with the moment computed from the weights of the counterweight and cableway tower and their moment arms around the same point. This computed weight is then increased so that the tower will not overturn when an occasional heavy load occurs on the cableway. The counterweight is usually located directly over the outside cableway tracks. The center of gravity of the tower and the counterweight must fall on the center line of the outside track or fall between the two tracks so that the tower will be stable before the track cable is erected and when the track cable is being replaced. It is important to check this weight resultant when cableways are being altered for reuse at locations other than their original ones. Photograph 13, in this chapter, shows the shoring required to support the forms for concrete counterweights used for two movable head towers.

The cableway span should be kept as short as possible. The longer the cableway span, the more wear there will be on the cables, the more frequently the cables must be replaced, and the more the work will be delayed due to lost time during cable replacements. Cableways have been used with long spans such as the 2,912-ft span used to pour Guri Dam, Venezula, and the 2,900-ft span used on the cableways for Dworshak Dam in Idaho, but frequent replacements of the track, traveling, and hoisting cables were necessary.

If more than one cableway is required, their towers may be located on the same or separate runways. If separate runways are used, they should be located so that one cableway will not cross in the path of another. Separate tower runways are warranted when the slope of the ground makes the construction of long runways a major construction problem.

After the type of cableway is selected and its runways are located, the track used to transfer the concrete from the mix plant to the pickup point, underneath the cableway, can be located. The cableway-layout drawing in this chapter, shows that as the loaded carriage travels along the track cable, it forms a load path that steepens in slope at both ends. Formulas for computing this load path are given in publications of cable manufacturers.[1] If the concrete transfer track is located under the steep part of the load path, there will be a sharp deflection in the track cable when a loaded bucket is handled in this vicinity. This will cause severe wear on the track cable and cause additional wear on the travel line. Therefore, the preferred location of the transfer track is where the load path levels out. A rule of thumb that can be used to determine the minimum horizontal distance the transfer point should be from the top of the cableway tower is 10 percent of the span; preferably, it should be a greater distance than this. The length of the transfer track should be sufficient to service the cableways when its movable towers are at each end of their travel. The elevation of the transfer track should be as close to the center of the mass of concrete as possible, since this location will result in the least amount of vertical bucket travel when concrete is being placed. Depending on the cableway layout and the type of dam, a transfer-track location at this elevation may be impractical when it interferes with the placement of concrete in a major dam monolith. If this interference occurs, the concrete of this block cannot be poured until the remainder of the dam concrete has been placed. Then, since a portion of the transfer track must be removed to pour this block, its placement may be costly and time-consuming because every bucket of concrete must be transferred from pickup point to the pour by movement of the cableway towers.

The mix plant should be located so that the elevation of its loading point is the same as that of the transfer track and as close to the cableway location as possible. This will allow short and level train travel from the mix plant to the transfer point under the cableway. Because of these different requirements, trial locations for the transfer track and mix plant must be made to determine which location will be most advantageous.

Cableway Capacity

The number of cableways required for a concrete dam is determined by computing the mass-placement rate required to keep the job on schedule and dividing into this the pouring capacity of one cableway. The pouring capacity of each cableway depends on the bucket size, the cycle time required to pour one bucket of concrete, the percentage of time the cableway will be pouring concrete, the hours worked per day, and the days worked per week. The cycle time of a cableway is the time required for vertical and horizontal travel and the fixed time required at the transfer and dumping points. Cycles will vary between dams and between different areas of the same dam due to the variations in travel distance. The use of a computer for determining cycle time saves considerable drudgery. To arrive at the average cycle without using a computer, it is necessary to compute individual cycles for different dam elevations and then find the average by weighing them against the yardage poured for each cycle. To determine the cableway-cycle time, the traveling and hoisting speeds of the proposed cableway must be known.

The following is an example of how cableway-cycle times are computed. This cycle is for a cableway with a horizontal travel speed of 2,100 ft per min and a vertical travel speed of 950 ft per min. It is assumed that the center of mass of the dam is located 1,500 ft horizontally and 225 ft below the bucket pickup point and that the accelerating and retarding time for both horizontal and vertical travel is 5 sec.

Cableway-bucket Cycle

Load bucket			30 sec
Hoist clear of bucket dock			5 sec
Horizontal travel:			
Accelerate	5 sec	121 ft	
Travel	36 sec	1,258 ft	
Retard	5 sec	121 ft	
Total	46 sec	1,500 ft	
Vertical travel:			
Accelerate	5 sec	38 ft	
Lower	10 sec	149 ft	
Retard	5 sec	38 ft	
Total	20 sec	225 ft	

Therefore, horizontal travel controls:

Horizontal travel from bucket dock to pour	46 sec
Spot and dump bucket	40 sec
Return to bucket dock	46 sec
Land bucket	10 sec
Total	177 sec
Change to min and sec	2 min 57 sec
Adjust to	3 min

The percentage of time that the cableway will be available for concrete pour depends on the amount of lost cableway time and the amount of cableway hook time required for other operations. Cableway lost time occurs when the cableway is being serviced, when cable replacements are necessary, and when the pour is switched from one block to another. As an example of this cableway lost time, the time required to change the track cable of a cableway is approximately one week. Much less time is required for replacement of the travel or hoisting cables. Cableway hook time is used for foundation cleanup, for concrete cleanup, for placing forms on blocks and moving them between blocks, for placing reinforcing steel in the blocks, and for setting gates, penstocks, and other materials.

The use of the cableway hook for applications other than concrete pouring becomes critical on jobs where only one cableway is provided. To relieve the cableway of this type of service,

Photo 17 Truck crane on a Bailey Bridge used to assist the cableway, Guri Dam, Venezuela.

provisions may be made for the use of other hooks. Photograph 17 shows a truck crane operating on a Bailey bridge located on the upstream side of Guri Dam. This truck crane was used to relieve the cableway of some of the hook time required for servicing operations.

When two or more cableways are provided for a project, one may make all the service lifts and pour concrete part of the time, while the others are used solely for concrete pouring. If this is done, the average nonpouring time for the cableway system will still be approximately the same as that for one cableway. For estimating purposes, it can be assumed that cableways will be placing concrete only 75 percent of the time.

After determining the average cycle time, the percentage of time the cableway will be pouring concrete, the number of hours that will be worked per day, the number of days that will be worked per week, and the bucket capacity of the cableway, the cableway's pouring capacity and the required number of cableways can be computed. Using the previously computed 3-min cycle time and assuming that a monthly placement capacity of 100,000 cu yd is required, the engineer can determine the number of cableways that must be installed in the following manner:

Bucket size	8 cu yd
Average cycle time	3 min
Cycles per hr	20 cycles
Average hourly pouring rate	160 cu yd
Hours worked per day	21 hr
Pouring hours per day (75% of time)	15.75 hr
Daily average pour (15.75 hr)	2,520 cu yd
Weekly average (5-day week)	12,600 cu yd
Monthly average (4.33 weeks)	54,600 cu yd
Number of cableways required	2

The assumed 75 percent availability on concrete pour is an average figure to use for cableway operation. Similarly, 50,000 cu yd per month for a 5-day week is a realistic figure to use for average concrete placement for one 8-cu-yd cableway. Pouring capacities of 8-cu-yd cableways have exceeded this when travel distances were short, mechanical bucket dumping was used, high speed cableways were installed, and cableway control was good. The following record pours were made at Bullards Bar Dam, California, where two high-speed cableways were employed.[3] These cableways had the same traveling and hoisting speeds as those used in the previous computations.

Calendar-month maximum production (7-day week)	Over 180,000 cu yd
Weekly maximum production (7-day week)	Over 40,000 cu yd
Daily maximum production (3 shifts)	Over 8,000 cu yd

If this production is adjusted to give the production of one cableway for a 5-day week, the results are:

Calendar-month maximum production	64,290 cu yd
Monthly average hourly pour	138 cu yd (approx.)
Weekly maximum pour	14,385 cu yd
Weekly average hourly pour	138 cu yd (approx.)
Daily maximum pour (21½ hr)	4,000 cu yd
Maximum hourly pour	186 cu yd

This monthly maximum production is 1.18 times the average production computed in the sample computation previously shown. This checks quite closely since the average monthly production should be less than the maximum monthly production.

Production during record months is always greater than the average monthly production, since record months are achieved when the concrete has been placed in each block in the dam's foundation, when there are no major cableway shutdowns, and when the cableway is not needed for handling gates, penstocks, or other structural items. Similarly, the average monthly production rate is achieved only when enough dam monoliths have been started so that adequate pouring area is available, and prior to starting the placement of the narrow pours required to top out the dam. Start-up monthly production should be reduced for several months to allow time to place concrete in all the dam monoliths in the base of the dam. This is known as "getting off the rock." Production at the end of the job will be reduced for several months when only small pours are left in the top of the dam. This slow production at the start and end of the concrete-placement period will cause a concrete-placement curve to take an S shape.

Effect of Bucket Cycle on Mixing-plant Capacity

To determine the required capacity of the mixing plant, it is necessary to review the cableway-bucket cycle. By examination of the previously computed bucket cycle one readily sees that when pours are made adjacent to the transfer track, 36-sec horizontal travel in each direction can be deducted from the cableway cycle, thus giving a cycle time of 1.80 min and a pouring rate for one cableway of 267 cu yd per hr, or an increase in production of 67 percent. When pours are made at the other end of the dam, and if one assumes that 1,258 lin ft must be added to the carriage's horizontal travel, the bucket cycle would be increased by 36 sec in each direction, giving a total cycle time of 4.20 min or a pouring rate for one cableway of 114 cu yd per hr, which is a decrease in production of 29 percent. To maintain the average production rate of 160 cu yd per hr for one cableway, the rate of pours made adjacent to the transfer track must exceed 160 cu yd an hr to compensate for the reduction in pouring rates when distant blocks are poured. In order to pour at this faster rate, the mixing plant and concrete-transportation method must be of capacities suited to accommodate this faster rate of pour. Also, the mixing plants must have sufficient capacity to permit a mixer to be shut down occasionally when servicing or relining is required.

The capacity of the mixing plant to be installed is a matter of engineering judgment. Except in special applications, the engineer can seldom justify installing a mixing plant of such capacity that the placement equipment can place concrete at its maximum rate when pours are adjacent to the transfer point. Mixing plants are often installed with a maximum capacity that permits the placement equipment to place concrete at 25 percent over the average hourly pouring rate. Then the concrete superintendent may complain that there is not enough mixing-plant capacity to allow him to pour at maximum capacity. However, a 25 percent increase is enough to permit the cableway to maintain the average placement rate, for the percentage decreases in production when making distant pours seldom equal or exceed this. If this 25 percent factor is used to determine mixing-plant capacity in the cableway production example, the mixing plant for the two cableways with an average pouring rate of 320 cu yd per hr would have a maximum capacity of approximately 400 cu yd per hr. This checks closely with the mixing plant installed for two cableways at Bullards Bar Dam, where six 4-cu-yd mixers were installed in one plant. Since it took $3^1/_2$ min to charge and discharge the six mixers, the total installed capacity was 411 cu yd per hr. A rule of thumb for determining mixer capacity is that the bucket capacity of the placement equipment should be multiplied by $1^1/_2$ to determine mixer capacity. If this rule is applied for two 8-cu-yd cableways, the mixing plant should have a capacity of 24 cu yd, or six 4-cu-yd mixers would be required. When both cableways and cranes are used to place concrete simultaneously, combined bucket capacities should be used to determine the required mixing plant capacity.

TRESTLE-CRANE SYSTEM

The trestle-crane method of dam concrete placement consists of pouring concrete from buckets which are swung into position over the concrete pour by cranes supported on the deck of a trestle. The trestle is located parallel to the concrete dam or the concrete section of any type of dam with an alignment and deck elevation that allows the cranes to have boom coverage of all the concrete pour areas. Photograph 18 illustrates this method of mass-concrete placement.

Three types of cranes are used: gantry cranes traveling on rails laid on each side of the trestle, rail-mounted hammerhead cranes, or self-contained cranes such as crawlers or truck cranes supported by the trestle deck. The use of crawler- and truck-mounted cranes is increasing because larger-capacity units are continually being developed. The cranes not only pour concrete, but, like cableways, they perform service lifts such as handling forms, handling lifts required for concrete cleanup, placing reinforcing steel, and handling embedded material. The trestle- and crane-placement method has a flexibility in placement capacity, since that capacity can easily be increased by placing more cranes on the trestle. Rail-mounted cranes are preferred, since they travel on a level track and maintain a constant roller-track position so that the swing of the boom is always level. Futhermore, their booms are more rigidly supported than those of crawler units and provide steadier control of the bucket,

Photo 18 Trestle and cranes for concrete placement, Libby Dam, Montana.
(Washington Iron Works.)

which allows rail-mounted crane units to pour concrete faster than rubber-tired or crawler units.

Rail-mounted cranes used for pouring dam concrete from trestles have been gantry cranes, with a few exceptions. Hammerhead, T-shaped cranes were used to pour concrete at Grand Coulee Dam, and the pouring rates for these cranes were excellent. However, because of their high initial cost and the large cost involved in their transportation, erection, and dismantling, new units of this type have not been purchased for use on other dams. Upon completion of Grand Coulee, some of the hammerheads were used in industrial plants. Others were used on Friant Dam, and later these were shipped to India for use on Bakra Dam and are no longer available for dam construction.

Crawler or rubber-tired cranes are used on trestles to pour small dams when such cranes can first be used for other purposes and only later for concrete pouring. For small concrete yardages, the increased placement cost caused by using these dual-purpose cranes is less than the increased capital write-off that would result if gantry-type cranes were purchased. Irrespective of the type of crane used, it is the practice to limit the maximum concrete bucket size to 4 cu yd. Cranes can handle larger buckets at shorter pouring radii, but whether the use of larger buckets can be justified depends on the particular job condition. Concrete buckets can be hooked and unhooked from these cranes manually or automatically. If done automatically, the hook is opened and closed by the pressure of compressed nitrogen, which is activated when the bucket bail contacts the crane hook. The compressed nitrogen is supplied from cylinders carried on the hook, and these cylinders are changed when the nitrogen in them is consumed. Concrete is dumped from the buckets by manually operated gates or by gates that open automatically when pressure is applied. If air-operated gates are used, air is supplied by a compressed-air hose, which is attached to the bucket at the pour area. Pouring procedure requires the crane to drop an empty bucket onto the bucket car, disengage the hook, hook onto a loaded bucket, and move it into pouring position. The concrete is dumped from the bucket, and the empty bucket is returned to the bucket car. This pouring cycle is repeated until the pour is completed.

Layout

If the trestle concrete-placement method is selected for a particular dam, the engineer must select the proper location and elevation of the trestle, the length of the trestle, the number of cranes, the size of cranes required to give adequate boom coverage, the location of the mix plant, and the method of concrete haul. To aid in this selection, a trestle-crane layout drawing is prepared similar to Drawing 2.

Prior to preparing a layout, it is necessary to secure the gantry widths, gantry heights, and boom capacities of different sizes of gantry cranes. Many sizes of gantry cranes suitable for concrete placement are available. As an example of the sizes available, following is a list of basic data on five gantry-crane models available from American Hoist and Derrick Co. Each of these cranes is equipped with a Model 180A three-drum hoist, which furnishes a concrete-bucket hoisting speed of 220 ft per min. Travel speed of the cranes is 125 ft per min on a level track. Swing speed is 1 rpm. Counterweights consist of rails to form the counterweight tank bottom, and the remainder consist of steel punchings or billets. The radius given is from the center line of the crane. The maximum corner load varies with track gauge and gantry height. The counterweight varies with the length of the crane's tail. Electrically powered cranes are usually used, but in some cases, when electricity has not been available, diesel-driven cranes have been used.

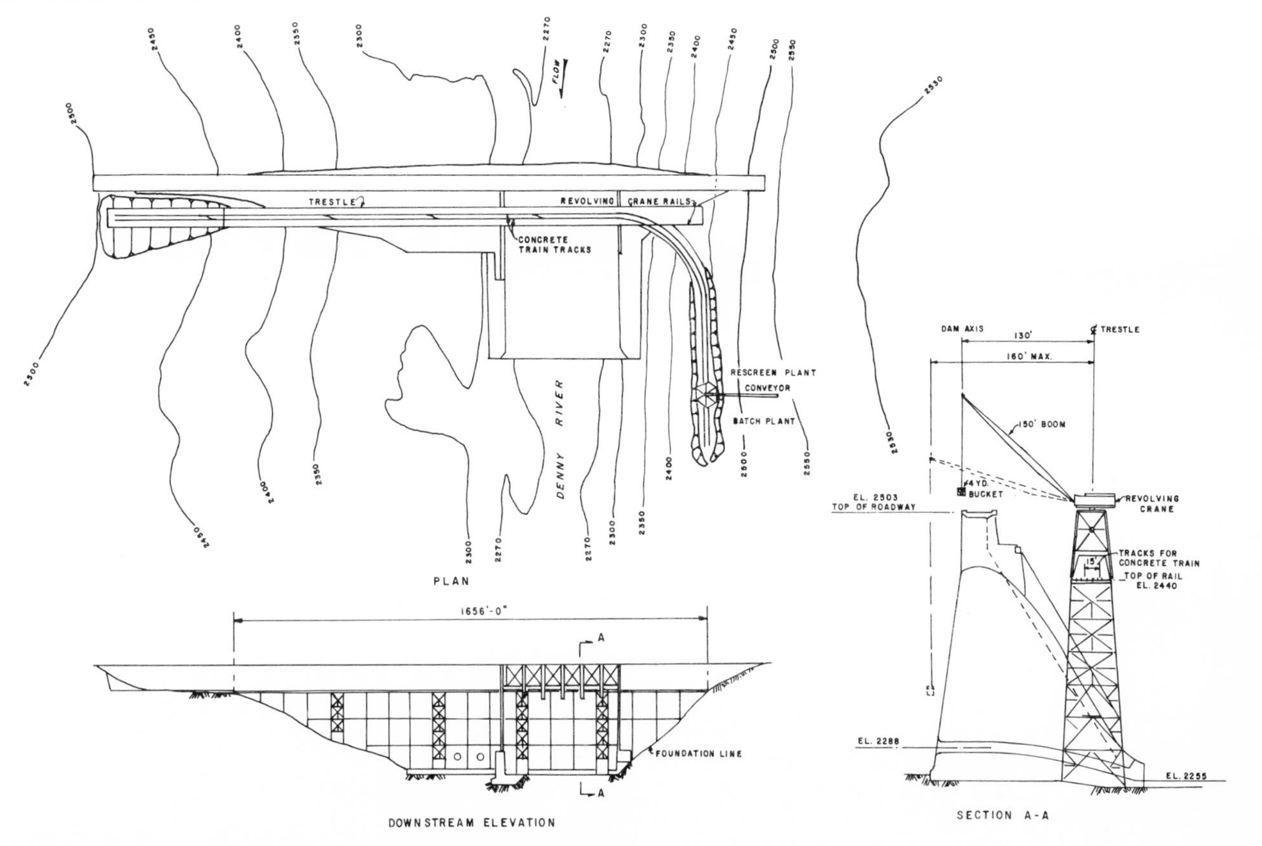

Drawing 2 Layout of a trestle-crane concrete-placing system for Denny Dam.

	Crane model				
	101	152	203	254	305
Maximum placing radius, ft, using a 4-cu-yd concrete bucket plus 25% impact	50	90	150	180	210
Maximum boom length, ft	120	140	160	180	200
Gantry height, ft	39–66	42–82	42–95	46–117	49–120
Track gauge, ft	16–32	20–36	24–40	28–44	32–48
Counterweight, tons	32–38	50–65	75–100	100–130	120–175
Maximum corner load, tons	46–52	70–82	110–118	142–164	202–228
Weight of gantry rail, lb/yd	100	175	175	175	175

A trestle-crane layout should give consideration to trestle cost compared to crane cost. If trestle height and length can be saved with gantry cranes equipped with high gantries and long booms, it is more economical to use the higher gantries. Their cost will generally be less than the cost of raising the trestle and increasing its length to allow the use of smaller cranes.

The most desirable location and elevation of the trestle deck places the concrete buckets, when they are on the bucket car, as close to the center of mass of the concrete as possible. This location involves the least amount of crane boom swing and the shortest bucket-hoisting distance required to position the concrete buckets over a pour. When narrow concrete dams are constructed, crane coverage can be secured without locating the trestle within the area to be covered with concrete. With narrow dams, when water is not to be stored behind the dam, a trestle located upstream of the dam with a height level with the center of mass of the concrete will result in the least amount of crane swing and hoisting. If, water is to be stored behind the dam before the dam is completed, the use of this area for a trestle is impractical, and the trestle must be located downstream from the dam.

Wide dams often cannot be poured with cranes located on trestles either upstream or downstream of the dam. The trestle must then be located above the dam's sloping downstream face, with its center line as close to the center of the dam's bottom width as possible. The concrete beneath the trestle's deck can be poured by drifting the concrete bucket or by pouring through openings in the trestle's deck. These openings may be framed into the deck, or the deck may have removable trestle deck sections.

On an exceptionally wide dam, it may be necessary to provide two trestles in order to furnish crane coverage for the full dam width. This was the case at Grand Coulee Dam where the trestle decks were located at different elevations. When the trestle legs are located within the dam area, they are encased in the dam's concrete as the dam is raised. After the dam is completely poured, the legs can be burned off in recessed pockets in the concrete formed around the legs for this purpose. The remainder of the trestle can then be removed and the pockets filled with concrete to furnish a smooth concrete surface.

Trestle length is controlled by the dam's length and the topography of the abutments. Often the trestle will intersect the abutment banks before sufficient trestle length is secured to allow the cranes to service the pours in the end blocks of the dam. Solutions to this problem are to excavate trestle-track extensions in the dam abutments, to raise the trestle, or to pour the end monoliths with other equipment. The topography and the volume of concrete in the abutment blocks will determine the solution to use for any particular dam. Trestle width is controlled by the width of the gantry, or, if gantry cranes are not used, the trestle must be wide enough to permit concrete-haulage equipment to pass by the placement cranes.

To properly estimate the cost of a trestle and crane system, it is necessary to know the volume of excavation required for the trestle foundations, the concrete yardage in these foundations, the weight of steel in the trestle, the board feet of wood decking, the quantity of rails and ties needed for the gantry and transfer tracks, the cost of the electrical distribution system, the cost of "baloney cable" (electric feed cable) needed for the cranes, the required number of cranes and their cost, and the cost of crane erection and crane counterweights. For preliminary estimating purposes, the steel in a trestle will run approximately 1.25 tons per lin ft of trestle, when cranes with 44-ft gantries and 150-ft booms are used. This tonnage factor increases with crane size until it becomes approximately 1.35 tons per lin ft of trestle when cranes with 84-ft gantries and 180-ft booms are used.

Similar to a cableway, the most economical method of transporting concrete from the mix plant to the placement cranes is by rail, since rail equipment causes less traffic and less congestion on the trestle than if trucks or truck-trailer units are used. Therefore, concrete is transported from the mixing plant to the placement cranes in rail-mounted bucket cars. The maximum size bucket cars used to date have five bucket spaces.[4] A rail-haulage system requires that the mix plant be located with its base at trestle height. If such a location is impossible, then truck-trailer units or trucks should be used to haul the concrete.

Capacity

The capacity of a trestle-crane system is determined by the number of cranes provided for the concrete pour. Small dams are often poured with one gantry, but these are exceptional cases, and hook coverage is supplemented with other cranes, such as truck cranes, to take care of the service lifts. Dams containing large volumes of concrete may require a large number of placement cranes in order to meet the scheduled production. As an example, for concrete placement in Grand Coulee Dam, enough cranes were installed to set a daily concrete-pouring record of almost 22,000 cu yd. This high rate of pour was made under less favorable conditions than those now existing; since refrigerated concrete was not used, the concrete had to be placed in small 50-ft-square blocks. To accurately determine the number of cranes necessary to meet schedule requirements, their bucket cycle time must be computed and used to establish the crane's pouring rate. Bucket cycle time is composed of the time required to hook the bucket, swing and hoist the bucket, spot the bucket, dump the bucket, and return the bucket to the hauling unit. For preliminary estimating purposes, one can assume that a crane can place approximately 100 cu yd of mass concrete per hr and that the crane will be pouring concrete 80 percent of the time. These capacities are slightly higher than the placement rates achieved at Libby Dam, Montana.[4] However, since only 24 cu yd of mixing capacity was installed, the placement capacity of the six cranes would be limited to the maximum production achieved from the mixing plant. These placement yardages are conservative compared to those given by T.G. Tripp in the *Handbook of Heavy Construction* as 144 cu yd per hr.[2] Placing capacity will be reduced when the crane is placing other than mass concrete. Like a cableway, a gantry crane cannot place concrete continuously since it must be serviced and make other lifts, but it can pour concrete a greater percentage of the time because there will not be any track-cable replacement delays.

After computing the bucket time cycle for the particular operating conditions to be encountered, the number of cranes required to meet a scheduled rate of pour can be determined in the following manner:

Size of bucket, cu yd	4
Average bucket cycle time, min	$2^1/_2$
Average pouring rate, cu yd per production hr	96
Percent of time crane will be pouring concrete	80
Average pouring rate, cu yd per elapsed hr	77
Number of hr worked per day	21
Cu yd placed per crane per day	1,617
Number of days worked per week	5
Cu yd placed per crane per week	8,085
Cu yd placed per crane per month	35,007
Number of cranes required to pour 100,000 cu yd per month	3

As explained when cableways were discussed, concrete-placement equipment cannot maintain its computed average hourly placement rates unless it can place some pours at much faster rates. To accomplish this and to furnish sufficient mixed concrete to maintain production when a mixer is being repaired or serviced, it is recommended that the mixing plant contain enough mixers to supply concrete at an hourly production rate 25 percent greater than the average pouring rate of the placement equipment. For example, if a mixing plant must furnish concrete to three gantry cranes that have an average pouring rate, when placing concrete, of 96 cu yd per hr, the mixing plant should be able to produce 360 cu yd per hr. Since one 4-cu-yd mixer can produce 80 cu yd of concrete per hr (Chap. 5), five 4-cu-yd mixers will be required. This is the same size mixing plant that would be selected if the previously stated rule of thumb for determining mixing-plant capacity were applied; for example, the cubic yards of installed mixer capacity should be $1^1/_2$ times the total bucket capacity of the placement equipment.

Bucket-car requirements consist of one car at each crane, one car at the mixing plant, and cars in transit between the mixing plant and the placement cranes. The number in transit depends upon the length of haul.

As previously discussed, the rate of concrete placement will be reduced at the start and completion of a concrete dam, resulting in a S-shaped concrete-placement curve. At the start of concrete placement and until the major part of the dam foundation is covered with concrete, there will not be enough dam blocks on which pours can be made to maintain maximum production. Later, when the dam is topped out, production will again decrease because only small volumes of concrete are needed for each block. Finally, the high blocks, which will be every other block, will be completed, leaving only one-half the blocks available for concrete placement.

CRANE SYSTEM

When a low concrete dam, or the concrete portion of any other type of low dam, is to be constructed, it is often possible to pour the concrete with gantry or other types of cranes operating from ground level. Gantry cranes can now be obtained with 120-ft-high gantries; this gantry height makes them very adaptable to this type of concrete placement. Concrete-placing procedure with this method is similar to that used under the trestle-crane method, with the exception that the trestle is eliminated. The placement cranes may place concrete from various ground elevations, so that trucks or truck-trailer units may be required to haul the concrete from the mixing plant.

The use of a crane-placement system is illustrated in Photograph 19, which pictures large gantry cranes operating from ground level

Photo 19 Cranes for concrete placement at The Dalles Dam, Columbia River. (Washington Iron Works.)

for placement of concrete in The Dalles Dam and Powerhouse on the Columbia River.

Layout

Similar to the trestle-crane system, before a drawing can be prepared of this method of concrete placement, crane dimension and boom-coverage charts should be obtained for the various-sized cranes. The layout should illustrate how crane coverage may be secured by gantry cranes located on tracks laid at ground level or whether space is available to operate crawler or rubber-tired cranes. Layouts should show how crane coverage will be provided for all concrete structures, since mental pictures of crane coverage may be misleading, and there may be concrete areas that would be extremely difficult or impossible to place in this manner. Gantry cranes can be obtained with different leg lengths to suit specific ground or foundation conditions. After the crane coverage is checked, locations must be selected for the mixing plant and the concrete-haulage tracks or haul roads from the mixing plant to the placement cranes. If possible, the mixing plant should be located near the center of the pouring area to reduce the concrete-haul distance.

Capacity

The number and capacity of the placement cranes and concrete-haulage units and mixing plant size should be computed in the same manner as for the trestle-crane method of concrete pour.

SELECTING THE MASS-CONCRETE-PLACING EQUIPMENT

The selection of the mass-concrete-placing equipment for any concrete dam is governed by the type of dam and damsite. In this respect, proposed dams and their damsites fall into four groupings.

1. ***High Concrete Dams Located in Narrow Canyons.*** Damsites in narrow canyons have steep side walls and most of the canyon floor is occupied by the river. Typical damsites of this type were those for Hoover, Hungry Horse, Shasta, Flaming Gorge, and Bullards Bar Dams.

Cableways are advantageous to use at these damsites because their towers and all other major construction facilities can be located outside the dam area and can be erected without interfering with, and before the completion of, the diversion facilities and foundation excavation. This allows concrete placement to start as soon as the dam's foundation is excavated. Often the cableway can be operated early enough to assist in the preparation and cleanup of the foundation. Another advantage of a cableway system is that most of the construction facilities will be located above the canyon floor where they will be free from any possible damage from flood flows.

Concrete placement in high dams entails a large amount of vertical bucket travel. Fast concrete pouring requires that the concrete bucket be hoisted or lowered at high speeds and that the placement equipment have sufficient drum capacity to contain the large amount of hoisting cable required. A cableway is particularly suited to this type of operation since it has fast hoisting speeds and an unlimited drum capacity.

When these high, narrow dams contain a large volume of concrete, the concrete-placement equipment must have a large capacity to meet schedule requirements. A cableway system has limited placement capacity, since cableways must span the canyon about as the dam does, which limits the number that can be installed at any damsite. However, this placement limitation can often be overcome, as was done at Shasta Dam where the terrain permitted the operation of seven movable cableway towers which radiated from one central fixed tower. At Glenn Canyon Dam the number of cableways that could be used at one time was increased by providing a double set of parallel tracks with two cableway units located on the inside tracks and operating underneath a third cableway with higher towers located on the outside tracks. Pouring capacity was increased also by using cableways capable of handling 12-cu-yd buckets (see Photograph 20).

Photo 20 Cableways at Glen Canyon Dam. Three traveling towers on two parallel sets of tracks. (Washington Iron Works.)

When the damsite meets the majority of the conditions suitable for cableway placement, the stability of the abutments may make its use impractical. One example of this was at Green Peter Dam, Oregon, which was a typical cableway site. However, the stability of the right abutment of the dam was indeterminable. The dam contractor elected to use gantry cranes operating from a trestle. Subsequent excavation of the right abutment proved that the choice of this placement method was correct, even though damage was caused to the trestle by a 100-year flood which plugged the diversion tunnel, overtopped the placed concrete, and damaged other construction-plant installations.

A perfect damsite for a cableway system occurs when the steep side walls of a canyon flatten above the top of the dam to furnish a level area for the runways for movable cableway towers. One of the best examples of this type of damsite was at Glenn Canyon Dam, where the Colorado River cut a narrow channel with almost vertical walls through a flat plain, providing an excellent location for runways for the movable towers of parallel cableways.

In comparison, trestle-crane-placement systems have disadvantages in their use for placing concrete in high dams located in narrow canyons. Before a trestle can be erected, the river must be diverted and the foundation completely excavated. The erection of the trestle becomes part of the critical path and lengthens the time required for the dam's construction. The trestle, the cranes, and other construction facilities will be located within the canyon where they will be exposed to flood-flow damage and will congest the limited work areas. Very high dams are often built at damsites in narrow canyons so that a large amount of vertical bucket travel is required for their concrete placement. Cranes are less suitable for making these high pours than cableways since their hoisting speeds are slower and they have limited cable capacity.

2. *Relatively High Concrete Dams Located on Wide, Flat Damsites.* The trestle-crane concrete-placement system is used for the construction of high concrete dams, or concrete portions of combination dams, at long, flat damsites that have low abutments. Typical dams and damsites of this type are Grand Coulee, Libby, Friant, and Noxon Rapids.

Trestle erection will not lengthen the time required for the dam's construction since the damsites are long enough to permit one portion of the dam to be excavated, a section of trestle erected, and concrete placement started while the remainder of the dam is being excavated and the stream is being diverted.

These broad damsites can only be spanned by long dams where concrete placement requires that concrete be hauled long distances between the mixing plant and the pours. A trestle-crane system can economically adapt to long hauls by increasing the number of concrete-haulage vehicles.

The large volumes of concrete contained in these long dams require that the selected concrete-placing system have high production rates. A trestle-crane concrete-placement system is preferred for large dams as any required production can be obtained for a minimum capital expenditure by increasing the number of placement cranes.

Other cost advantages resulting from a trestle-crane system include the use of the trestle to support the water, compressed air, and electrical distribution lines that service the pour areas, and the use of the trestle deck as a work platform. Compared to a cableway, there will be fewer pouring delays with a trestle-crane system since there will be no track and operating cable replacement delays. Finally, a trestle-crane system has great flexibility in providing hook coverage to dams of irregular shape.

Conversely, a cableway concrete-placement system is costly to install at wide, flat damsites, and often not enough cableway units can be installed to provide the desired pouring capacities. Cableways must operate across the width of a dam, as compared to a trestle-crane system that operates along its length; therefore, the number of cableways that can be installed to pour long, narrow dams is limited, and the pouring capacity of a cableway system is restricted. If enough cableways can be installed to meet pouring requirements, there will be a high acquisition and installation cost for each as a result of long cable spans and the need for high towers; the latter compensate for low abutment heights and the large cable sags which occur when there are long cableway spans. Acquisition and erection costs for the cableway system will also be high because the long horizontal distances that the cableway must transport the bucket, between the bucket pickup point and the concrete pour, will increase the cableway cycle time, decrease its production, and require that a large number of cableways be installed to meet the scheduled rates of production. This increase in cycle time and decrease in production for each cableway will cause a proportionate increase in the direct cost of concrete placement.

3. *Low Concrete Dams.* When the topography at the damsite permits and the dam is low enough that concrete can be poured with cranes operating from ground level, this is the most economical concrete-placement method. This method minimizes acquisition cost and operating cost. Gantry cranes, crawler cranes, truck cranes, or tower cranes are used for this purpose. Typical of the concrete

dams that have been poured in this manner are the low-head spillway dams on the lower Columbia and Snake rivers. This system was used to pour concrete in the powerhouse at Chief Joseph Dam. A large walking dragline was removed from its base and was mounted on a low, specially built gantry which traveled on rails laid on the excavated surface of the intake channel. Because of the dragline's large size, it could utilize a 200-ft boom to make all the powerhouse pours.

4. *Other Dams and Damsites.* Some dams and damsites will not readily fall into any of the three foregoing groups. For these dams and damsites, it is often necessary to prepare comparative estimates that evaluate the acquisition and operating cost of each type of concrete-placement system before the preferable system can be selected. If these comparative estimates do not show a definite cost advantage to a particular system, then the system often selected is one that makes use of available equipment owned by the contractor or one that is preferred by the future contract manager.

MISCELLANEOUS-CONCRETE-PLACEMENT EQUIPMENT

The term miscellaneous concrete is used to describe concrete located separate from mass concrete and of such a small quantity that the purchase of large-capacity equipment for its placement is not justifiable. Specifically, miscellaneous concrete is concrete that cannot be placed with mass-concrete-placement equipment. Included in this category are the small end blocks of concrete dams, small concrete dams, slab facing of some rock-fill dams, diversion-tunnel linings, the backfill of diversion conduits in dams, training walls, spillway floors, fishladders, intake structures, and any small isolated concrete structure connected with dam construction.

Only minor amounts of miscellaneous concrete are included in dam contracts. Individual pours are small and there is usually a large amount of standby and delay time connected with each pour, resulting in low hourly pouring rates. This restricts the expenditures that can be justified for the purchase of concrete-placing equipment. Because of the great variety in the types of structures containing miscellaneous concrete, there are many ways that it can be placed, and new placement methods are constantly being developed. Methods commonly used for placing miscellaneous concrete are:

1. Direct placement from agitator or dumpcrete trucks
2. Placement with truck or crawler cranes
3. Conveyor-belt placement
4. Placement with air guns

5. Placement with pumps
6. Slip-form placement
7. Tower-crane placement
8. Placement with guyed derricks
9. Placement with stiffleg cranes

When the concrete occurs in low, narrow pours and when trucks have access to the pour, depositing concrete in the forms directly from the trucks is the most economical pouring method. Dumpcrete trucks are preferred over agitator trucks, since they are lighter, have fewer moving parts, require less capital investment, and have a lower operating cost. Agitator trucks are used when the job specifications prohibit the use of dumpcrete trucks or when the concrete contains 6-in. aggregate.

When concrete pours are too wide or high to permit direct placement from trucks, such as pours for concrete training walls, the end blocks of dams, and spillway floors, then concrete is usually placed by buckets swung over the pour area by truck cranes or crawler cranes. Concrete may be delivered to the point of use by agitator trucks, dumpcrete trucks, or trucks hauling concrete buckets. Often the concrete-placement crane is one originally purchased for other work, such as foundation excavation; when this work is completed, it is then used for pouring concrete. Placement with cranes is mandatory when the job specifications prevent the transfer of concrete from one container to another. Under this type of specification, the concrete must be placed in buckets at the mixing plant, the buckets must be transported to the pour area, and final placement must then be done by cranes.

Conveyor-belt placement can be used when it is impractical to provide portable-crane coverage of the concrete pour and when specifications permit their use. This method of pouring concrete is shown by Photograph 21. Conveyor belts can be obtained which have a pouring capacity of 200 cu yd per hr, but because of the limited belt width available, their use is limited to the placement of concrete containing 3-in. or smaller aggregates. The belts can receive the concrete from dumpcrete or agitator trucks or from buckets swung over the belt by cranes. In some cases, belt transportation may be used directly from the mixing plant to the point of pour. The belts come in lengths that vary between 17 to 40 ft. The end discharge of the belts permits the addition or deletion of individual belt units as pour lengths change. Belts can be supported from crane booms to pour isolated and relatively tall structures. Belts have great flexibility of alignment for pouring thin slabs over large areas, such as spillway floors. Concrete placement with belts is rapidly increasing because of their flexibility, low capital cost, and high pouring rate. Trade publications contain many articles describing the use of conveyor belts for pouring concrete.[5]

The most economical way to place the concrete lining of diversion tunnels is with air guns, when they can be located close to the pour and when the lining is continuous. Since air guns have fewer moving parts than pumps, they have lower operating and maintenance costs.

Concrete pumps are used to place tunnel lining when the use of air guns is restricted or when a more versatile placer is required. Concrete pumps are preferred over air guns for short pours, intermittent pouring, and when the concrete placer must be located a distance from the pour. Pumpcrete machines can pump small-aggregate concrete as far as 1,200 ft in a horizontal direction. This distance must be reduced if vertical pipe runs are necessary or when bends must be installed in the pipeline. Pump manufacturers' literature will give the pumping limits for combinations of lengths of vertical

Photo 21 Pouring concrete with portable conveyor belts.

and horizontal pipe, which will help determine the pumping limit for any pour. Typical pours made with pumps are intermittent lining in a diversion tunnel, placement of tunnel concrete when the pump must be located at the tunnel portal, placement of backfill concrete in diversion openings in the dam, and isolated pours that cannot be placed with other placing equipment. Placement of tunnel concrete is a specialty item; if descriptions of this procedure are desired, they are available in other publications.[6]

The use of slip forms for concrete placement is rapidly increasing, primarily because structures are now designed for this method of concrete placement. For slip-form placement, the main structure must be consistent in shape, and bolts must be located in blockouts to prevent their interference with the slip-form movement. Slip-form placement has many advantages for concrete placement in tall, regular-shaped concrete structures such as intake structures or shafts. Pouring rates depend upon the volume of concrete in the cross sections of the structure, and unless large quantities are involved, the forms can be moved at a rate between 12 and 16 in. per hr. Slip-form pouring is economical for forming and placing the concrete in steep spillway slabs and for pouring concrete facings for rock-fill dams. The slip forms can be jacked up steep slopes from rods embedded in the poured concrete or can travel down the slope controlled by rods anchored at the top of the slope. Concrete can be supplied to the distribution belt located over the slip forms by cableway, conveyor belt, or skips running on rails up and down the spillway slope. Use of a conveyor belt for this purpose is limited to installations not exceeding a grade of approximately 20 percent.

Tower-crane placement is economical for high pours and for pouring small concrete arch dams. The larger the tower crane, the larger the capital investment, the larger the concrete bucket can be, and the higher the pour.

Guyed derricks are sometimes used to place concrete in small concrete dams. On large jobs they are seldom used for pouring concrete but are often used to handle heavy lifts connected with the concrete pour. Lifts of this type include the placement of large penstock sections and the installation of powerhouse equipment.

In the past, stiffleg cranes were often used to pour miscellaneous concrete. They are now seldom used because they are tied to one location, their capital cost is high, and pouring rates are low, resulting from slow boom travel and the number of parts of line used in the hoisting operation. One continuing use for stiffleg cranes is the pouring of shaft concrete since they can also be used to handle lifts and serve as all-purpose cranes during shaft excavation.

REFERENCES

1. *Wire Rope Handbook for Western Rope Users*, copyright 1959, United States Steel Corp.
2. Frank W. Stubbs, Jr., ed., *Handbook of Heavy Construction*, copyright 1959, McGraw-Hill Book Company.
3. California's Bullards Bar Dam, *Western Construction*, July, 1968, page 29.
4. Libby Dam Sheds Its Electric Blanket, *Contractors & Engineers Magazine*, April, 1970, page 56.
5. Versatile Belts Place Mix for Maze of Bridges, *Construction Methods and Equipment*, December, 1968, page 82.
6. Albert D. Parker, *Planning and Estimating Underground Construction*, copyright 1970, McGraw-Hill, Inc.

CHAPTER FIVE

The Concrete-construction System

Concrete-placement equipment, its production capacities, and its selection were discussed in Chap. 4. Since this equipment controls the selection of the other units that complete the concrete-construction system, this knowledge is a prerequisite for understanding this chapter.

This chapter describes the remaining components of the concrete-construction system. The primary purpose of these remaining components is to supply sufficient mixed concrete, formed areas, placement crews, cleanup crews, curing facilities, and concrete finishers to permit the placement equipment to continually place concrete at scheduled production rates.

CONSTRUCTION-SYSTEM SELECTION

That a concrete-construction system is complex is shown by Figure 4, a flow diagram for concrete dam construction.

Selection of the components of a concrete-construction system is accomplished by establishing work quantities, preparing a construc-

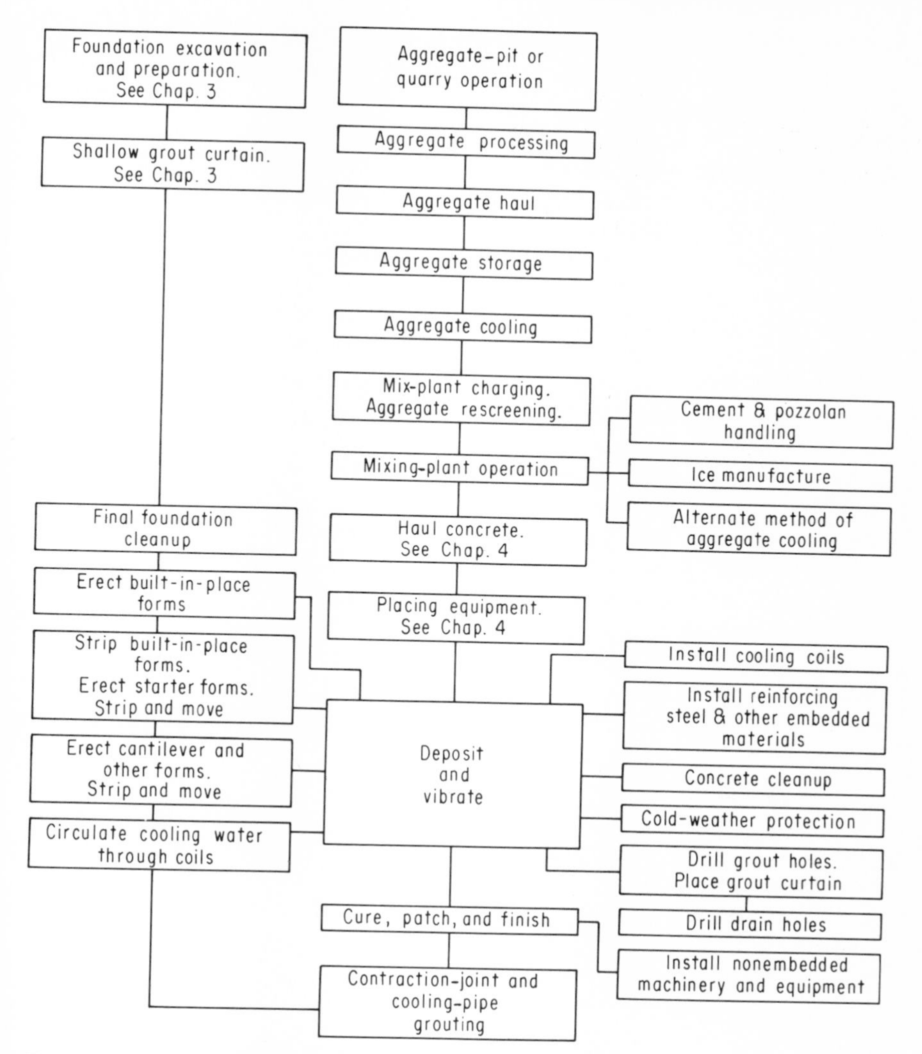

Fig. 4 Work-flow diagram of a concrete system for dam construction.

tion schedule, and then choosing system components that will operate efficiently within the system and that will have the capacity to meet the scheduled rate of production.

If the construction of the project requires the placement of large quantities of concrete, the system concept is to reduce labor cost by the use of equipment. If only small quantities of concrete must be placed, the system concept is to reduce equipment cost by performing more work with hand labor. When equipment and labor cost are balanced properly, a low cost estimate will result.

After all units of the concrete-construction system have been selected, a final review should be made to determine whether the essential facilities have been provided. Then the capacities of all plant and equipment units should be checked against the capacities required to keep the work on schedule.

To provide background for the selection of the components of the concrete-construction systems, descriptions are given of the preparation of quantity takeoffs, the scheduling of concrete construction, the performance of each phase of concrete construction, and the equipment and plant units used to perform this construction. The order followed is that shown in Figure 4: aggregate production, haul, rescreening and mix plant charging; cement and pozzolan handling; aggregate and concrete cooling; concrete mixing (concrete-haul and concrete-placement methods were described in Chap. 4); foundation and concrete cleanup, placement of embedded items; concrete forming; concrete depositing and vibration; cold-weather protection; curing; patching and finishing; installation of permanent facilities; and drilling and grouting.

After construction equipment and plant types have been selected, the required number of operating units is established by dividing their production capacity into the maximum scheduled rate of concrete placement.

For large plants and equipment units, equipment availability is not used to increase the number of operating units to arrive at the total equipment requirements because it costs less to shut down operations when a large equipment or plant unit, such as a cableway, is out of service than to provide spare units. The converse is true for smaller equipment units. As previously discussed, equipment availability on heavy-construction projects is between 70 and 80 percent, with the remainder of the equipment under repair or being serviced, and therefore unavailable for productive use.

QUANTITY TAKEOFFS

Quantity takeoffs are made to check the work quantities listed in the bid schedule and to establish work quantities which are not listed in the bid schedule but which form a part of the cost of the listed work items. Such work quantities commonly include: the square feet of rock foundation that must be cleaned prior to being covered with concrete; the square feet of concrete that must be cleaned prior to the placement of the next pour; the square feet of each type of surface finish; the square feet of construction joint treatment; the surface contact of each type of concrete form; the square feet of any required form lining; the square feet of each type of form that must be built or purchased; the amount of blocking or shoring that must be used to support soffit or other types of forms; the trusses that may be required to support forms; the quantities of walkways, ladders, and stairs that are required to provide access to and from the pour and forming areas; and the cubic yards of *overbreak concrete.*

Overbreak concrete is designated as nonpay concrete in the specifications. It occurs when the plans show concrete pay lines and when the specifications state that the concrete required to fill any excavated area that extends past these pay lines will not be measured as pay quantities. This necessitates that the quantity takeoff list the estimated yardage of anticipated overexcavation, which is the same as the anticipated overbreak concrete. The estimator places the cost of this overbreak concrete against the pay concrete bid items, as this is the only way the contractor will be reimbursed for this work. Specification writers insert this clause in the specifications to force the contractor to use careful blasting techniques and to avoid excavating more rock than is necessary.

Form quantity and concrete takeoffs should be made simultaneously, using the same personnel, in order to eliminate errors in establishing *form ratios.* (Form ratio is the square feet of form contact for 1 cu yd of concrete.) If concrete and form quantities are computed from the same concrete dimensions and an error is made, this error will not change the form ratio, since it will be a compensating error for both quantities. Estimators use form ratio to establish form cost for each concrete bid item.

Form cost is such a large portion of the total cost of concrete that the determination of the form ratios is necessary for proper concrete-cost estimating. The form ratio is a measure of the form cost and placing cost of each concrete bid item. The thinner and more complicated the structure, the greater will be the number of square feet of forms per cubic yard of concrete, the greater will be the form ratio, the greater will be the form cost for each yard of concrete, and the longer it will take, and therefore the more costly it will be to place a cubic yard of concrete. Also the ratio of the form cost to total direct concrete cost increases with an increase in form ratio, since form cost is the major cost for most concrete structures. When mass concrete is poured with a form ratio of between 1 and 2, the form cost will be between one-eighth and one-fourth of the direct cost of concrete. This proportion increases as the form ratio increases and may constitute four-fifths of the direct cost of concrete when concrete structures with a form ratio of 80 are poured.

The form takeoff should show the square feet of forms that should be purchased, the square feet of forms that must be job-made, and the number of times each type of form will be reused. The form takeoff should list the different types of forms: built-in-place forms, lift-starter forms, cantilever upstream face forms, cantilever downstream face forms, transverse and longitudinal cantilever bulkhead forms, collapsible gallery forms, curved-panel forms, straight-panel forms, soffit forms, stairway forms, special-shaped forms, and block-out concrete forms. Listings should be made of the square feet of form contact that will be handled by each type of form and the number of times each form will be reused.

Before the form takeoff is started, the minimum number of form reuses that will justify the use of purchased forms should be established. If steel forms are economical when a form can be used four times, this should be reflected in the form takeoff. The form takeoff should make maximum use of purchased steel forms. This includes using special steel forms for special shapes, using high panel forms for wall pours, and using collapsible forms for galleries and other interior forms. Slip-form usage is economical for spillway slopes, for towers, for shafts (both vertical and inclined), and for concrete facing on rock-fill dams. Sonotube forms are economical for circular shapes. Plastic forms are being developed for many applications.

Many points require special attention when preparing the form takeoff for a concrete dam. Among these is the amount of *built-in-place* forms that will be required to make the pours that are placed on the rock surface of the dam foundation. This number should be determined as accurately as possible, since the construction of these forms is costly. Another item is the *block-out concrete*, the forms used to block off these areas, and the forms later required to contain the block-out concrete when it is poured. Block-out concrete is concrete that encases steel gate guides or other steel parts that must be accurately placed and aligned. This concrete is not placed at the same time as the major pours, since at that time the space it occupies is formed off. After the major portion of the concrete structure is poured, the gate guides or other steel parts are set and aligned, and the block-out concrete is then formed and poured. Because of the small quantities involved, the difficulty in placing, and the high form ratio, the cost of placing a cubic yard of block-out concrete is very high. However, since each project has only a few cubic yards of block-out concrete, its total cost is not a major item.

The concrete-pouring schedule and sequence should be established before the form takeoff is made so that maximum form reusage can be planned. This is of prime importance when it is planned to complete one section of a dam before the remainder is started, since forms from the first section can then be reused on the remainder of the dam. A concrete-pouring diagram will help in form scheduling. The preparation of this diagram will be discussed under scheduling of concrete construction, which is the next subject in this chapter. By using this placement diagram and the dimensions of each dam monolith, the required number of dam forms can be determined. Since the time available for estimate preparation is limited, the form takeoff often must be started before the concrete schedule has been prepared. In this case the form takeoff is made using an assumed schedule and corrected later to fit the final schedule.

To properly prepare a form takeoff, it is necessary to understand form types and how these forms are used in dam construction. Dam monoliths are poured by starting the pour in alternate blocks across the bottom of the dam. The first lift of concrete placed on the rock surface is contained by built-in-place forms that are braced against the rock and anchored to steel hooks placed in holes drilled into the rocks. These forms must be built in place since they must be hand-tailored to the rock surface.

The next few pours are formed with *lift-starter forms.* These are noncantilevered forms anchored at the bottom to she-bolts placed in the previously placed concrete lift. The tops are anchored to inclined tie rods fastened to anchors placed in the surface of the concrete during the previous pours. At the top of the forms, she-bolts are installed so that they will be imbedded in the pour and will be available to anchor the bottom of the forms when they are raised to contain the next pour. Lift-starter forms are used one or two times for each dam monolith until the concrete surface is raised sufficiently in elevation that anchorage space is available for the cantilever legs of the regular forms. Only enough lift-starter forms are required to form one or two monoliths, because after they have been used on one monolith, they can be moved to the next one. These forms can be shop-built wood forms or purchased steel forms.

Cantilever forms are then moved onto the block and are anchored at both the top and bottom of the cantilever by she-bolts placed in the previous two pours. She-bolts are located at the proper spot near the top of the form and encased in the concrete pour to provide anchorage for the cantilevers when the forms are raised. These forms are used for the remainder of the concrete in the monolith. Each dam monolith is formed with three types of cantilever forms: upstream face forms, transverse and longitudinal bulkhead forms, and downstream face forms. If the dam has variable curvature of its downstream face, the cantilever downstream face forms must have adjustable surfaces to fit this change in curvature. When 7.5-ft or higher lifts are used and the dam has an inclined downstream face, the downstream forms are hinged at the middle. This allows the top section of the forms to be swung back during the placement of the first half of the concrete pour, which permits the concrete bucket to be spotted nearer the downstream face of the pour.

High-gravity dams may be so wide at the base that the lower section of each monolith must be divided into an upstream and a downstream pour. These pours must then be separated by longitudinal bulkhead forms. To provide space for cantilever forms, a 10- to 15-ft differential in elevation between the concrete surfaces of the two pours must be maintained.

After the alternate dam monoliths have been poured to a height that gives clearance for the cantilevered bulkhead forms, pours can be started in the remaining monoliths. They are formed like the first group of monoliths, except that transverse bulkhead forms are not required. These remaining monoliths must be kept from 10 to 15 ft below the lowest section of the first group of monoliths so that cantilever form clearance can be maintained. If the alternate block system of pouring is not practical, monoliths can be poured in a stepped fashion to provide cantilever anchorage clearance. The difference in block elevations and the use of cantilevered steel panel forms is very well illustrated by Photograph 22 of Glen Canyon Dam. Concrete forms, form hardware, and dam-forming techniques are amply illustrated in form hardware catalogs.[1]

When a pour is completed, small hydraulic cranes are used to support and raise the forms while they are being disconnected from their position and raised and connected to the next set of she-bolts. These small cranes are moved from pour to pour by the concrete-placing equipment. Forms must be designed so that form stripping, raising, cleaning, and alignment can be done with a minimum crew. The forms must be strong enough to stand the stress of many reuses, to withstand the shock of occasional contact with the loaded concrete bucket, and to support walkways. The labor required to anchor and align cantilevered steel panel forms has been reduced by using forms with quick and efficient anchoring devices such as the coffin-handle anchorages used on the forms at Libby Dam in Montana[2] or by using forms that contain self-raising jacks such as those used at Dworshak Dam in Idaho.[3] The extra cost of purchasing self-raising forms is justifiable when they are used on a high dam which permits many form reuses.

Photo 22 Concrete forming at Glen Canyon Dam. (Washington Iron Works.)

The number of steel dam forms required to keep concrete forming ahead of concrete placement can be readily computed if concrete placement is started in the lowest dam block and pours are made across the total width of the dam as the dam is raised in elevation. Face forms are the controlling factor in form purchase since a greater area of face forms than of bulkhead forms is usually required. The length of face forms necessary to maintain a constant pouring schedule and to allow placement in each block increases as the dam is raised in elevation. Often the pouring schedule can be arranged so that end abutment blocks are poured after the center blocks are completed. Then the upstream and downstream face forms from the center blocks can be used for forming these blocks. This scheduling problem may require a study comparing the savings in form purchase to the additional cost of prolonging the schedule before the best solution can be determined. After the quantity of face forms has been determined, then sufficient bulkhead forms should be purchased so that bulkhead forming can match dam-face forming.

A greater number of bulkhead forms are required per dam monolith near the dam's foundation than are required per monolith when the dam approaches its final height. As the dam approaches its final height, a larger number of monoliths can be poured at one time, a fact which must be taken into account when bulkhead form requirements are reviewed. The number of monoliths that can be poured at one time depends on the lineal feet of face forms purchased. To find the maximum length of bulkhead forms required, the amount needed at each dam elevation to balance the length of face forms purchased can be determined and a curve can be drawn, with dam elevation as one ordinate and lineal feet of bulkhead forms as the other ordinate. The maximum length of bulkhead forms required can be taken from this chart.

If sectional concrete placement is used for pouring the dam, it is necessary to examine a pouring diagram to determine the form requirements. Pour diagrams are discussed in the following section.

When the dam concrete has been placed to the elevation of the dam galleries, gallery forms must be erected on the pour. Gallery forms should be collapsible so that they can be readily stripped and reused. When forms must be hand-tailored to fit surfaces, such as vertical form surfaces with projecting embedded materials spaced at irregular intervals, then forms must be built in place.

The extent to which the other types of forms are used varies with each project. If concrete pouring is scheduled in a stepped fashion, special steel-shaped forms used on one monolith can be stripped and used on the next monolith and reused as required along the length of the dam. Models that can be disassembled by pours have been used to solve complicated forming problems encountered in miscellaneous structures.

When the forms and concrete quantities are to be computed for large regular-shaped structures, a considerable amount of takeoff time can be saved by a computer takeoff. A computer takeoff will save in time and manpower when the forms and concrete yardages must be computed for a thin, double-curvature arch dam. The computer can be programed to tabulate the number of forms and amount of concrete for every pour in the shape of a pour diagram, with each pour in its proper location and elevation to the dam's cross section. The programing of a computer to take off forms for interior openings such as galleries, etc., is so complex that this work is done by hand. Using a computer to take off form quantities for irregular or special structures is seldom justifiable, since more time may be spent programing than is saved on the takeoff.

After the form takeoff has been completed, it should be reviewed to determine if forms have been used to maximum advantage. Form manufacturing companies will review proposed projects, give a quotation on the forms, and make recommendations on the number and kinds of forms required. This information can be used to check the contractor's takeoff.

WORK SCHEDULING

Before the components of a concrete-construction system can be selected, it is necessary to prepare a concreting schedule so that the scheduled hourly concrete-placement rates will be available. Then the maximum placement capacity of concrete in cubic yards per hour can be established, which will control the capacity of many of the system components and can be used to establish the tons of aggregate required per hour, which in turn controls the capacity of the aggregate plant and the aggregate-handling facilities.

The construction schedule for a concrete dam must allow time for work access, for the mobilization of plant and equipment, for the construction of diversion facilities, and for the excavation of the dam's foundation. Initiating concrete placement differs with different diversion methods and sequences of performing the dam's excavation. These subjects were discussed in Chaps. 2 and 3.

Scheduling the dam foundation excavation, the construction of the diversion facilities, and the time required to mobilize and erect the concrete construction equipment will establish the date that concrete pouring can start. The date that concrete pouring must be completed is established by allowing time at the end of the schedule for the completion of any work that must be done after all concrete is placed. After these control dates have been determined, the best method of scheduling concrete placement is to use a pour diagram. A typical concrete-pouring diagram consists of an elevation of the dam showing each pour, the concrete yardages in each pour, and the date that the pour can be placed. Instead of showing dates of pour placement, the pours made in every month can be colored. Drawing 12 in Chap. 11 is a typical concrete-pouring diagram.

Pour scheduling must allow for a slow start of concrete placement, since, at the start of each monolith, time must be spent on rock cleanup and considerable time is needed to construct built-in-place forms. Alternate blocks are poured until they are approximately 15 ft above the intermediate blocks. This was discussed under form takeoff in the previous section.

Pouring will be slow and limited in volume until enough monoliths have been started to provide sufficient pouring area. With a five-day work week, seldom can more than one lift per week be placed on the same monolith. Time between lifts is required for concrete cleanup, for raising and setting forms, for installing embedded material, and for allowing the concrete to develop enough strength to support the raised form.

After sufficient monoliths are started, the number of pours that can be made in one day is controlled by the maximum pouring rate that can be achieved and not by the one-lift-a-week criterion. The rate that concrete can be raised in each monolith is delayed when gates, penstocks, gallery forms, etc., must be installed.

As the concrete is raised in the dam, the amount of concrete in each pour will decrease until the criterion of one lift a week controls the rate of concrete placement. This will continue until the top pours are reached.

Topping out of the dam is slow since special forms must be installed for the dam's parapet and sidewalks and the quantities of concrete finish required will increase. After each high alternate block has been topped out, the low alternate blocks must be raised and topped out.

Completing the pouring diagram establishes the desired maximum pouring rate to use when computing the required number of concrete-placing units. The production capacity of the different types of placement-equipment units was covered in Chap. 4; also, it was stated that the mixing plant's capacity could be computed by multiplying the bucket capacity of the placing equipment by 1.5. The number of tons of aggregate that must be produced in one day is computed by multiplying the maximum daily production of mixed concrete by 1.85. These production rates are then used in selecting the equipment and plant units required to complete the concrete-construction system.

AGGREGATE PRODUCTION

Aggregate-production facilities should be selected and arranged to produce aggregates to specification requirements at a production rate that will equal the mixing plant's demands. Depending on the specifications and the suitability of the aggregate source, concrete aggregates may be natural aggregates produced from gravel deposits or manufactured aggregates produced by crushing and processing quarried or excavated rock. Specification requirements for aggregates differ because contracting agencies and owner's engineers differ in their concrete technology.

One of the major differences in concrete technology that is of importance to a contractor is whether the specifications permit the use of natural aggregates that contain reactive elements. If concrete is made from reactive aggregates, the alkalies released by hydration of the cement will have a chemical interaction with siliceous-reactive aggregates to form alkali silica gel. Over a time period, the concrete may be broken by the osmotic swelling of this gel. When concrete is made from reactive aggregates, formation of this gel can be prevented by using low-alkali cement or by including certain pozzolanic materials in the mix that neutralize the alkali produced during the cement's hydration. This type of pozzolan can be produced from certain volcanic materials, from pumice, and from Monterey shale. Fly ash is also a suitable type of pozzolanic material to use for this purpose. Some concrete technicians will permit the use of reactive aggregates when this type of pozzolanic material is added to the mix. Under this type of specification, when suitable local gravel deposits are located adjacent to the projects, they can be used as an aggregate source. Such gravel deposits are the most economical source of aggregates, because it costs less to excavate gravel and install and operate a natural-aggregate processing plant than it costs to operate a quarry and install and operate a plant for manufacturing aggregates from quarried rock. Aggregates produced from stockpiled rock from the dam's excavation are less costly than aggregates produced from quarried rock, since the cost of stockpiled and reclaimed excavated rock is less than the cost of quarrying rock. Rock from the dam's foundation is seldom suitable for aggregates, so this must be classed as a special application.

Other specifications prohibit the use of reactive aggregates in the concrete mix. Under this restriction, when any local gravel deposits contain reactive materials, they cannot be used as an aggregate source. Instead, aggregates must be manufactured from approved rock quarries containing nonreactive rock. At many damsites, these approved quarries have been located a long distance from the dam or on top of steep, inaccessible hills. This required use of manufactured aggregates has often greatly increased the cost of concrete construction.

Plant Layout

A comprehensive knowledge of aggregate-processing equipment, its application, and its production rates is a prerequisite for the layout and design of an aggregate plant. Because aggregate production is a specialized field, the prime contractor often subcontracts this production to aggregate contractors; thus allowing him to concentrate his efforts on the major work items. If an aggregate plant is required and the estimator lacks experience in aggregate-plant design, either a consultant should be retained or an aggregate equipment distributor should be contacted to design the plant, advise on its capital and operating cost, and establish operating procedures. Engineers who are interested in aggregate-plant design should review the references listed at the end of this chapter.[4-8]

Equipment manufacturers' literature contains tabulations on every phase of aggregate production. These include the horsepower required, the tons of aggregate crushed per hour, and the crushed-products screen analysis for different sizes and types of crushers; the size, horsepower, and capacity of conveyor belts, vibrating screens, log washers, scrubbers, heavy-media equipment, rod mills, classifiers, dehydrators, conical separators, etc.; the tons of each size of aggregates required for a typical yard of mass concrete; and the quantities of aggregates that can be stored in various-shaped stockpiles. Manufacturers will lay out and quote on stationary aggregate plants, on portable aggregate plants that have all major equipment mounted on rubber-tired dollies, or on any equipment component of the plant. The use of portable plants for aggregate production is rapidly expanding because of the excellent resale values based on their mobility.

The specification restrictions on the concrete aggregates should be reviewed prior to preparing an aggregate-plant layout. Particular emphasis should be placed on determining whether the specifications call for the performance of unusual items of work so that these requirements can be incorporated into the plant's design. For example, some specifications have required that minus 200-mesh rock dust be produced for use in the concrete mix. Requirements such as this deserve special attention since this rock dust must be ground as fine as cement, and large quantities of this material cannot be produced in a regular aggregate manufacturing plant but only in plants that have grinding mills similar to those used for cement manufacturing.

The first step in aggregate-plant design is to determine the number of tons per hour of sand and of each size of aggregate that must be produced to meet schedule requirements. The total amount of sand and aggregates used in one cubic yard of concrete varies between 1.65 and 1.95 tons, depending on the specific gravity of the aggregate and the maximum size of aggregate used in the mix. When the mass concrete contains 6-in. aggregate, the average consumption of sand and aggregates per cubic yard of concrete will be approximately 1.85 tons. If the aggregate plant is operated the same number of daily shifts as the mixing plant, its required production in tons per hour is 1.85 times the cubic yardage of concrete scheduled to be produced per hour by the mixing plant. If the aggregate plant is to be operated a fewer number of shifts than the mixing plant, its production should be increased proportionately. This total aggregate production in tons per hour should be divided into the tonnages required for sand and each size of aggregate by applying mix proportions to tonnage requirements.

After the aggregate tonnage requirements are determined, an

aggregate-plant flow chart can be prepared showing the tons of plant feed, the relative location of each piece of equipment, the tonnages that will be processed by each piece of equipment and transported by each belt, and the tonnages of each size of aggregate produced. If the flow chart indicates that the production in the individual aggregate sizes differs from the mix-plant requirements, equipment must be added to crush and process the extra production in one size to make up the shortage in production of a smaller size. A flow chart for a manufactured aggregate plant is shown in Figure 5.

The aggregate plant should be located strategically between the

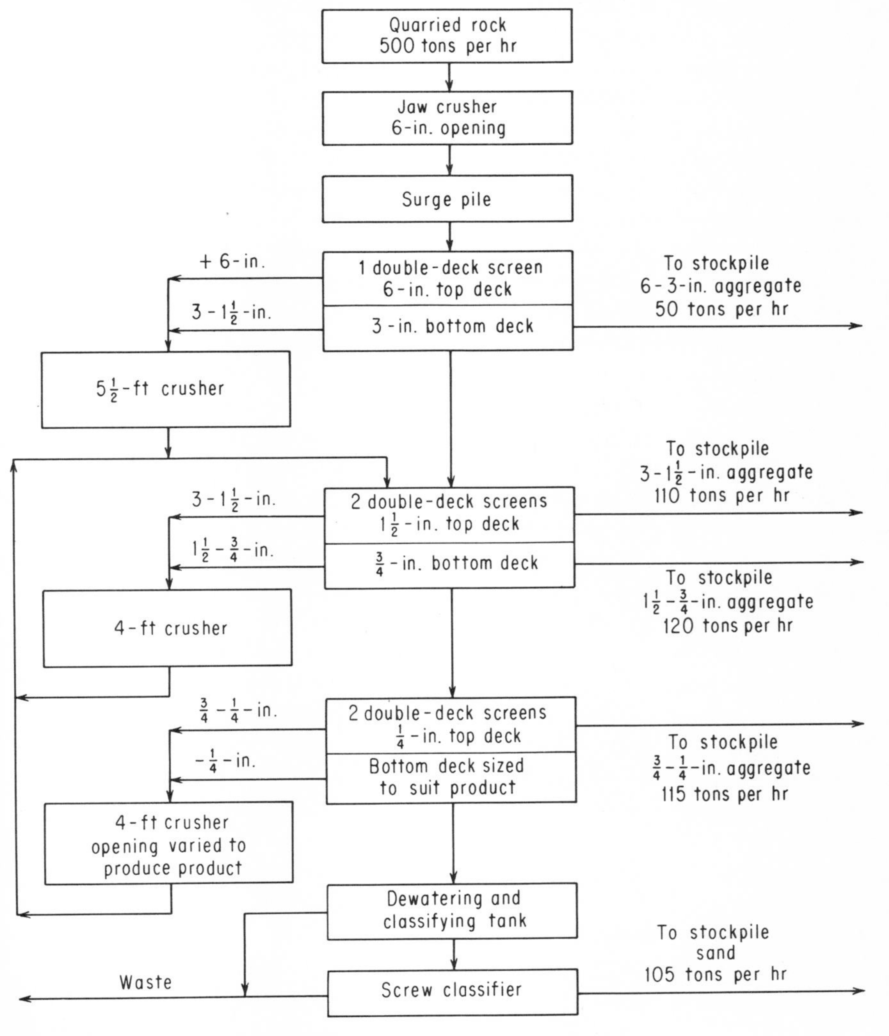

Fig. 5 Flow diagram of a manufactured-aggregate processing plant.

aggregate pit or quarry and the mix plant, so that hauling cost will be held to a minimum. Unique conditions concerning the location of aggregate source and plant have resulted in equally unique layouts. At Detroit Dam, Oregon, the source of aggregates was a quarry located high above the left abutment of the dam. Since there were no level areas between the quarry and the mix plant, shelves were excavated in the hillside, and the aggregate plant and the timber-crib bins used for aggregate storage were located on these shelves. The only alternate location would have been on level ground across the river. This alternate location would have resulted in a long haul from the quarry to the aggregate plant, with a return haul from the aggregate plant to the mix plant almost as long.

At Dworshak Dam, Idaho, the quarry was situated at a location that allowed the primary and the secondary crushers to be placed in an underground chamber that was excavated directly below the quarry.[9] Quarried rock was shoved from the quarry floor into a 20-ft-diameter vertical shaft. An apron feeder located at the shaft's bottom was used to feed the primary crusher. The discharge from the primary crusher was then crushed to 6 in. minus in two secondary crushers. This crushed product was small enough to be transported by conveyor belts through a tunnel to the left abutment of the dam where the remainder of the aggregate plant was located. The glory-hole shaft between the quarry and the primary crusher eliminated the need for a truck load and haul operation from quarry to crusher, with a resulting savings in hauling cost.

At locations that have unrestricted areas, the aggregate-plant layout may encompass several acres. Separated stockpiles are often used for sand and each size of aggregate. When large stockpiles are desired, they can be readily obtained by using a truss-supported stockpiling conveyor. The truss for this conveyor is hinged on the charging end, while the other end of the truss is framed down and supported on wheels that travel on curved rails. The truss travels and pivots on these wheels, forming a semicircular stockpile.

To give some idea about the cost of providing and operating these plants, examples of both natural-aggregate and manufactured-aggregate plants are described below. However, these descriptions are not accurate enough to be used as a basis for aggregate-plant design.

Natural-aggregate Plants

A natural-aggregate plant must be capable of producing aggregates that will meet specification requirements either from an approved gravel source or from a source that has had enough development work done to it to justify its use as the basis of a bid. If the specifications designate approved aggregate deposits, the deposits have been explored by the owner's engineers and this information furnished to the bidders. Exploration consists of drill holes, test pits, exploratory trenches, screen analysis of samples, and physical and chemical testing to determine aggregate suitability.

When the contractor plans to use aggregates from a source that has not been approved by the engineers, then enough exploratory work should be done on the deposit to prove its extent, to establish the size gradations, and to determine whether the aggregates meet specification requirements. The simplest way to determine if the aggregates will meet specification requirements is to have the physical and chemical testing done by private material-testing laboratories. Detailed descriptions of methods of exploring and testing aggregate deposits are very well presented in a Bureau of Reclamation publication.[7] If the contractor wishes to base his bid on the use of an unapproved aggregate source, then he must accept the risk that the unapproved aggregate source may not be accepted by the owner.

The gravel deposit, test pits, drill logs, and screen analysis should be examined, and the resulting data should be tabulated and analyzed. The total aggregate tonnage in the deposit should be computed and subdivided into tonnages of sand and each size of aggregate. These tonnages are computed by establishing the relationship of screen analysis to gravel thickness in each test pit and drill hole, and multiplying these results by applicable areas.

The *gradation* and *fineness modulus* of the sand should be reviewed to determine if they meet specification requirements. The gradation is determined by the amount of sand retained on standard screens. The fineness modulus is a measure of the fineness of the sand. This is determined by dividing by 100 the sum of the cumulative percentages retained on U. S. Standard sieves nos. 4, 8, 16, 30, 50, and 100. Specifications often require that the fineness modulus shall be between 2.30 and 2.90 and that it shall not vary from the average by more than 0.15. To give an indication of fineness modulus with respect to grain size, coarse sand has a fineness modulus of 2.50 to 3.50, fine sand from 1.00 to 2.50, and very fine sand from 0.50 to 1.50.

The exploratory data should be studied to determine the quantities of clearing, grubbing, and stripping required to develop the necessary tonnage; whether the top layer of aggregate is contaminated by organic materials and must be wasted; whether unsuitable materials are present in lenses through the deposit necessitating that selective excavation be performed; whether there is a shortage or overage of sand or any size of aggregate; and whether there is oversized material in the deposit that must either be scalped out at the gravel pit or crushed at the processing plant.

It is necessary to provide loading and hauling equipment to excavate material in the aggregate deposit and transport it to the plant. The equipment used to excavate the dam's foundation may be available for aggregate-pit excavation, provided that the dam-foundation excavation is completed before aggregate production is started. A dragline will be required if the gravel lies below the water surface.

If the gravel deposit contains aggregates suitably sized to meet the concrete mix demands and sand of the proper gradation and with the right fineness modulus, then the aggregate plant only needs to be a screening plant. A layout of this type of plant is illustrated by Drawing 3. This plant was laid out in a straight line with single-deck screens which were located on towers over stockpiles and provided with conveyors to transport the material that passes through each screen to the next screen. The flow of materials can be readily followed in this type of plant. The other type of plant contains a central screening tower supporting multideck screens. Conveyor belts fan out from this tower and transport the sized material to stockpiles arranged around the tower. The free fall of the large aggregate into stockpiles is restricted by the use of rock ladders which limit the vertical drop of the aggregates to between 2 and 3 ft, thereby eliminating size separation and aggregate breakage.

The capacity of the plant must be sufficient to meet the mixing plant's demand. Also, the surge pile should have sufficient live-storage capacity to permit the aggregate plant to operate continuously and be independent of the receipt of plant feed from the haul trucks. Wide conveyor belts are required to handle 6-in. aggregate; therefore, the capacity of these wide belts often establishes plant and screening capacity. The plant may be sized so that it must operate as many shifts as the mix plant or so that one or two shifts a day furnish enough aggregate for three shifts a day of mix-plant production.

If a large tonnage of aggregate is to be handled, reclaiming from stockpiles is done by conveyor belts located in reclaim tunnels underneath the stockpiles. Batch feeding of the sized aggregates onto the reclaim belts is controlled by feeders located under gates placed in the roof of the reclaim tunnel. These belts can discharge either directly into aggregate-haulage trucks or into bins equipped for automatic loading of trucks. When the aggregate tonnage is small, the cost of installing a belt reclaiming system seldom can be justified; consequently, front-end loaders are used for aggregate reclaiming.

If the gravel deposit does not contain well-graded suitable mate-

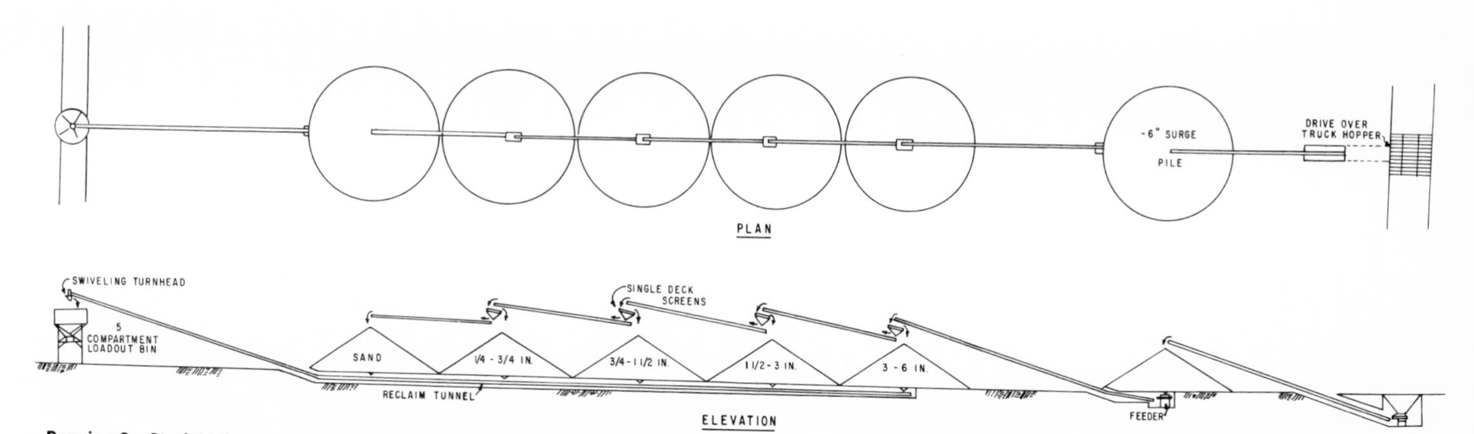

Drawing 3 Straight-line arrangement of an aggregate-screening plant.

rials, if there is a deficiency of some aggregate sizes, or if material is lacking in other qualities, additional equipment must be added to remedy these situations. If roots, sticks, and other similar materials are present in the gravel, then a log washer may be required. When the aggregates are coated with objectionable material, it may be necessary to add a scrubber to the plant to remove this coating.

When the gravel deposit does not contain sufficient 6- × 3-in. aggregates, one method of securing more quantities of this size aggregate is to crush the oversized material in the deposit. If there is not a sufficient quantity of oversized material to make up this deficiency, then it may be possible to make up this size shortage from other aggregate sources. If this is impractical, excess gravel must be processed through the plant until sufficient 6- × 3-in. aggregate is produced. This increased plant feed will result in an overproduction of the other aggregate sizes which must be wasted.

When there is a shortage of smaller-sized aggregates, extra production of the smaller sizes can be achieved by crushing larger-sized aggregates. This requires the addition of a cone crusher and of recirculating conveyors which feed the crusher. When specifications do not permit crushing of aggregates and if shortages occur in any size of aggregate or sand, enough additional gravel must be processed to make up this shortage, and the corresponding excess of production in other sizes of aggregates must be wasted.

The deposit may contain small-sized particles of unsuitable, lightweight materials with specific gravities of less than 2.50. Then the aggregate plant must be equipped with *sink-float* processing equipment to remove this unsuitable material. Sink-float equipment is drum-shaped and contains a *heavy-media* mixture of water, magnetite, and ferrosilicon. The light, unsuitable aggregates float on the media and are discharged over the lip of the drum and wasted. The heavier, suitable aggregates sink and can be removed and used. Other methods can be used for removing lightweight aggregates, such as other types of heavy-media separation, hydraulic jigging, and *elastic fractionation.* (Separation of the lightweight aggregates by elastic fractionation is accomplished by a process that makes use of the principle that denser aggregates will rebound a greater distance when they are dropped on a steel plate.)

If the gradation and the fineness modulus of the sand do not meet specification requirements, often this can be corrected by splitting the sand into several sizes and reblending the sizes to the desired gradation. This necessitates the addition to the aggregate plant of sand classifiers, sand pumps, settling ponds, additional belts, additional stockpiles, and reblending facilities. Depending on the complexity of the required splitting, sand can be separated into different sizes by the use of screw classifiers, rake classifiers, hydraulic classifiers, or hydraulic sizers. Water settling ponds are required since specifications seldom permit the discharge of the processing water into drainage systems.

If gradation and fineness modulus requirements cannot be met by separation and reblending of the sand, then crushing or grinding of the sand may be required. This can be done by adding a rod mill or a sand crusher.

When there is a shortage of sand, when there is a lack of coarse sand, or when the sand that is present in the deposit is unsuitable and must be wasted, then sand must be manufactured from any excess in production of any size of aggregate. This necessitates the addition of sand-manufacturing equipment to the processing plant. This sand-manufacturing equipment consists of Cone or Gyrasphere crushers and recirculating belts to furnish the plant feed, rod mills or sand crushers to produce the sand, sand classifiers, sand pumps, and settling ponds. Since these facilities are similar to those required for the manufacture of sand from quarried rock, more detailed descriptions of their use will be given under manufactured-aggregate production.

From this description, it can be seen that natural-aggregate processing plants may vary from relatively simple screening plants, like the one in Drawing 3, to very complex aggregate-processing plants.

Manufactured-aggregate Plants

Concrete aggregates are manufactured either from quarried rock or, in special cases, from rock excavated for the dam's foundation. Quarried rock is used when the specifications make its use mandatory or when the haul from the closest suitable gravel deposit to the jobsite is so long or difficult that it makes the use of natural aggregates uneconomical.

If aggregates are to be manufactured from rock excavated from the dam's foundation, a large stockpile is a necessity since aggregate usage occurs after the major portion of the dam's foundation has been excavated. If the haul to the rock stockpile is greater than to a disposal area, the extra haul cost should be charged to aggregate production. Selective loading of the excavated rock may be required since the stockpiled rock should be of sizes that can be handled by the primary crusher; otherwise secondary drilling and blasting will be required.

If aggregates are to be manufactured from quarried rock, the quarry site should be inspected so that the quarry's development can be planned and suitable equipment can be selected. The development of a rock quarry and the production of broken rock will be described in Chap. 7.

The shovel used in the quarry should be sized to fit the opening of the primary crusher, so that the shovel cannot pick up larger rocks than the crusher can handle. Equipment manufacturers list the size of shovel that should be used with each size of crusher. To obtain efficient loading and haulage operation, truck size should also be matched to shovel size, using the criterion that only two or three shovel passes should be required to load a truck to capacity. Since the development of the large, front-end loaders, economics often dictate the use of loaders instead of shovels in the softer rock quarries.

An aggregate-manufacturing plant is composed of two plants: a rock plant and a sand plant. Quarried rock is the feed for the rock plant; the sand plant's feed is minus $^{3}/_{4}$-in. material that is produced by the rock plant. If an insufficient amount of minus $^{3}/_{4}$-in. rock is produced in the primary and secondary crushing of the quarried rock, additional sand-plant feed can be secured by using Cone or Gyrasphere crushers to crush an excess production of any size of aggregate. The design of both plants is established by preparing flow charts. Figure 5 in this chapter shows the flow chart for the aggregate-manufacturing plant used in the example estimate in Chap. 11.

A typical rock plant starts with a drive-over truck dumping bin equipped with grizzly bars to reject oversized material. This oversized material must be drilled and shot or broken with a headache ball before it can be placed back into the system. From this bin, fines fall through grizzly bars directly onto a conveyor belt and the coarse material is fed into a primary crusher, or in some instances, both the fine and coarse material are fed to the primary crusher. The primary crusher can be either a jaw or a gyratory type. The feed to the primary crusher is controlled by a feeder equipped with a magnet to remove shovel-bucket teeth or other pieces of iron. The larger the crusher, the larger the rock that the crusher can handle and the less the amount of secondary drilling and shooting required in the quarry. At Detroit Dam, Oregon, where aggregate was manufactured from quarried rock, the primary crusher was a 62- by 84-in. Blake-type jaw crusher. This crusher had a large enough opening to permit the use of coyote blasting in the diorite-rock quarry and to minimize the amount of secondary drilling and shooting, thus reducing the quarrying cost and, correspondingly, the aggregate cost.

If the aggregate plant is designed for low production, the

discharge opening of the primary crusher is set to produce minus 6-in. material and secondary crushing is not installed. When high-production aggregate plants are installed, the primary crusher is set with a larger discharge opening, i.e., from 12 to 15 in., and a secondary crusher or crushers are installed to reduce the rock to minus 6 in. These secondary crushers are usually of the gyratory type.

The crushed rock is conveyed from the crushing unit to a raw-material surge pile of sufficient size to permit the remainder of the aggregate plant to operate independently of the rate of rock production in the quarry. From the surge pile, the rock is conveyed to vibratory screens, where the aggregate is separated into sizes. The screened products are then placed in separate stockpiles.

The screening section of the plant can be laid out in a straight-line pattern, or with a central screening tower. The difference between these two was explained in the former section on natural-aggregate processing plants. Similar to a natural-aggregate plant, rock ladders are used beneath the discharge end of each large-aggregate stockpiling conveyor to limit the free fall of the rock into the stockpile, so that size separation and aggregate breakage will not occur.

The aggregate stockpiles are often equipped with recirculating belts which reclaim any excess production of any size of aggregate and feed it to tertiary crushers to produce a finer-sized aggregate or feed for the sand plant. If excess production is not reclaimed from the stockpiles, it may be split at the screens' discharge or distributed between the stockpile and the tertiary crushers from a storage bin located beneath the screens. A typical installation of a tertiary Gyrasphere crusher is shown in Photograph 23.

Two types of sand-manufacturing plants are in common use, a rod mill plant and a sand crushing plant.

If a rod mill is used, its feed is minus 1/4-in. material produced by Cone or Gyrasphere crushers. A rod mill can produce sand to meet specification requirements by varying the maximum size of the feed to the rod mill, the speed of rotation of the rod mill, the length of time that the material is in the rod mill, and the rod charge in the mill. Not only does the rod mill control the sand gradation, but it controls the rate of sand production. The history of most rod-mill-type sand-manufacturing plants is that insufficient rod-mill capacity was initially installed; consequently, in order to make sufficient sand, additional rod mills were required.

A recent development in sand manufacture is to use a vfc sand crusher (Barber-Greene) or a gyradisc crusher (Nordberg) in a closed-crushing and screening circuit to produce sand. These

Photo 23 View of an aggregate plant showing screening towers and a Gyrasphere crusher. (Smith Engineering Works.)

crushers have fine adjustable-discharge openings; however, fine sand particles are produced by the layers of particles acting upon each other to accomplish reduction. Feed for the plant is the excess in production of $^3/_4$- by $^1/_4$-in. material and a portion of the sand that passes a no. 4 screen and is retained on a no. 12 or 14 screen (see Figure 5 in this chapter). By varying the discharge opening of the crusher, the size of the top sand screen—no. 4 or 6, the size of the bottom screen—no. 12 or 14, sand of required gradation can be produced from many types of rock. The feed should have less than 4 percent total moisture and less than 7 percent surface moisture. This sand-manufacturing process has produced concrete sand successfully at many concrete dams, including Green Peter, Oregon, and Dworshak, Idaho. When its use is applicable, the cost of sand production will be less compared with the use of a rod mill.

After sand is produced, it must be dewatered and sized. Often dewatering and sizing can be done in single-stage screw classifiers. In some instances, it is necessary to separate the sand into different sizes in a series of screw classifiers, or in hydraulic sizers, or in hydraulic classifiers. These sizes of sand are separately stockpiled; when reclaiming is done, they are blended on the reclaim belt to produce sand of the required gradation.

The waste water from sand plants must be clarified by using thickeners or by retaining the water in large settling ponds. Multiple sand stockpiles may be necessary when specifications require that the sand remain in free-draining stockpiles for a specified time period before usage.

Similar to a natural-aggregate plant, the method used for reclaiming the aggregate and sand from the stockpiles and for truck loading of this material depends on the tonnages to be handled. Conveyor-belt reclaiming is done when large tonnages are involved, and front-end loader reclaiming when only small tonnages are handled.

The foregoing descriptions of aggregate and sand manufacturing plants are very general. If the plant feed has special characteristics, other types of equipment may be required, such as roll crushers, reversible impactors, hammer mills, etc.

AGGREGATE HAUL

The processed aggregates must be transported from the aggregate plant to the mixing plant and charged into the mixing plant's batching bins. When the aggregate plant is located adjacent to the mixing plant, the aggregates are transferred between the two plants on a conveyor belt. When there is a considerable distance between the two plants, the most economical method of making this transfer is with bottom-dump trucks, since their ratio of payload to truck weight is excellent. At Bullards Bar Dam on the Yuba River, the concrete aggregates for a plant with six 4-cu-yd mixers were hauled with one primary mover pulling two 60-ton bottom-dump trailers.[10]

When there is a long distance between the aggregate plant and the mix plant, it is impractical to attempt to schedule the proper arrival of aggregates sized to comply with the mixing plant's demands. Instead, aggregate storage and reclaiming facilities are installed adjacent to the mixing plant to provide a surge storage for recharging the mixing plant's batching bins. These aggregate storage and reclaiming facilities usually consist of a drive-over truck receiving bin and conveyor belts to transport aggregates to the stockpiles. The elevated section of the conveyor belt that is located over the stockpiles must be equipped with a traveling tripper or a plow so that each size of aggregate can be discharged into the proper stockpile with rock ladders limiting the free fall of the large aggregates into the stockpiles. Similar to aggregate plants, reclaiming is done by conveyor belts if large tonnages must be handled and by front-end loaders if the tonnages handled are minor. The reclaimed aggregates pass through a rescreening plant if one is located near the mixing plant; otherwise, they feed the mixing plant's charging conveyor.

RESCREENING AND MIX-PLANT CHARGING

Some specifications require that the concrete aggregates be rescreened to eliminate any size degradation before they are charged into the mixing-plant bins; if this is specified, a complete screening plant is required, equipped with spray nozzles to reduce dust, with dewatering equipment, and with settling ponds to clarify the water used in the screening process. This rescreening plant can be located on top of the mixing plant or on the ground adjacent to the mixing-plant charging conveyor. The latter location usually results in the least capital expenditure.

Charging of the mixing plant's aggregate storage bins is accomplished by repetitive batch loading of the sand and each size of aggregate on an inclined conveyor belt. The belt extends from ground elevation to the top of the batch plant or rescreening plant. The required capacity and method of operation of this belt are described when mixing plants are discussed.

CEMENT AND POZZOLAN HANDLING

Since cement is a required ingredient in all concrete, it is necessary to provide facilities to handle its receipt, storage, and transfer from the storage silos to the mixing plant's batching silos. Similar facilities are a necessity for pozzolanic materials when they are used in the mix. Pozzolanic materials are often a required ingredient in dam concrete since they can replace part of the cement, act as a cement dispersing agent, and improve the workability of the concrete. When the concrete's workability is improved, the water-cement ratio can be decreased, which increases the strength of the concrete. Also, some types of pozzolanic materials will combine with the free alkali in the cement to neutralize its reaction with reactive aggregates.

If interground cement and pozzolan can be purchased and this mixture will satisfy the specifications, its use is economical, since only one set of handling facilities will be required.

If the cement and pozzolan are delivered by truck to the mixing-plant area, the installations necessary to store and handle them will be of a minimum. Modern cement and pozzolan delivery trucks contain pressure equipment for truck unloading. The pressure system forces the cement out of the truck, through pipes, and into the storage silos. With this type of delivery, the facilities required at the site are storage silos; screw conveyors beneath the silos, which transfer the material from the silos into bucket elevators; and bucket elevators, which discharge the cement into the top of the mixing plant's batching silos.

Separate silos and handling facilities are required for cement and pozzolan unless they are interground and delivered in a mixed condition. Sufficient storage silo capacity should be installed so that the mixing plant will operate independently of the delivery of cement and pozzolan. The minimum storage to be installed varies with the distance from the project to the location of the cement source, the job location, the ease of delivery, and whether weather conditions will affect delivery.

If the dam is at a location where the cement and pozzolan delivery trucks discharge the load by gravity through a bottom gate and are not equipped with a pressure discharge system, then a receiving hopper, screw conveyor, and bucket elevators are required to receive and place the cement and the pozzolan into storage silos.

If the cement and pozzolanic materials are delivered by rail, then the contractor must supply and operate pumps to transfer these materials from railroad cars into trucks for haulage to the mixing-plant site.

The mixing plant may be in such an inaccessible location that cement-delivery equipment will not have access to the mixing-plant site. In this event, the cement is often pumped from the point of receipt, through a pipeline, to silos at the mixing plant. At Detroit Dam, Oregon, cement was received on railroad cars at a siding located on the right abutment, directly across the canyon from the mixing plant which was located on the left abutment. The railroad cars were unloaded with pumps that pumped the cement through a pipeline into storage silos located on the left abutment, adjacent to the mixing plant. The pipeline was suspended from a cable stretched across the river canyon.

AGGREGATE AND CONCRETE COOLING

If cold concrete is placed in a dam and, when necessary, if the temperature of the concrete is controlled after placement, then the heat of hydration of the cement will not raise the temperature of the concrete above its final temperature, which is approximately the same as the average temperature of the locality. If the temperature of the concrete never rises above its final temperature, it will not drop in temperature and will not be subject to shrinkage; if it does not shrink, grouting becomes a minor consideration. Also, the concrete will reach its final temperature in a relatively short time, as compared to the years required when its temperature is not controlled. If the temperature of mass concrete is controlled, then concrete can be placed in large blocks and the contraction joints need not be grouted, or they can be grouted shortly after concrete placement of each grouting lift has been completed. Whether contraction-joint grouting is necessary depends on the type of dam and the requirements specified by the contracting agency.

The placement temperature of the concrete and the amount of temperature control that must be maintained after the concrete has been placed vary with the amount of heat generated by the hydration of the cement and the rate at which heat is dissipated from the pour. The heat generated in the pour varies with the quantity and type of cement used in the mix. Heat dissipation from the pour varies with the pour height, pour dimensions, time interval between pours, type of aggregate, type of forms, length of time the forms are in contact with the pour, concrete curing method, and temperature of the locality. Concrete should not be placed at such cold temperatures that freezing will occur. When concrete pours are massive (preventing rapid heat dissipation) or when placed in extremely hot weather, the temperature rise of the poured concrete cannot be prevented solely by the use of cold concrete. Under these conditions, additional concrete cooling must be done after the concrete is placed in the form. This additional cooling is achieved by pumping chilled water through cooling coils located on the top of the rock foundation and between each pour. The average length of time that this chilled water must be circulated through each cooling coil to control the temperature rise varies but often is approximately 12 days. Near the dam foundation where heat generation is greater than heat dissipation, pour heights are reduced so that the cooling coils can be installed at closer intervals. When the dam concrete reaches an elevation that contains smaller pours, more heat will be dissipated, so that cooling coils may no longer be required and the use of cold concrete will control the temperature rise in the concrete.

At some locations the river water may be cold enough when circulated through the cooling coils to cool the concrete; otherwise, refrigerated water is required, as is the refrigeration equipment needed to produce it. Control of the concrete temperature with cooling coils is based on the use of cooling-water velocities of 2 cfs (5 gal per min) through 1-in.-diameter cooling coils. The tonnage of refrigeration required to chill this water is determinable by establishing the maximum amount of water that must be chilled. This can be computed using the construction schedule to establish the maximum number of pours that must be chilled during one time period.

Specification restrictions on concrete placement during hot weather differ depending on the contracting agency, the size of the concrete structure, and the temperature at the locality. Specifications may require that aggregate piles be shaded from the sun, that they be sprinkled to permit evaporative cooling, that aggregate conveyor belts be shaded, that concrete placement not be done during the summer months, that concrete placement be restricted to nighttime pouring during the summer months, that concrete be placed at a specified temperature, and that river water or chilled water be circulated for a defined time period through cooling coils placed in the concrete. The specifications may limit the maximum and minimum time for lift raising and may designate the time for form removal. Specifications should always be reviewed to determine the restrictions on concrete placement because of temperature conditions and how these restrictions will affect the construction schedule.

The temperature of mixed concrete can be controlled using many different methods. The simplest methods are those that cool the large aggregates and utilize ice in the concrete mix. The cooling of sand, cement, and pozzolan is undertaken only when it is a necessity, since their cooling is difficult and expensive. Sand can be cooled by using the vacuum process to provide effective evaporative cooling. Sand, cement, and pozzolan can also be cooled in commercial-type coolers.

If the aggregate storage piles are shaded, this will prevent them from rising in temperature. Evaporative cooling of aggregates to wet-bulb temperatures can be accomplished by spraying water on the aggregate storage piles. Large aggregates can be cooled by spraying chilled water on them while they are transported by a covered conveyor belt. Photograph 24 shows an installation of this type.

The simplest and easiest method of reducing the temperature of concrete is to replace part of the free water in the mix with ice. This method provides enough cooling action when the concrete temperature must be lowered only a few degrees. Since the ice that can be used in the mix is limited to approximately 80 percent of the free mixing water, its cooling action must be supplemented by other methods of cooling when large temperature reductions are required. The required temperatures can often be obtained by cooling aggregates and by using ice as a varying mix ingredient to control the final temperature. A simpler explanation is that the hotter the day, the greater the amount of ice used in the mix.

If ice is used in the mix, it is necessary to provide ice-making machinery, storage bins, and a method of conveying ice to the mixing plant, ice bins in the mix plant, and an ice batcher.

Large aggregates can be cooled by immersing them in chilled

Photo 24 Cooling concrete aggregates. Two enclosed 60-in. belts with chilled water sprays, Dworshak Dam, Idaho. (Lewis Refrigeration Co.)

water. This is a batch-type process. The aggregates are placed in watertight bins and then the bins are sealed and flooded with chilled water. The aggregates are retained in these bins until their temperature is lowered to that of the chilled water. The water is then withdrawn, the bottom of the bin is opened, and the aggregates discharged onto a conveyor belt for transportation to the mixing-plant charging bins. With proper scheduling of cooling operations and mixing-bin charging, aggregate cooling can be accomplished with one bin for each size aggregate. This method of aggregate cooling was used at Detroit Dam, Oregon, constructed in 1953.

Large aggregates can be cooled in the mixing-plant batch bins by the upward circulation of cold air through the aggregates. When this cooling method is used, the mixing-plant batching bins should have capacity to allow the aggregates to be cooled for approximately 3 hr. Since each bin is charged in succession, to achieve this cooling capacity the bins must have a capacity sufficient to run the mixing plant from 4 to 5 hr. The ammonia compressors required for this cooling process are installed adjacent to the mixing plant. Ammonia piping, pressure controls, and ammonia expansion chambers connect the compressors to heat exchangers that are located on the plenum chamber erected around the mixing plant's batching bins. These heat exchangers cool the air in the plenum chamber. Fans are located in this chamber to force the cold air upward through the aggregates in the batching bins and back to the plenum chamber where it is again cooled. This type of cooling has been used on many projects, including Mossyrock Dam in Washington and Glen

Canyon Dam on the Colorado River. Photograph 25 shows the installation of heat exchangers on a mix plant at Dworshak Dam, Idaho.

Another method of cooling aggregates is to use a partial vacuum to expedite evaporative cooling. Aggregates are placed in a large, airtight bin connected to a chamber containing a steam jet. The action of this jet draws air and water vapor from the bins, producing a partial vacuum and causing the moisture on the aggregate to vaporize, thus cooling it. Approximately from 30 to 45 min are required to cool the aggregates in each bin. The bins are baffled inside and these baffles serve as rock ladders, protect the walls of the bin, and distribute the aggregate throughout the height of the bin. Bins may be located adjacent to the mix plant or may be part of the mixing-plant structure. Before being charged into the bins, the aggregates and sand should be dampened until the minimum moisture content of the aggregates is 1.5 percent and of the sand is 3.0 percent. Photograph 1 in Chap. 2 and Photograph 26 picture mixing plants that have vacuum-cooling bins mounted on the mixing-plant structure.

When sand is cooled by the vacuum process, the sand bins are equipped with special baffles that distribute the sand so that air spaces are provided in the centers of the bins and next to the bin walls. Another method of constructing a sand-cooling bin for this cooling process is to separate the bin into lower and upper compartments. When the bin is sealed and placed under vacuum, a sand

Photo 25 Heat exchangers mounted on the plenum chamber on the exterior of a mixing plant, Dworshak Dam, Idaho. (Lewis Refrigeration Co.)

Photo 26 Mixing-plant and vacuum-cooling aggregate tanks, Green Peter Dam, Oregon. (C. S. Johnson, Division of Koehring Company.)

meter opens and lets the sand flow at a measured rate from the upper half of the bin into the lower half. This sand flow permits the moisture to boil off and cool the sand by evaporation.

The number of aggregate bins required for vacuum cooling can be reduced by mixing the various sizes of aggregates in the same proportions as they are used in the concrete mix before they are charged into the cooling tanks. After the mixed aggregates are cooled, they are resized by an aggregate rescreening plant located on top of the mixing plant. This premixing of the aggregates makes it possible to cool all the aggregates in three cooling bins. While one bin of premixed aggregates is being cooled, another aggregate bin can be charged, and the cooled aggregates can be drawn from the remaining bin. Two additional bins are required when the sand must be cooled. In addition to the bins, steam generators, steam jets, and conveyor belts are required. The vacuum-cooling method has been used on many projects, including Derbindi Kan Dam in Iraq, Hartwell Dam in Georgia, Allegheny Dam in Pennsylvania, and Green Peter Dam in Oregon.[11,12]

The selection, design, and layout of a suitable refrigeration and cooling plant is such a complex problem that it is recommended that the planning engineer contact either refrigeration-plant equipment suppliers or competent consultants to assist in laying out the cooling plant.

When a consultant or a refrigeration-plant manufacturer recommends the method and extent of cooling, the planning engineer should always check the results. The computations that are made to check refrigeration requirements are illustrated in Chap. 11, in the discussion of estimating task XII.

CONCRETE MIXING

The concrete-mixing plants used for dam construction differ in design because of differences in specification requirements and differ in their capacity because of differences in project requirements. Concrete mixers must have a capacity of 4 cu yd or greater when 6-in. aggregate is used in the concrete mix. The hourly production rates of the mixing plants will vary from 60 cu yd per hr, when they are equipped with one 2-cu-yd mixer, to 410 cu yd per hr, when they are equipped with six 4-cu-yd mixers. Their control boards become complex when many different concrete mixes are required, and supports often must be enlarged to take care of the weights of rescreening plants placed on top of their bins or of aggregate cooling equipment hung on the side of the plant.

In Chap. 4 it was explained that the total cubic yard capacity of the concrete mixers should be 1.5 times the bucket capacity of the placement equipment in cubic yards. This means that if the total bucket capacity of the placement equipment is 16 cu yd, then six 4-cu-yd mixers should be installed. The hourly capacity of the mixing plant can be determined by computing the number of cycles per hour made by each mixer and multiplying that by the mixer capacity and number of mixers. The number of cycles per hour depends on charging time, mixing time (which is a specification requirement), and mixer discharge time.

Typical hourly production rates for mixers of different size are as follows:

	2-cu-yd mixer	3-cu-yd mixer	4-cu-yd mixer
Charging time	$^1/_4$ min	$^1/_4$ min	$^1/_4$ min
Mixing time	$1^1/_2$ min	2 min	$2^1/_2$ min
Dumping time	$^1/_4$ min	$^1/_4$ min	$^1/_4$ min
Mixing cycle	2 min	$2^1/_2$ min	3 min
Cycles/hr	30	24	20
Cu yd production/hr	60	72	80

The mixing plants cannot be scheduled to constantly mix concrete at maximum-capacity production rates, since there are times when one or more mixers have to be removed from service for repair or relining.

When six mixers are arranged in a circle in one plant, the time required to go around the circle and charge each mixer may be greater than each mixer's cycle, and this factor may control production. As an example of how hourly mix-plant production is determined, a 4-cu-yd mixer will produce 20 batches per hour when it is operated in conjunction with five mixers or less; under these conditions it has an hourly production of 80 cu yd per hr. When it is operated with six mixers, the charging cycle controls the number of mixing cycles, reducing them to 17 and the hourly production to 68 cu yd per hr.

Specifications for different projects may differ in mixing-plant requirements, such as whether accumulative or individual weighing and measuring equipment must be used; what kind of sampling equipment, and what amount of control equipment should be installed; whether rock ladders are necessary; and what type of bins is preferred. The number of different types of concrete mixes that are required influences the size and number of mixers. For rapid plant operation, push-button control for a rapid change of mix ingredients is desirable.

Aggregate cooling requirements may necessitate the use of large storage bins and the installation of plenum chambers, heat exchangers, fans, etc.

Specifications often require that aggregate rescreening facilities be placed adjacent to or on the top of mixing plants. When they are located on top of the mixing plants, this, of course, increases the height of the plant, necessitates the use of a longer charging conveyor, and increases the load on the bin supports.

When there is a small quantity of concrete to be placed, the specifications may permit the use of a *minimum mixing plant*. A typical minimum plant consists of low, open-type aggregate bins, a cement silo equipped for truck delivery of cement, a weigh batcher that successively accumulates the charge for the mixer as it travels under the aggregate and sand bins and the cement silo, a conveyor belt that receives the charge from the weigh batcher and charges the mixers, and a small mixer of either the tilting type or the rotary type. Since the aggregate bins are of a low, open type, they can be charged with aggregates by using a rubber-tired, front-end loader. When only one type of mix is required, when belt conveying of the batched material to the mixers is permissible, and when concrete must be produced in large quantities, then this type of mixing plant can be secured with a tilting mixer that has a capacity up to 12 cu yd. However, because of restrictions concerning the method of charging mixers, because different types of concrete mixes may have to be produced almost simultaneously, and because individual weigh batching may be required, this type of plant is seldom used in concrete-dam construction. When many mixes are required, when individual weigh batching is needed, and when mixer charging must be done without producing dust, then a stationary plant with more complex weighing, measuring, and batching equipment is required.

Dams containing large volumes of concrete of many different mixes may require mixing plants that contain multiple mixers. The mixing plant used for the construction of Glen Canyon Dam contained six 4-cu-yd mixers. It also included a rescreening plant, ice batching facilities, and facilities for air cooling of large aggregates. Six 4-cu-yd mixers is the greatest number of mixers that can economically be located in one plant. Two mixing plants have been installed on dam projects that required high rates of concrete production. At Grand Coulee Dam, two mixing plants were installed, each containing four 4-cu-yd mixers. At Dworshak Dam, two mixing plants were used, one containing six 4-cu-yd mixers and the other four 4-cu-yd mixers.[9]

A schematic layout of a mixing plant that includes an aggregate charging conveyor, bucket elevators and chutes for charging the cement and pozzolan silos, aggregate rescreening facilities, batch bins, aggregate-cooling facilities, ice batching, six 4-cu-yd tilting mixers, and two wet-batch hoppers is shown in Drawing 4. This type of plant will meet most specification requirements since it is equipped with self-cleaning batch bins, individual automatic-measuring and weighing equipment, equipment for automatic measuring of concrete consistency, push-button control of the scales so that any one of the multiple of mixes can be weighed simultaneously for mixer charging, batch-type tilting mixers, and automatic recording of each operation. This schematic layout is of an open-type plant, but contractors place siding over the lower part of the plant to protect the operators and control equipment and to provide protected areas for concrete testing and office space.

The flow of concrete materials through mixing plants of this type is started by loading the charging conveyor with successive charges of sand and each size of aggregates, leaving empty spaces between each size so that there cannot be size intermixing. Because the charging conveyor must operate as a batching conveyor, the maximum tonnage it can handle will be approximately 70 percent of its

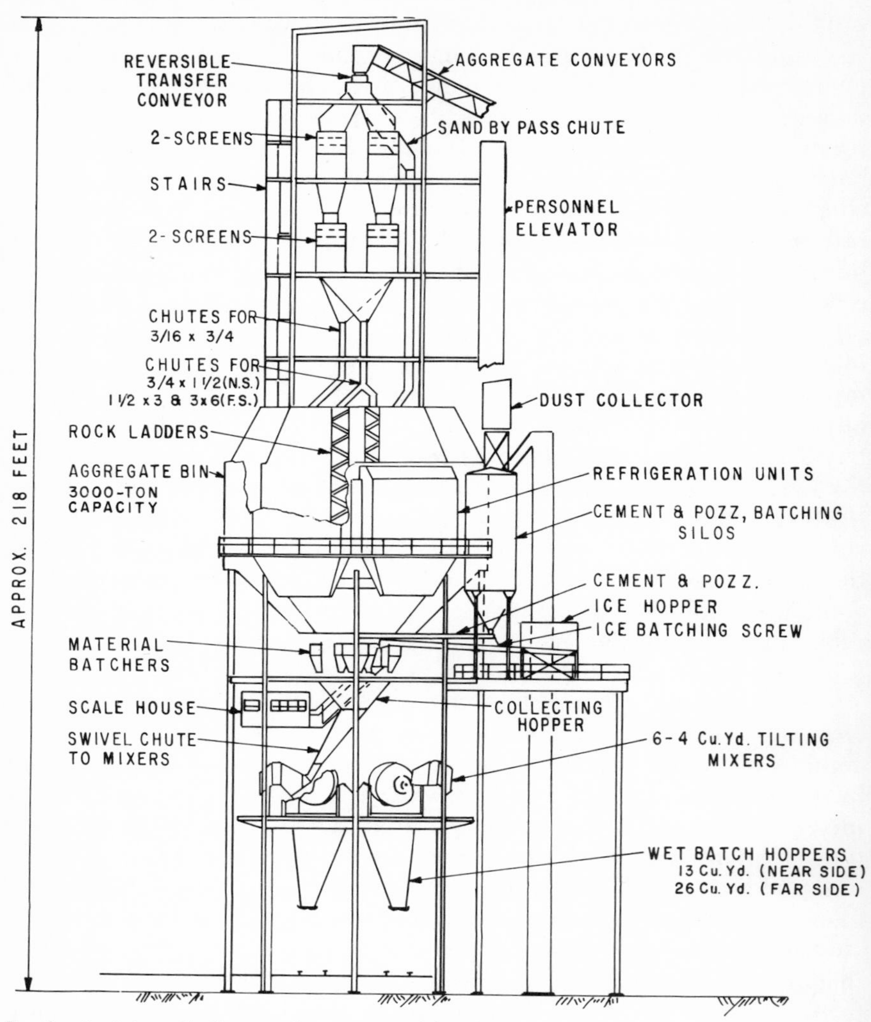

Drawing 4 Schematic diagram of a concrete-mixing plant.

theoretical capacity. To illustrate, six 4-cu-yd mixers will mix 410 cu yd of concrete per hr requiring 760 tons of aggregate per hr. At 70 percent efficiency, this will increase to 1,090 tons per hr. This establishes a charging conveyor size of 42 in. or greater, and it must operate at a speed of 250 ft per min.

The charging conveyor transports the sand and aggregate to the top of the mix plant where it passes through a rescreening plant. The purpose of this plant is to eliminate any size degradation in the aggregate. The rescreening plant must be sized to take the production of the charging conveyor; for a plant containing six 4-cu-yd mixers, it should have a screening capacity of 1,090 tons per hr.

After passing through the screens, the aggregates are chuted into the batching bins. These bins are equipped with rock ladders to restrict breakage and prevent segregation. Five sand and aggregate bins are arranged around center silos for cement and pozzolan. Bins are sized in accordance with the percentage of sand and each size of aggregate used in the concrete mix. Cement and pozzolan are charged into their respective batching silos by bucket elevators and chutes. The silos and bins should contain enough cement, pozzolan, sand, and aggregate to supply the mixers for 2 or 3 hr. This permits the mixing plant's charging conveyor to transport one size of aggregates for 15 to 30 min.

For the mixing plant containing six 4-cu-yd mixers, 3 hr of bin storage would require sand- and aggregate-storage bins to have 2,200 tons of storage capacity. If cold air is circulated through the mixing plant's storage bins to cool the aggregates, it often takes 3 hr to cool them to the desired temperature. This necessitates increasing the bin sizes to 4 or 5 hr capacity, which is a capacity of from 3,000 to 3,600 tons.

The required pounds of aggregates, sand, cement, and pozzolan for charging the mixers are released from the bins, individually weighed, and dumped into a collecting hopper, and then the required amount of ice, water-reduction agent, and water are added. This collecting hopper has a swivel chute that allows it to progressively charge the tilting mixers arranged in a circle beneath it. At the end of the mixing cycle, the mixers discharge the mixed concrete into one of two wet-batch hoppers, and these hoppers discharge the concrete into concrete buckets or into transfer cars, depending on the type of haulage equipment used.

Mixing-plant manufacturers will quote and supply complete mixing plants that will comply with various specifications. When requested, they will quote on a complete mixing plant, including charging conveyors, cement- and pozzolan-storage silos and bucket elevators, and aggregate rescreening and cooling plants. When a request is made for a mixing-plant quotation, the request should specify the number of mixers, include the specification requirements, designate whether pozzolan is to be used in the mix, list the number of additives that will be used in the mix, and specify if ice batching is required or whether cooling is to be incorporated into the plant. The estimator can then use the manufacturer's quoted price for the complete mixing plant and storage silos in his cost estimate.

To this he must add the cost required for site preparation, foundation construction, distribution of utilities to the plant, construction of a bucket cleaning yard, plant erection, and furnishing and installation of the housing around the mixing plant and of any required insulation.

FOUNDATION CLEANUP

All rock foundation surfaces must be cleaned prior to being covered with concrete. Cleaning consists of the removal of unsound rock, debris, mud, and running or standing water. The equipment required for this work consists of air and water jets, hand tools, and skips to contain the excavated material. Cranes may be required to handle and dump these skips. Air and water must be supplied to the cleaning equipment.

CONCRETE CLEANUP

The surface of all concrete pours must be cleaned, and standing water must be removed before the next concrete lift can be placed. All *laitance* (loose or defective concrete) must be removed, and the larger aggregates partly exposed to provide good concrete bonding. This can be done from 4 to 14 hr after the concrete has been placed, using high pressure air and water cutting jets. Recently, air and water cutting equipment with pressures as high as 6,000 psi have been developed. This high-pressure cutting is so effective that it can replace the more expensive sandblasting procedure; sandblasting equipment may, however, be required by the specifications.

Equipment used by the cleanup crews are hand tools, air and water jetting equipment, and sandblasting equipment. Utility distribution to each pour area is also required.

PLACEMENT OF EMBEDDED ITEMS

The reinforcing steel and all other embedded materials must be placed in a pour before concrete placement can start. The other embedded materials may be steel penstocks, air-vent piping, pipe drains, concrete cooling coils, grout pipe, metal grout seals, water pipe, trashrack anchors, anchor bolts, electrical conduit, electrical outlet boxes and fixtures, pipes through which drilling is performed, manhole and hatch-cover frames, ladder rings, grating guides, etc.

So that pours will not be unnecessarily delayed, it is essential that all the embedded materials be installed and checked before form erection and concrete cleanup are completed. Pour drawings showing the dimensions and survey control points for each pour are prepared prior to concrete placement. These drawings also show the location and quantities of embedded materials. All embedded materials must be transported to and placed inside the formed area. A cableway or crane hook is needed for lifting and placing penstock sections. Pours may be delayed for a week or more when penstock or other large embedded materials are installed and fabricated. Other equipment required are rod benders, welding machines, and hand tools.

Often the erection of reinforcing steel, embedded conduit, penstocks, etc., is done by subcontractors, but the contractor usually must supply the hook required to move the material into the pour.

CONCRETE FORMING

Concrete forms are required to maintain freshly placed concrete in the desired shape until it develops enough strength to be self-sustaining. The cost expended on the erection of these forms is one of the largest components of the total direct cost of the concrete. Form materials, form oil, *form hardware,* and form labor are the major items of form cost. Form hardware is a general term designating form anchors, form ties, tie clamps, and form fasteners. Form oil and form hardware are expenses per square foot of form contact. The cost of form materials and forming labor has increased so sharply that economical form cost can only be achieved when the expenditures for these items are minimized. The cost of these two items can be reduced by using purchased metal forms instead of shop-built wood forms and by scheduling concrete forming in such a way that the greatest reuse of forms is obtained while consistently providing sufficient formed areas to permit the scheduled rate of concrete placement.

The facilities required to construct and erect forms consist of a carpenter's shop complete with saws, a staging area for form construction, form storage areas, a truck to move forms, small hydraulic cranes for lifting dam-monolith forms, and powered hand tools.

CONCRETE DEPOSITING AND VIBRATION

After the concrete or foundation cleanup has been completed, the forms have been set, the embedded materials and reinforcing steel have been placed, and the pour has been approved, then concrete can be deposited in the pour.

Mass-concrete placement is started at one end of the pour by covering the foundation or the previously placed concrete with a $^{3}/_{8}$-in. layer of mortar that is worked into any irregularities. Next, a horizontal layer of concrete (not exceeding 22.5 in. in thickness) is placed across the downstream end of the pour. Then successive horizontal layers of concrete are placed, bringing a short length of pour to its full height across the width of the block. The upstream face of the pour is stepped with the bottom layer extending farther into the pour and each succeeding layer extending a shorter distance into the pour. Concrete placement is continued in an upstream direction in this stepped manner until the pour is complete, keeping the unconfined upstream edge of the successive layers of fresh concrete as steep as practical. As a bucket of concrete is deposited, vibrators are used to level and compact the concrete except at the upstream edges of the concrete layers. Gravel pockets are disbursed and vibrated into the concrete. This is performed using large vibrators with vibrating speeds of 6,000 rpm and heads 4 in. or more in diameter. Compressed air, water, and power must also be distributed through the pouring area.

Placing structural concrete is slow, since pours are small and accurate bucket spotting is required to discharge the concrete into the narrow form openings. More time is required for concrete vibration in this type of concrete placement which necessitates that more time elapse between the dumping of buckets.

COLD-WEATHER PROTECTION

Concrete placed during cold weather must be protected from freezing. The amount and cost of providing cold-weather protection increases as the temperature decreases. Mass pours do not require as much cold-weather protection as do thin reinforced pours, because the heat of hydration of the cement is not readily dissipated from mass pours and helps control their temperature. When the concrete-placement schedule is prepared, mass pours can be scheduled in cold weather, but structural concrete should only be scheduled in warm weather. If it is necessary to place structural concrete when temperatures are below freezing, the cost of pouring and protecting this concrete may almost double its warm-weather placement cost.

If concrete is placed when the mean daily temperature is 40° F and above, the specifications often require that it be protected against freezing temperatures for at least 48 hr after placement. Specifications also generally require that when the mean daily temperature is below 40°F, concrete be placed at a temperature not less than 50°F, be maintained at this temperature for 72 hr, and be protected from freezing temperatures for an additional three days.

During cold weather, steam jets are used to remove all ice, snow, and frost from the interior of the forms, from the reinforcing steel, and from other embedded materials before concrete placement is started. If concrete pours are on the rock forming the dam's foundation, the rock must be in an unfrozen state. Foundations can be covered with straw or other materials to help prevent their freezing. Steam may be required, however, to thaw the foundation if it becomes frozen. If cooling coils have been installed between concrete lifts, steam can be passed through these coils to remove snow and ice before pouring and to maintain the proper temperature in the concrete after it is placed. Steam is also used to heat the aggregates prior to their use in the mix. With preheated aggregates and warmed mixing water, mixed concrete can be produced with the desired temperature. When mass pours are made, pour protection can be provided on the sides of the pour with form insulation, and the forms may be left on the pours until temperature protection is no longer required. The top of the pour may be protected from a temperature drop by using heat lamps, or if the area is covered, the temperature of the covered area can be controlled by the use of piped steam or salamanders (oil heaters). During cold-weather pouring, the water-cement ratio of the concrete may often be lowered, and if the concrete is not subject to sulfate attack, its freezing temperature can be lowered by using 1 percent calcium chloride in the mix. When concrete is placed during extremely cold temperatures, steam curing is often substituted for water curing. When thin structural pours are made, either a greater amount of form insulation is required or steam must be applied to the form surfaces. The amount of insulation required for different thicknesses and types of pours is tabulated in various publications.[7]

Concrete placement during the winter months may be suspended if extremely low temperatures are anticipated. This occurred at Libby Dam where concrete was not placed from mid-November to the first of March since during this period temperatures reached as low as −20°F. To protect the concrete, the contractors installed $1^1/_2$ to 2 in. of urethane insulation over the tops of the monoliths, on the upstream and downstream sides of the top two pours, on exposed bulkhead joints, and over openings in the dam concrete. The tops of the 19 highest blocks had heating cables installed between the concrete and the insulation that maintained temperatures of 50 to 56° F throughout the winter. A total of $12^1/_2$ acres of insulation, 100 electric heating units, and 50 miles of heating cable, wire, and conduit were installed. The power consumption for heating was 1,900 kw.[13]

CURING

The amount of water contained in fresh concrete is more than enough for hydration of the cement. This contained water must be maintained or replenished in the concrete during the early rapid stages of cement hydration. This control of the contained water is called *concrete curing,* and is accomplished by keeping the concrete moist, by sealing exposed surfaces with a sealing compound, or by using steam when concrete is placed during periods of cold weather.

The curing required for each type of concrete and the length of the curing time is delineated in the specifications. Requirements for moist curing will differ in regard to both the required application and the length of the curing time, depending on the type of concrete surface, the type of forms, and the length of time that the forms are left on the pour. Flat concrete surfaces are often cured by ponding or by being covered with wet sand, wet earth, or wet burlap. Vertically formed surfaces are often cured by using a perforated pipe or soaker hose system that trickles water down the face of the forms or concrete. Another moist curing method used on both flat and vertical surfaces is the use of sprinklers to continually wet the concrete surfaces.

PATCHING AND FINISHING

The amount of patching and concrete repair that must be done to the cured concrete is controlled by the adequacy of the concrete placement. All imperfections in the concrete surface must be corrected. Any concrete that is damaged, honeycombed, or fractured, that has excessive surface depressions or is otherwise defective, must be excavated and built back up to the desired surface with dry pack mortar or concrete. Ridges and abrupt protruding irregularities must be removed by bush hammering or grinding. The permissible amounts that the concrete surface can vary from a true plane are set forth in the specifications. The owner's engineer uses a template to check these variations. Unsightly stains on exposed surfaces may have to be removed by grinding, sandblasting, washing with a solvent, or using additional sealing compounds. Holes greater than $^1/_4$-in. in diameter left by the removal of form hardware must be filled with dry pack mortar.

The kind of finishing required for different concrete surfaces is defined in the specifications. The amount and kind of finish should be tabulated when the form and concrete takeoffs are made. Finishing of reinforced surfaces will vary between screened surfaces, floated finish, steel trowel finish, or special finishes such as hardened floor topping or terrazo. Finish of formed surface is seldom required. The work involved after form stripping is concrete repair and patching, consisting of the repair of defective concrete, the removing of excessive irregularities, and the filling of the holes left by form hardware. If the concrete-formed surfaces are not up to specification requirements, sack rubbing of the surface may be necessary.

When the formed concrete surface is subject to special wear such as hydraulic erosion on a spillway slope or the concentrated wear on stair risers, a special stoned finish may be required. Special finishes on formed surfaces may also be required for architectural reasons.

INSTALLATION OF PERMANENT FACILITIES

Dam construction includes the installation of equipment and facilities required to control the water at the damsite or to service these water-control facilities. Typical items of this nature are gate and bulkhead guides, gates and bulkheads, penstock sections, navigation lock gates and operating mechanisms, elevator guides, elevators, cranes, etc. The construction schedule must allow sufficient time to install, fabricate, and align these items. Installation time is extensive when large spillway gates or navigation lock gates must be installed. The installation of permanent machinery and equipment cannot commence until the surrounding concrete has been placed. Therefore, their installation must be correlated with the concreting schedule. A level plant area is required for site assembly, fabrication, and storage of these items. Construction equipment required consists of welding machines, cranes, and heavy-duty transports. Concrete-placement hook time is needed for erection and assembly of these items in their final position. The fabrication and installation of these items require that special crafts be employed, so prime contractors often purchase these items on an installed basis, or if the owner supplies these items, the prime contractor may subcontract their installation.

The concrete immediately surrounding the guides for gates, bulkheads, and similar equipment is not placed when the dam blocks are raised, but is blocked out for future placement. When all the surrounding concrete has been placed, the guides are erected and aligned in these blocked-out areas. After alignment is completed, the concrete encasing these guides is placed. Using this procedure, gate-guide alignment can be accomplished to the accuracy specified. A large amount of fitting and welding is required during erection of all these items, and the extent of this work can only be determined by making quantity takeoffs.

If the dam contains a navigation lock, the fabrication and erection of the lock gates and closure mechanism is a major problem, and large heat-treating ovens may be required. Another specialty work item is the fabrication and erection of the permanent cranes required to handle trashrack sections or stop log bulkheads.

DRILLING AND GROUTING

Drilling and grouting for concrete-dam construction must be started as soon as the dam's foundation is excavated and cannot be completed until all major concrete pours are made. The work that must be done after the foundation has been excavated is the placement of the shallow grout curtain as described in Chap. 3.

The main grout curtain is placed after the concrete has been poured above the dam's foundation galleries. When the foundation galleries are constructed, $1^1/_2$- and 3-in. sections of pipe are embedded in the gallery floor, through which the grout holes and drain holes are later drilled. Drilling and grouting are started by drilling $1^1/_2$-in. grout holes to the prescribed depth. Then a piston-type grout pump capable of operating at high pressures is used to place the grout. Photograph 12, Chap. 3, depicts a grouting setup. After the grout curtain has been placed, 3-in. drain holes are drilled on the downstream side of the grout curtain; these holes remove any water that passes through the grout curtain and discharge the water into drainage ditches located along one side of the foundation gallery. Grouting is also required to fill the contraction joints and the embedded cooling pipe upon completion of the chilled water circulation. Grout piping and grouting connections to the contraction joints are embedded in the concrete as the dam is constructed. Metal grout seals, which retain the grout, are placed near the downstream and upstream edges of the contraction joints, and at approximately 50-ft intervals, horizontal grout seals are installed to form separated grout lifts. After concrete placement has been completed in each grouting lift and the concrete has been cooled, the contraction joint and the embedded cooling pipes can be filled with grout.

Drilling and grouting are specialty items usually subcontracted to drilling and grouting contractors. When the prime contractor does this work with his own forces, he must purchase or rent diamond drills and grout pumps and distribute grouting sand, cement, water, compressed air, and electricity to this equipment.

UTILITY DISTRIBUTION AND SERVICE FACILITIES

Water, power, and compressed air must be supplied and distributed to the various plants and construction areas. If any work is performed on a two- or three-shift basis, it is necessary to illuminate the work areas. Dam areas may be illuminated by stringing floodlights on a cable spanning the damsite or by placing banks of floodlights on elevated towers.

The service facilities required are compressor plants, water-pumping plants, a machine shop, a rigging loft, craft buildings, toolboxes, service cranes, service trucks, and pickups. Some of these facilities may be available from other construction operations, such as the service facilities provided for the dam's excavation.

REFERENCES

1. *Forms and Rock Bolt Engineering Catalog No. 1968–69*, copyright 1968, Williams Form Engineering Corporation, Grand Rapids, Michigan.
2. First Concrete at Libby Dam, *Western Construction*, October, 1968, page 33.
3. Dworshak Dam Concreters Crank Up for 200,000 cu yd per month, *Engineering News-Record*, December 5, 1968, page 32.
4. Donald D. Barnes, ed., *Aggregate Producers Handbook*, 2d ed., copyright 1960, Smith Engineering Works, a division of Barber-Greene Co., Milwaukee, Wisconsin.
5. *Crusher Slide Rule*, Smith Engineering Works, a division of Barber-Greene Co., Milwaukee, Wisconsin.
6. *Heavy Media Separation*, Wemco, a division of Western Machinery Company, San Francisco, California.
7. *Concrete Manual*, 7th ed., 1963, U. S. Department of the Interior, Bureau of Reclamation, Denver, Colorado.
8. Frank W. Stubbs, Jr., ed., *Handbook of Heavy Construction*, copyright 1959, McGraw-Hill Book Company.
9. Concreting Starts at Dworshak Dam, *Western Construction*, August, 1968, page 34.
10. Truck Takes Over as Concrete Plant's Lifeline, *Construction Equipment and Materials*, February, 1969, page 34.
11. New Way to Cool Aggregates by Steam, *Engineering News-Record*, April 9, 1959, page 38.
12. When Snow Blows at Alleghany—Rock Chiller Warms the Aggregate, *Engineering News-Record*, April 9, 1959, page 38.
13. Libby Dam Sheds Its Electric Blanket, *Contractors and Engineers Magazine*, April, 1970, page 56.

CHAPTER SIX

Earth-fill Embankment Construction

This chapter describes the construction of earth-fill embankments. Systems used in the construction of this type of embankment differ since individual embankments have different slopes, a different number of fill zones, and different types of fill. For economic reasons, earth-fill embankments are designed to suit each foundation condition and to utilize fill from adjacent borrow pits. This chapter describes the construction of a typical earth-fill embankment, containing a zone of impervious fill to prevent water passage through the embankment; a downstream pervious drainage zone, to remove any water seepage; a downstream zone of random fill, for dam stability; an upstream random-fill zone, to provide dam stability when the level in the reservoir is lowered; and an upstream surfacing of riprap, to protect the fill from wave action.

Some of the work required to construct an earth-fill embankment has been described in previous chapters. Briefly, the planning and the construction of any required diversion facilities were described in Chap. 2. The scheduling and performance of the foundation excavation and preparation were discussed in Chap. 3, as was the proper synchronization of fill placement in the embankment with the sequence of performing the foundation excavation. The planning and the construction of any associated concrete structures are described in Chaps. 4 and 5.

The facets of earth-fill embankment construction presented in this chapter include the planning required for the excavation of fill materials from borrow pits and the processing of these materials, transporting the fill to the embankment and spreading it on the embankment, control of the moisture content of the fill and compaction of the fill, placement of any required riprap surfacing, and the installation of settlement- and pressure-recording instruments.

Other aspects of earth-fill embankment construction are dealt with elsewhere. The development and operation of quarries that may be required to produce pervious material and riprap will be described in Chap. 7. When earth-fill dams contain random zones of coarse gravel, transition zones are located between the impervious zone and the gravel fill. (This transition zone material is also required on rock-fill dams with impervious cores.) The method of producing transition-zone material is likewise described in Chap. 7.

CONSTRUCTION-SYSTEM SELECTION

Economical construction of earth-fill embankments is accomplished by using construction equipment to perform the major work and limiting hand labor to the performance of a few minor tasks. For this reason, the cost of constructing earth-fill embankments mainly comprises the operating, maintenance, and depreciation costs of the construction equipment. To minimize the equipment's operating and maintenance costs, a construction system should be selected that utilizes large equipment units. The larger the equipment units, the greater their productivity and the less the amount of operating and maintenance labor required for each work unit. The depreciation resulting from the large capital expenditures required to purchase these large equipment units will be small per work unit since it can be amortized against the large fill quantities required for most embankments.

Many variables must be taken into consideration prior to the selection of the construction system. Earth-fill embankments are constructed using many different types of equipment since they are composed of different types of fill; borrow pits vary in depth and in their characteristics; haul roads are of different lengths and grades; embankment zones vary in number and widths; and there are differences in the lift thicknesses of embankments.

When the hauls are short, and unless excessive haul grades are encountered, transportation cost is low and loading cost controls equipment selection. As haul distances increase, the transportation cost also increases until it becomes the major consideration. Since long haul distances are required for the construction of most embankments, the method of haulage becomes the critical item in selecting the construction system.

Construction-system selection cannot be made until quantity takeoffs of embankment quantities are completed and tabulated in the order that they will be placed in the embankment. The borrow pits must be investigated, and a determination made whether processing of the borrow-pit materials is required. A flow diagram and a layout of any required processing plant must be prepared, and the placement of the fill into the embankment must be scheduled. Finally, the haul roads must be laid out and the haul distances and grades determined. When all these data have been assembled, haulage cycles can be computed and hourly production figures established. These figures will control equipment selection.

QUANTITY TAKEOFFS

Before quantity takeoffs are started, the relationship between bank yards, loose yards, and embankment yards should be established. Payment is received for fill measured in the embankment, termed *embankment cubic yards*. The yardage hauled from borrow pits to the embankment is *loose cubic yards*. The yardage loaded by the loading equipment is *bank cubic yards*. There are such wide differences in the relationship of these units of measurement for most fill materials that accurate determinations of their relationship can only be established by performing field tests. A very rough approximation of the relationship among these three units of measurement is given in the following table.

Comparison of Bank Cubic Yard, Loose Cubic Yard, and Embankment Cubic Yard

	Bank cu yd	Loose cu yd	Embankment cu yd
Clean, dry sand or gravel . . .	1.00	1.14	0.88
Clean, wet sand or gravel . . .	1.00	1.16	0.86
Loam and loamy soil	1.00	1.20	0.83
Common earth	1.00	1.25	0.80
Dense clay	1.00	1.33	0.75

To illustrate the use of this table, 1 cu yd of common earth in the borrow pit will produce 1.25 cu yd of loose earth in the haul trucks and 0.80 cu yd of compacted earth in the embankment.

Quantity takeoffs of the materials in an embankment are prepared to check the bid quantities and to determine the required quantities of each type of material in each foot of elevation in the dam embankment. Graphs should be prepared with embankment elevations as one ordinate and tons or cubic yards of fill as the other. These curves are helpful in determining the amount of each type of fill that must be placed between any two embankment elevations. When different longitudinal sections of the dam are scheduled to be placed at different times, the quantity takeoffs should be computed so that separate tabulations can be prepared for each embankment section. Embankment fill curves should be prepared in a similar manner.

BORROW-PIT INVESTIGATION

Borrow pits should be investigated to determine whether they contain sufficient quantities of fill for each embankment zone. The drill logs and inspection pits should be examined to determine whether fill materials lie in stratified layers, which may necessitate that they be blended during excavation. When blending is required, the loading equipment selected must be of the type that can operate in this manner. It should be determined whether the borrow-pit materials require processing to comply with specifications for fill materials. Processing may consist of scalping, screening, washing, blending, or any combination of the four. At some locations it may be necessary to open quarries to produce riprap; at other locations field stones must be gathered for this purpose. When gravel deposits are not present, it may be necessary to open quarries to produce this material.

When specifications state that the inspection personnel will judge the suitability of each load of material before it is placed in the embankment, production will drop, cost will increase, and claims may be presented by the contractor.

The moisture content of the impervious borrow pits should be reviewed. Some impervious borrow pits are so wet that the material must be dried by harrowing, aerated by passing it through pug mills, or dried in gas-heated dryers. When the impervious borrow pits contain dry material, then it may be necessary to raise the moisture content of this material by extensive sprinkling. If there is any indication that there will be difficulty in maintaining the moisture or in handling the impervious fill material, a qualified consultant should be retained to study this problem. Sometimes these problems become so acute that specifications must be changed before embankment placing can be performed.

Impervious materials that may be difficult to handle are those containing appreciable amounts of the mineral halloysite or cohesionless silts and fine, uniform silty sands. Even good impervious material such as a well-graded mixture of sand, gravel, and claylike or silty fines may be subject to coarse-particle separation when it contains too many gravel-sized particles.

If gravel is to be excavated below water level, it should be examined or tested to determine whether it contains so many fines that it will not readily drain. If it is of this character, it may be necessary to windrow the material and permit it to drain before it is placed in the embankment. At Navajo Dam, New Mexico, the gravel that was excavated below the water surface and hauled directly from the borrow pit to the embankment had such fluid characteristics that equipment flotation on the fill could not be maintained.

PROCESSING BORROW-PIT MATERIALS

If process of fill materials is needed, a flow diagram and a proposed layout of the processing plant should be prepared. Processing varies from the scalping of large rocks from impervious materials to screening, washing, and reblending of fill materials. The most suitable location for a processing plant is determined by evaluating all site conditions. If wet processing is required, the separation or settlement of fines from the wash water will influence the location of the processing plant. Specialists should be retained to assist in the design of extensive processing plants when required. Detailed discussion of processing plants will not now be presented, but since the layout of a processing plant is similar, the aggregate-plant layouts in Chap. 5 should be referred to.

One of the largest earth-fill embankment processing plants was installed at Portage Mountain Dam, Canada. This plant processed fill materials for five of the zones in the dam embankment. Portage Mountain Dam contained 57,500,000 cu yd of fill, of which 27,500,000 cu yd required processing for the dam's impervious core. The prime source of fill material was pit-blended sand and gravel from a glacial moraine. This material did not require processing for use in the main random shell zone but had to be processed and blended for use in the other zones. The local silt material was too wet and had the wrong characteristics to be used for impervious fill. Impervious fill was produced by blending 89 percent minus $^{3}/_{8}$-in.

moraine material with 11 percent of the local silt. Other processed material was 3- by $^3/_8$-in. washed gravel and two sizes of washed sand. Washed gravels and sands were blended in different proportions to produce the specified fill materials for the transition, filter, drain, and pervious shell zones.[1]

The processing plant consisted of a dry-screening plant, a wet-screening plant, and a washing plant. Part of the sand and gravel from the glacial moraine was feed for the dry-screening plant. This plant contained twelve 8- by 20-ft double-deck screens which had a screening capacity of 7,200 tons per hr and which produced minus $^3/_8$-in. material. Most of the minus $^3/_8$-in. material was blended with the local silt to produce impervious core material, and the remainder was used as part of the feed for the washing plant. The plus $^3/_8$-in. material was used as feed for the wet screening plant.

The basic screens for the wet-screening plant were two 8- by 20-ft double-decked screens. The top screen had 3-in. openings and the bottom screen had $^3/_8$-in. openings. This plant produced washed 3- by $^3/_8$-in. gravel. Oversized material and excess production of washed 3- by $^3/_8$-in. gravel was used in the random shell zones. Minus $^3/_8$-in. material from this plant and minus $^3/_8$-in. dry-screened material from the dry-screening plant were the feed for the washing plant. The washing plant produced two sizes of washed sand that were used for blending purposes.

Another large material processing plant was installed at Tarbella Dam, Pakistan. This plant used a computer to control the flow of material.[2]

HAUL-ROAD LAYOUT

Haul roads should be laid out from the borrow pits to each height of embankment so that haul distances, road grades, and road-construction quantities can be determined. Errors made in computing truck turnaround cycles seldom result from poor judgment in regard to equipment performance but are made because haul roads had to be constructed differently from those initially planned.

The haul roads are often planned for travel of loaded trucks up one of the dam's abutments, across the dam as the trucks dump their loads, and down the other abutment. This permits a circular flow of the haul trucks back to the borrow pits. If more than one type of haulage unit is used, the haul roads should be wide enough to permit haul trucks to overtake and pass each other; otherwise, all haulage units will be restricted to the maximum speed of the slowest unit.

Haul roads should have a well-maintained, dust-free surfacing so that maximum truck speed can be obtained and tire wear can be minimized. Grades should be held to the practical minimum. Continual grading and sprinkling of the haul roads are of prime importance. At San Pedro Dam, California, where 120-ton bottom-dump trucks were used to transport the dam fill, the contractor laid out the haul roads so that grades were under 3 percent. Rainbird sprinklers were used in place of water trucks to keep road surfaces free of dust.[3]

EMBANKMENT SCHEDULING

The scheduling of the placement of fill into the embankment is of maximum importance to embankment construction-system selection, because maximum equipment requirements are controlled by the fill quantities scheduled for daily placement.

The first consideration in embankment scheduling is determining the date that fill placement will start and the date that it must be completed to allow all project work to be finished prior to the specified completion date. The embankment construction schedule must allow time for providing access to the project; for plant, equipment, and personnel mobilization; for the construction of any required diversion facilities; and for foundation excavation. (Scheduling of foundation excavation and grouting was discussed in Chap. 3.)

Other factors that control the initiation of embankment placement are first whether the specifications allow sectional construction of the embankment and second whether the embankment is long enough to warrant sectional construction. If there is a long section of embankment on one or both abutments and if specifications permit, then considerable yardages can be placed in these sections while the diversion facilities are being constructed and the bottom of the dam is being excavated.

The dates when embankment placement can start and when it must be completed will establish the number of calendar days available for embankment placement. These calendar days must then be reduced by the number of calendar days when weather will be unsuitable for embankment placement. If the embankment is wide, if the random-fill zone is pervious, and if specifications permit fill placement at different elevations across the embankment width, then fill placement in the embankment can be scheduled so that it can be performed during part of the rainy seasons. This is accomplished by expediting the placement of the impervious zone during dry weather so that it is raised in elevation above the surrounding zones, thus providing space for placement of the pervious zones during the wet weather. Pervious-zone placement can continue during the wet weather until the surface of the pervious zones approaches the elevation of the previously placed impervious zone.

The method selected for placing fill into the embankment establishes what control weather will exert on embankment placement. When the method is selected, the weather records should be reviewed to determine how many calendar days each year are suitable for embankment placement. If other earth-fill projects have been constructed in the surrounding area, research of their construction will provide data on this subject. In some heavy rainfall areas, only 100 days out of the year have suitable weather for impervious placement. In other heavy rainfall areas, the contractor has had to erect tents over the embankment so that he could finish the embankment within the contract period.

Another weather control over embankment placement is the number of days that heavy freezing may occur. Frozen material should not be placed in an embankment, nor should fill material be placed on the frozen surface of an embankment. When the surface of the embankment is frozen to a shallow depth, embankment placement can continue if this frozen material is removed by scrapers or other equipment. Similarly, frozen material must be removed from the surface of the borrow pits so that unfrozen fill material can be delivered to the embankment.

After a determination has been made of the total number of calendar days available for embankment placement, the maximum number of working days when maximum fill placement can be achieved should be determined. Calendar days are reduced to working days by establishing whether the project will be worked five or six days a week. If a five-day workweek is used, overtime payments will be held to a minimum, but construction equipment requirements will be at a maximum. If a six-day workweek is used, overtime payments will be at a maximum, but construction equipment requirements will be at a minimum. To properly select the number of working days per week, comparative cost estimates may be required. At some locations the supply of skilled labor may be short and it may be necessary to work six days a week so that labor will be attracted by the overtime premiums.

When the total number of working days has been established, the number of maximum productive working days is determined by reducing the total workdays by the number of low productive working days that are necessary at the start and at the completion of the embankment. When work is started, fill placement will be slow since the rock foundation under the impervious zone's surface must be covered with hand-compacted fill until a sufficient area has been developed to provide working space for mechanical compactors. Embankment placement will continue to be slow until enough length of fill is constructed to allow rapid and continuous dumping by the haulage vehicles. Fill placement will be slow while the embankment is being topped out, since the embankment will be so narrow that it will restrict the rapid and continuous dumping of the haulage units. Also, the impervious zone will occupy most of the embankment width and the more easily placed random zone will have become very narrow. Therefore, maximum embankment-placement rates can only be scheduled when fill is placed in the major part of the embankment, since the random fill zones will then be wide and there will be unlimited area to dump haulage vehicles and for rapid fill spreading and compaction.

The maximum daily rate of fill placement is computed by subtracting the number of low productive starting and finishing working days from the total available working days, by subtracting the fill placed during these low productive days from the total fill, and then by dividing the remaining fill by the remaining working days.

After the maximum daily production is determined, the number of shifts that will be worked per day and the number of hours that will be worked per shift must be determined before hourly production rates can be established. Two different methods of scheduling daily working hours are used for embankment construction. One method is to work two 10-hr shifts per day and to service the construction equipment between shifts. If overtime pay is time and one-half and if multiple-shift work is seven hours' work for eight hours' pay, then 18 hr will be worked a day for an equivalent of 22 hr of straight-time pay, or a work-to-pay ratio of 1.22.

The other method of scheduling daily working hours is to work three 8-hr shifts and to service the construction equipment while it is being used. This results in 21 hours' work for an equivalent of 24 hr of straight-time pay, or a work-to-pay ratio of 1.14. This second method of scheduling will reduce overtime payments, but hourly production will also be reduced because of lost time while equipment is being serviced. On the other hand, roadside servicing of haulage equipment has become so fast and efficient that very little equipment time is lost and many large earth-fill projects are constructed in this manner.[3]

After the hours that will be worked have been established, the maximum hourly fill-placement rates are computed by dividing the working hours a day into the maximum daily production requirements. This hourly placement rate is then used to establish the required number of operating construction-equipment units.

The detailed scheduling of embankment construction should be planned so that all excavated material to be used in the dam embankment is hauled directly from the point of excavation to the embankment without being placed in storage stockpiles. The use of storage stockpiles is expensive, since reclamation of material from these stockpiles costs almost as much as it does to load the fill material into haulage units in the borrow pits. Direct placement of fill cannot always be done when conveyor belts or trains are used to transport the fill, since surge piles are often required by these transportation systems. However, surge-pile storage is not as expensive as stockpile storage, since surge-pile reclaiming is usually done with belts and therefore is not expensive.

An embankment-placing schedule should permit the piezometer tubes and settlement instruments to be installed without delaying embankment placement. Piezometers are often required in the dam foundation and in the embankment, with piezometer tubing running to a central station. Vertical- and horizontal-movement measuring installations may be required in the fill. It may be necessary to install foundation settlement base plates on the embankment foundation. Specifications may call for surface settlement points to be installed on the embankment and for temporary control points outside of the embankment area. This work is not difficult and the total cost of installation is relatively minor. These installations necessitate the use of a large amount of hand labor.

HAULAGE EQUIPMENT

Prior to discussing the selection of embankment-construction equipment, descriptions will be given of the available equipment and the advantages that apply to the use of each type of equipment. Because the equipment used for transporting the fill material often controls the selection of other embankment equipment, it will be discussed first.

Bottom-dump Off-highway Trucks

This is the most versatile method of hauling fill material. This type of truck has the best payload-to-truck weight ratio of any off-highway type of haulage vehicle. Most earth-fill embankments have been constructed using bottom-dump haulage of fill material. These trucks can be loaded by any top-loading method and they are economical for haul distances exceeding 7,000 ft, provided that excessive grades are not encountered. They are suitable for use on all but the smallest of embankments, and they are particularly advantageous for transporting fill material from scattered and widely separated borrow areas.

Bottom-dump off-highway trucks can be secured in sizes up to 120-ton capacity, and manufacturers are constantly producing larger units. Photograph 27 shows a 15-cu-yd front-end loader loading a 120-ton bottom-dump truck. Maximum truck size is controlled by tire capacity and not by engine or truck size. On long hauls, tires heat and therefore the tire wear increases. This tire heating can be reduced by using smaller dual tires. Large single tires are advantageous to use when truck flotation on the embankment is a problem.

As an example of bottom-dump truck application, they were used on the construction of San Luis Dam, California. This was an earth-fill embankment-type dam. The 65 million cu yd of fill in the embankment was hauled from adjacent borrow pits by 100-ton bottom-dump trucks. These trucks were loaded by towed belt loaders during the process of preparing borrow pits for a wheel excavator. Fill from the prepared borrow pits was loaded into the bottom-dump trucks by a large Bucyrus Erie wheel excavator which could load them in 45 sec.

Scraper Haulage

Scraper haulage of fill material is more expensive than bottom-dump transportation because scrapers have a lower payload-to-equipment weight ratio. However, scrapers are economical to use on hauls of under 7,000 lin ft because the low cost of scraper loading outweighs the more expensive haulage cost.

Photo 27 A 15-cu-yd front-end loader and a 120-ton bottom-dump truck. (K. W. Dart Truck Co.)

Photo 28 A 30-cu-yd elevating scraper. (Caterpillar Tractor Company.)

Scrapers are powered by wheeled tractors that are securable with either one or two axles and come in many size ranges with different power units. The advantages of using one- or two-axle towing units depend upon haul conditions. One-axle tractor units have better traction because they have more weight on their driving wheels and because their single axle, controlled by hydraulic steering, allows them to duck walk through mud that would stop two-axle tractors. They are more maneuverable since they can turn in less than their length. The advantage of a two-axle tractor is that it has better riding qualities so it can be operated at higher speeds on well-surfaced, flat-graded haul roads. Scrapers can be powered by either single or dual engines and can be secured with a struck bowl capacity varying from $8^1/_2$ to 40 cu yd.

For single-engine scrapers, the loaded weight to horsepower ratio increases as the scraper size increases. With good haul roads at minimum grades, the lowest haulage cost is achieved by the use of large-capacity single-engine scrapers. When grades increase, it is often necessary to match smaller-size scrapers with more powerful tractor units. When extreme grades are encountered, the lowest cost is achieved by using two engine-driven scraper units, with one engine driving the scraper axle. This type of scraper is more costly to purchase and operate, but there will be such a time saving on steep hauls that lower unit cost will result.

Elevating, self-loading scrapers are suitable for embankment construction when only one scraper is required or when scrapers and pushers cannot be properly balanced. Since elevating scrapers contain equipment that allows them to self-load, they are less efficient transportation units due to the additional weight they must transport. The weight of the elevator will reduce the carrying capacity of the scraper from 2 to 3 yd. A 30-cu-yd elevating scraper is shown in Photograph 28.

The most economical type of scraper to use for any haul condition usually cannot be selected without making comparative cost studies between different types. Computers can be used to eliminate much of the detailed computation required for this type of study.

Off-highway Rear-dump Trucks

The off-highway rear-dump truck is a sturdy truck designed to take the shock of having a shovel dump large rocks into its bed or having a shovel hit the truck with its bucket. The truck's low payload-to-truck weight ratio results in high direct cost when it is used for transporting common fill material. Savings in equipment write-off may outweigh this high direct cost when rear-dump trucks have been used on rock excavation and are available for use on embankment construction. If these trucks have been secured for other uses and are available, they are often used to haul the small volumes of fill required for narrow-embankment zones. Typical zones of this type are those consisting of sand or other granular materials located between the impervious-core and coarse-gravel zones. Rear-dump trucks have more horsepower per ton of loaded capacity than other types of trucks; thus they may be economical for transporting common material over steep haul roads.

On-highway Bottom-dump Trucks

The standard on-highway bottom-dump truck has narrow tires, a low horsepower to weight ratio, and a high payload-to-truck weight ratio. They are designed for highway haulage, cannot maintain speed on grades, and have difficulty maintaining flotation on an embankment. Their use is economical when they haul fill over existing highways or over flat, well-maintained construction roads. The longer the haul, the more economical is their use. As an example, four 70-ton capacity on-highway bottom-dump trailers have been used as trailer trains hauling fill over a level roadbed. Each train was powered by a 735-hp tractor unit supplemented with a 318-hp convertible dolly inserted between the first and second trailers. The train was loaded by a portable-belt loader. Turnaround time for the train on a 9.1-mile haul was 57 min.[4]

Bottom-dump on-highway trailer units can be secured with wide single tires that give good flotation on fill. Often when scrapers are used to transport fill material on long, level hauls, one of these lightweight units is hitched behind the scraper since the scraper will have sufficient power to pull this extra load. To do this, it is necessary to have a system for top loading the trailer units in the borrow areas. These trailer units also have been used behind off-highway bottom-dump units to increase their capacity on long hauls. Since the tractor for the bottom-dump unit usually does not have enough power to pull both trailer units at high speeds, a remotely controlled power unit is inserted between the two trailers.

Rail Haulage

Rail transportation of fill has very limited application. It is suitable for use when the fill material is located in deep, continuous deposits; when hauls exceed 5 miles; when haul grades are suitable for rail haulage; when a large hourly placement capacity is required; and when there are sufficient quantities to economically amortize the rail system. When rail transportation is utilized, separate excavation and transportation systems must be used in the borrow pits to supply fill to train loading stations and a separate system is required at the damsite to transport the material from the train discharge point onto the dam embankment.

In recent times, rail transportation was used to construct one earth-fill embankment at Oroville Dam on the Feather River in northern California. The main fill in the dam was 60 million cu yd of dredger tailings located approximately 12 miles downstream of the dam. A single-track railroad line which passed the damsite and was located adjacent to the borrow pits was made available to the contractor because the Western Pacific had replaced it with a relocated main rail line.

A wheel loader was used in the borrow pit to load dredger tailings onto a 54-in. portable-collecting conveyor belt. This belt transported the tailings to a train loading station where it deposited the tailings on a large shuttle conveyor. The shuttle conveyor stockpiled the tailings over a reclaim tunnel of a diameter adequate to permit a train to pass through it. The tunnel was long enough to contain 10 cars, and loading gates were spaced so that all 10 cars could be loaded at one time. The existing rail line was extended to the train-loadout station and was double-tracked between the station and the dam site. Hauling from the borrow pit to the dam was done by four trains that operated at 20 min intervals. Each train consisted of diesel locomotives pulling 42 gondola cars of 70-cu-yd capacity.

At the dam site, each set of locomotives dropped its string of cars in front of a double-rotary car dumper and then hooked onto an empty string of cars for the return trip to the loadout point. A winch and cable were used to pull the parked loaded string of cars into the car dumper. The car dumper dumped two cars at a time on 60 sec intervals. The dumped tailings were discharged into a concrete hopper that fed a conveyor which transported the material to the main stockpile.

At the stockpile, a traveling stacker stockpiled the material over a 1,600-lin-ft reclaim tunnel. Reclaiming from the stockpile was done using a belt conveyor located in the reclaim tunnel. This conveyor belt transported the material up the dam's right abutment. As the dam was raised in elevation, the conveyor was successively extended up the abutment. The conveyor belt discharged the fill onto a portable conveyor located on the top of the embankment. Portability of this belt was achieved by supporting it with cables suspended from an A frame constructed on the undercarriage of a shovel. The portable conveyor belt discharged the fill into truck-loading bins that could load two 100-ton bottom-dump trucks simultaneously. These trucks distributed the fill across the embankment. As the embankment was raised in elevation, a higher foundation was prepared for the truck-loading bins. They were then skidded from their old location onto this higher foundation, and the portable belt was relocated so that it discharged into the relocated bins. Capacity of this system was 4,500-cu-yd maximum per hr with an average production of 3,500 cu yd per hr.[5,6]

Conveyor Belts

Conveyor belts provide another specialized method of fill transportation which is used when the fill occurs in large deposits, when high hourly production is required, when the yardage in the embankment is sufficient to economically amortize the cost of the system, and when haul distances are not so great as to result in too large a capital write-off for the cubic yards of fill in the embankment. Conveyor belts are economical to use when material must be transported over steep terrain. Two earth-fill dams where conveyor belts were used for transportation of fill material are Trinity Dam, California, and Portage Mountain Dam, Canada.

Trinity Dam contained 10 million cu yd of impervious material. The source of this material was on the top of hills surrounding the dam. Since truck transportation of this material down the steep haul grades presented a definite problem in truck braking capacity, the contractor elected to use conveyor belts for this haul. The material was excavated in the borrow areas by scrapers which transported the material to drive-over belt loading stations where they dumped their loads. Grizzly bars at these loading stations were set at 24 in. to reject oversized material. Material over 8 in. in size was crushed to minus 5 in. in roll crushers. A flight of conveyor belts transported this crushed material and the minus 8-in. material down the hill. The belts were 48-in. and 42-in. wide; the total flight length was 10,145 ft; and the drop in elevation was 1,000 ft. The capacity of the system was 1,850 cu yd per hr. The longest flight was 1,895 lin ft. The belts discharged into three truck-loading bins where water could be added to increase the moisture content of the impervious material. Trucks were used to transport the fill from the bins to the embankment.[7]

The largest-capacity conveyor system and the longest single-flight conveyor belt installed to date were used for the construction of Portage Mountain Dam, British Columbia, Canada.[1] Portage Mountain Dam contained 57 million cu yd of fill. The fill materials available for embankment construction were primarily sand and gravel from a glacial moraine located approximately 4 miles from the dam. The terrain between the moraine and the dam had considerable grade differences. Weather conditions permitted embankment placement during only six months of the year. To meet the construction schedule, fill had to be placed in the embankment at a rate of 2.5 million cu yd a month. In order to move this amount of fill material from the moraine to the damsite over the existing grades, the contractor installed a conveyor-belt system that had a capacity of 12,000 tons per hr.

Material was blended in the pit by using large bulldozers to shove the material down steep slopes into portable-belt loaders. The belt loaders discharged onto 48-in. shuttle conveyors; these shuttle conveyors fed 60-in. gathering conveyors; and the gathering conveyors fed a 72-in. main conveyor 15,000 lin ft long. Two belt loaders were used to feed each gathering conveyor, and two gathering conveyors fed the main belt.

Near the dam, the main belt discharged into two surge piles. One surge pile was used for storage of unscreened material, which was used in the random-fill zone. The other surge pile furnished the feed for the dry-screening plant. Part of the product of the dry-screening plant was wet-screened. Two 60-in. conveyor belts were used to transport the fill material from the unscreened surge pile and the screening plants to two truck-loading bins located on the dam embankment. From the bins, the material was transported and distributed on the embankment in bottom-dump off-highway trucks.

LOADING EQUIPMENT

Scraper Loading

Scraper loading is less costly than using loading equipment to top-load trucks because the scraper participates in the loading. The extent of this participation is dependent on the type of scraper. However, as haul distances increase, the additional cost of scraper haul makes their use uneconomical.

Crawler tractor-drawn scrapers are usually self-loading but are not used in embankment construction because they are slow and have limited capacities. Self-loading elevating scrapers are self-contained units that do not require pushers and operate individually. The elevator reduces the payload-to-scraper weight ratio and thus increases transportation cost. As haul distances increase, this increase in haul cost soon becomes more than the savings in loading cost. Self-loading scrapers have application for small-yardage embankment construction when haul distances from borrow pits are short, when a small number of scrapers are used, or when more scrapers are used than the pushers can load.

Scraper trains consist of a number of scrapers connected together and controlled by one operator. They are self-loading because the operator can use all the power in the train to load each scraper in succession. To date, scraper trains have had slow operating speeds, and if one unit breaks down, the whole train will stop. These units have not been used for embankment construction, but have been used on short-haul applications such as canal construction.

A recent development which permits scrapers to be self-loading is a quick-connection attachment that can be installed on scrapers. This permits scrapers to be self-loading for they can be automatically connected in pairs during the loading operation and can use the power of both towing units to load each scraper. After loading, they are automatically disconnected and operated as separate units. During the loading operation, the front scraper is loaded first and the rear scraper last. The rear scraper can be loaded with a greater load than the front scraper, since when it is being loaded, the power unit for the loaded front scraper can exert more tractive effort because of the load on its drive wheels. On one construction project, two quick-connected scrapers have been used to load a third scraper.[8]

The majority of scrapers are push-loaded by crawler-type tractors. The larger the scraper, the larger the pusher must be. For small scrapers in good material, a 180-hp crawler tractor may supply sufficient pushing power. For 40-cu-yd struck-capacity scrapers in tough loading, the equivalent of three 385-hp pushers may be required. Crawler-type pushers develop better tractive effort than wheel-type pushers. Dual crawler-type pushers, controlled by one operator, can engage scrapers quicker and will load approximately 25 percent more scrapers than two pushers operated as independent units. It is advantageous to use wheel-type pushers in abrasive sand (since track wear is eliminated) and for pushing applications where their greater speed can be utilized.

Front-end Loaders

Common materials can be rapidly loaded into trucks by wheel-mounted front-end loaders. The capacity of this type of front-end loader is constantly increasing, and they are now available with 15-cu-yd buckets. As an example of their loading capacity, a 15-cu-yd front-end loader has loaded 120-ton bottom-dump trucks with dredger tailings (large gravel) in five passes, taking approximately 3 min.[3] This type of loader is shown in Photograph 27 in this chapter. Both the capital cost and the operation cost of this type of loader are of a minimum.

Towed-belt Loaders

Older plow-type belt loaders are equipped with 54-in. belts. Their loading capacity is between 1,000 and 1,200 cu yd per hr when two large tractors are used to pull the loaders in good material on long, flat runs. Photograph 29 shows this type of loader. New plow-type belt loaders are now manufactured by Euclid, Inc. They are equipped with 72-in.-wide belts, and it is reported that they have a maximum loading capacity of 3,000 bank cu yd per hr. This is an economical method of loading bottom-dump trucks from suitable borrow areas. Production drops sharply if borrow areas are short, since frequent turning of the loader will be necessary. This type of load-

Photo 29 Towed-belt loader. (Earthmoving Equipment Division, General Motors Corporation.)

ing equipment should not be used when selective loading or blending is required in the borrow pit.

Elevating-belt Loader Attachments

Elevating-belt loader attachments can be secured to a grader or a large-type tractor. Production is less than one-half that of the older type of towed-belt loaders, but equipment cost is reduced.

Portable-belt Loaders

Portable-belt loaders are used for loading common material from deep deposits located on steep slopes. Photograph 8 in Chap. 3 shows this type of loader in use. The capacity of portable-belt loaders is dependent on the size of the belt, the number of bulldozers feeding the belt, and the distance the bulldozers must push the material. As discussed in Chap. 3, portable-belt loading has been done with production rates of 1,800 cu yd per hr. As the material is loaded out, the distance that the bulldozers must push their load is increased. To keep this distance to reasonable limits, the belt loader must be relocated as often as once a day. This takes approximately 1 hour for three bulldozers. The loaders are usually moved on nonshift hours at overtime rates so that they will always be available for truck loading during the regular shift.

Wheel Excavators

Wheel excavators have been used to load out borrow-pit material at Oroville Dam and at San Luis Dam. At Oroville Dam, a wheel excavator loaded dredger tailings onto a gathering conveyor belt. It had a maximum capacity of 4,500 cu yd per hr and an average capacity of 3,500 cu yd per hr. At San Luis Dam, a wheel excavator loaded random material from a long, previously prepared bank into 100-ton bottom-dump trucks, loading each truck in 45 sec. This wheel loader is shown in Photograph 30.

The continuous banks required for the operation of this loader were prepared by using towed-belt loaders to load the shallower material into the trucks. The first cost of large-wheel excavators is so large that economical equipment write-off can only be secured when large volumes are to be handled. This was possible at these two dams since Oroville Dam contained approximately 65 million cu yd of dredger tailings and San Luis Dam contained approximately 65 million cu yd of random fill.

Small-wheel excavators are being developed.[9] One manufacturer is producing a wheel excavator of a smaller size using a ditcher principle for the excavating wheels. This type of wheel excavator,

Photo 30 Wheel excavator used at San Luis Dam, California. (Bucyrus-Erie Company.)

pictured in Photograph 31, has achieved a loading capacity of 3,600 loose cu yd per hr.[10]

Power Shovel

Power-shovel loading is done when the shovel has been used for rock excavation and is available for use on common excavation. Or in some cases power-shovel loading may be used to break out and load hard-packed material in confined areas.

Photo 31 Wheel excavator. (Barber-Greene Company.)

Draglines

Large draglines are used to excavate large quantities of material from below the water surface. Truck loading can be expedited by having the dragline discharge its load into a portable truck-loading bin. When the dragline must raise the bucket high enough to dump into the bin, its cycle time increases and production decreases. This decrease in production can be eliminated by having the dragline discharge its bucket into a low hopper and by using a conveyor belt to transport the material from the hopper to the top of the truck-loading bin.

Sometimes materials that have been excavated in the wet contain enough fines to restrict their rapid drainage, and therefore, if they are placed directly in the dam embankment, it will be difficult to maintain equipment flotation. When this situation occurs, the dragline must windrow the material and it must be given time to drain before it is hauled to the embankment. After it has drained sufficiently, it can be loaded out by the dragline, by front-end loaders, or by other loading equipment.

Truck-loading Bins

Truck-loading bins, charged by either conveyors or draglines, are often used to rapidly top load haulage units. There is no reason why portable truck-loading bins cannot be used with many other types of excavating equipment. As an example, front-end loaders could dump on portable belts that discharge into loading bins. Then less truck spotting and loading time would be required and fewer trucks could be used.

Scraper Top-loading Trucks

Scraper top-loading of haul trucks has been done when the borrow pit is located a long distance from the embankment. The scrapers are used to excavate the material in the borrow pit and to transport their load to a loading bridge where they can discharge their load into trucks. The main disadvantage of this system is that it is difficult to key scraper capacity to truck capacity.

A good application of scraper loading and bottom-dump hauling occurred at a wet impervious borrow pit located a long distance from Hell Hole Dam on the American River, California. The material was dried in the borrow pit by digging drainage ditches, exposing extensive surface areas, and then aerating the material by harrowing. The top layer of dried material was then harvested with scrapers that dumped their loads adjacent to a portable belt loader. Two bulldozers then fed this material to a belt loader which top loaded bottom-dump haulage units.

EMBANKMENT EQUIPMENT

After the fill material has been transported to the embankment, it must be spread to the desired lift thickness, brought to nearly the optimum moisture content, and compacted to the required dry density.

Spreading Equipment

The depth that each layer of fill material can be placed in the embankment varies between 4 and 24 in. This depth is dependent on the type of fill material, on the type of compactor, and on the number of compactor passes. Fill spreading on the embankment may be done with either crawler-type or wheel-type bulldozers, with blades mounted on the front of compactors, and with motor graders. The amount of equipment required for spreading the fill varies with the type of haulage equipment. Since scrapers can discharge their load to any lift thickness, fill spreading will be at a minimum when they are used. Bottom-dump trucks discharge their loads in relatively deep, narrow windrows, and considerable fill spreading must be done when they are the haulage vehicles. Rear-dump trucks deposit their loads in relatively deep piles, which entails the maximum use of spreading equipment.

Equipment for Controlling Moisture

The moisture content of the fill material is controlled by the use of large sprinkler-tank trucks of from 8,000- to 12,000-gal capacity. Photograph 32 is of this type of truck.

These trucks are pulled by the same type motive unit as is used on scrapers and bottom-dump trucks. Some contractors have converted used bottom-dump trucks to this service. A separately driven

Photo 32 A 12,000-gal sprinkler truck. (Southwest Welding and Manufacturing Company.)

gas or diesel 6- to 8-in. pump is used to furnish pressure for the sprinkling system. In order to maintain the tank truck in continuous operation, its water tank must be refilled in as short a time as possible. This can be accomplished by having elevated storage water tanks of larger or of similar capacity outfitted with a large-diameter, drop-valve truck filling spout. There is always sufficient time between tank-truck fillings to replenish the water in the storage tanks with relatively small-capacity water pumps or water lines. At some dams where large amounts of water have been required in the embankment, water has been distributed on the embankment by fast-connecting pipelines equipped with sprinklers. This requires the continuous relocation of the distribution piping as the embankment is raised in elevation.

Compaction Equipment

Compaction of the embankment is handled in different ways by different specification writers. One method is to specify the type of compactor and the number of compactor passes. Another method is to specify the required dry density and place the resonsibility on the contractor to determine the type of compactor and the number of compactor passes that will produce this density.

The type of compactor used and the number of compaction passes required varies with the type of fill material and the thickness of lifts. The more pervious and granular the fill material, the thicker the lifts and the fewer compaction passes required. Since the impervious zone must be fine-grained plastic material to provide an efficient water barrier, this zone must be placed in the thinnest layers and receive the most compaction. If granular, gravelly material is used in the outer shell zones of the embankment, it often can be placed in 2-ft layers and compacted with crawler-tractor treads. Compaction is quite a complex subject on which a variety of information has been published.[11, 12]

There is a wide variety of compaction equipment available for embankment construction and a wide choice in motive power. Compactors may be of the towed-drum type; compaction-drum-type wheels can replace standard wheels on wheeled tractors; single drums can replace the rear wheels of wheeled tractors; and motorized units can be secured composed of two, three, or four compaction drums which are capable of movement in either direction. The wheeled-tractor units and some of the motorized-drum units can be outfitted with blades for spreading the fill material. The operating speed of compaction equipment varies between 3 and 10 mph, depending on the type of unit.

Following are listed the different types of compactors, the types of fill material best packed by each type of compactor, and the thickness of fill layers suitable for each type of compactor:

Sheepsfoot rollers are used for compacting clays, clay and silt mixtures, and gravel with clay binders. Lift thickness varies between 6 and 12 in.

Tamping-foot compactors are used on sandy clays and silts with clay binders. Lift thickness varies between 7 and 12 in.

Pneumatic, small-tired compactors are used on sandy silts, sandy clays, and gravelly sands. Lift thickness varies from 4 to 8 in.

Pneumatic, large-tired compactors are used on sandy silts, sandy clays, and gravelly sands. Lift thickness can be as much as 24 in. Photograph 33 is of a compactor of this type.

Vibratory compactors can be secured with smooth steel drums, sheepsfoot drums, and tamping drums. Lift thickness varies between 3 and 12 in.

Grid rollers are used on granular soils. Lift thickness varies between 5 and 10 in.

Steel-wheel rollers are used on sandy silts and other granular materials. Lift thickness varies between 4 and 8 in.

Segmented wheel-roller wheels are used on wheel-type tractors equipped with dozer blades. This type of compactor is useful in pulverizing material.

Tractor treads on large crawler tractors are used in compacting pervious gravels. Lifts may be placed as thick as 2 ft.

Hand-held compactors are required to compact fill material placed in thin layers (3 or 4 in.) on and adjacent to the rock foundation and adjacent to any structures. Hand compaction must be done until there is sufficient compacted working area so that machine com-

Photo 33 A 100-ton rubber-tired compactor, Portage Mountain Dam, Canada. (Southwest Welding and Manufacturing Company.)

paction can be used. Additionally, hand compaction must be continued up the rock surface of each abutment so that the impervious fill is compacted into all the irregular rock pockets and because mechanical compactors cannot operate adjacent to the abutment. The more frequent the occurrence of sharp irregularities in the dam foundation, the greater the amount of hand compaction that will be required. In order to estimate this work quantity, it is necessary to visualize the appearance of the final prepared rock surface.

Equipment Used for Placing Riprap and Rock-slope Surfacing

The riprap and rock surfacing on the embankment slopes can be placed as the embankment is raised or it can be placed after the embankment is virtually completed. If it is placed as the embankment is being raised, the easiest method to use is a backhoe or a crane to place this material on the slope. The crane can work off the top of the embankment and can secure the rock from small, truck-dumped stockpiles located on the embankment and adjacent to the slope. The placement crane or backhoe must have lifting capacity sufficient to handle the largest size of rock. Gradalls, as shown in Photograph 34, are often used for this purpose.

If riprap and rock surfacing are not placed until the embankment is completed, one of two methods is generally preferred for placing this material. One method is to use rear-dump trucks to dump the surfacing material over the edge of the embankment. Spreading and leveling of the surface material is done with bulldozers operating up and down the slopes. The other placement method is to construct haul roads in the embankment slopes so that trucks can haul surfacing material to cranes located on these haul roads. The roadways are progressively leveled out by the cranes as they place the surfacing material on the embankment slopes.

EQUIPMENT SELECTION

Before the construction equipment is selected, it should be determined if equipment requirements can be reduced by using equipment made available from other dam-construction operations, such as equipment used for excavating the dam's foundation. Priority in equipment selection should then be given to equipment that will be suitable to use on other project work, either before or after it is required for embankment construction.

Remaining equipment selections should be of types and units that will maintain the required rate of fill placement in the most economical manner. The advantages of each type of loading, trans-

Photo 34 A G-1000 Gradall. (The Warner and Swasey Company.)

porting, and embankment-placing equipment have already been discussed in this chapter. Final selection of equipment types must be based on judgment, experience, and, if required, on comparative cost estimates.

Comparative Cost Estimates

If comparative cost estimates are used to determine the most economical type of equipment, the comparative estimate should show the hourly write-off cost and the hourly operation cost of the equipment. To illustrate how comparative cost estimates are prepared, following is one that compares the cost of using two different loading methods to top load 100-ton off-highway bottom-dump trucks. One method of loading is to use a portable-belt loader fed by four bulldozers. The other loading method is to use two $8^1/_2$-cu-yd front-end loaders and to provide a bulldozer to clean up and push material to the front-end loaders. This cost comparison indicates that front-end loading will be the preferable loading method.

	Front-end loading system	Belt loading system
Estimated capital cost:		
Belt loader		$ 73,000
Bulldozers	$ 70,000	280,000
Front-end loaders	169,000	
Total	$239,000	$353,000
Useful life, hr	10,000	10,000
Write-off/hr	23.90	35.30
Estimated direct cost/hr:		
Operators	24.00	32.00
Truck spotter	5.25	5.25
Maintenance labor	8.00	10.00
Repair parts and supplies	16.00	20.00
Total direct cost per hr	$53.25	$67.25
Direct cost plus write-off/hr	$77.15	$102.55
Cost/bank cu yd at 1,000 cu yd/hr	0.077	0.103
Cost/embankment cu yd at 800 cu yd/hr	0.096	0.128

Selecting Borrow-pit Processing Equipment

If the borrow pits contain oversized material that is not permitted in the fill, this oversized material should be removed by screening and wasted in the borrow pits. When borrow pits contain minor amounts of oversized material, its removal can be accomplished by using portable screens. If belt loaders are used for borrow-pit loading, oversized material can be removed by screens installed on the discharge end of the loading belt. If other types of loaders are used, oversized material can be removed on grizzly-type screens mounted on portable frames, constructed to permit gravity truck loading of the screened material.

When the borrow pits contain large quantities of oversized material, a scalping screening station is installed at the borrow pits. These are arranged so that the pit trucks can dump their loads directly onto a grizzly, and from the grizzly the screened material is transported by conveyor belt to truck-loading bins or directly into haul trucks. The oversized, rejected material can be loaded into trucks in a similar manner.

If processing of fill material is necessary, then a processing plant must be designed. A flow diagram furnishes the required data for processing-plant design and the selection of the size of screens. The design of a processing plant is quite similar to the design of a natural aggregate plant, a subject discussed in Chap. 5.

Determining the Number of Loading Units

After the type of loader has been selected, the number of loading units required to maintain the scheduled production must be determined. This is accomplished by dividing the loader's hourly production capacity into the scheduled hourly rate of fill placement. The best method of determining loading capacity is from past job records or from work experience. If these are lacking, the production that can be achieved with different loading units can be found in tables published by equipment manufacturers.

The tables are based on the equipment working 60 min of every hour, on the theoretical capacity of the loading bucket, and on the maximum efficiency obtainable from the operator and from the equipment. This necessitates that production rates be reduced because such conditions will never occur. The 60-min hour should be reduced to a 50-min hour to take care of unavoidable delays. The bucket capacity must be reduced to that obtainable for each type of material. An evaluation must be placed on the loading cycle, since it varies with the shovel swing or the distance a front-end loader must travel. Production must be reduced because the machine and operator will not operate at 100 percent efficiency. All these adjustments are explained by equipment manufacturers in the fine print of the text accompanying their tables.

The following approximations of the production that can be achieved by different loading equipment are included as an aid to individuals who do not have access to other information:

Shovel Production in Bank Cu Yd per Hr;
70% Efficiency, 90° Swing

	Optimum bank height			
Bucket size, cu yd	Sand and gravel	Common earth	Hard clay	Wet clay
1/2	60	55	45	30
1	140	110	85	65
2 1/2	260	210	180	140
3 1/2	360	320	280	200
4 1/2	440	390	310	240
6	560	500	400	300
8	750	650	550	420
10	900	800	650	500
14	1,200	1,000	860	660

When the angle of swing increases to 180°, shovel production will be 70 percent of that shown in the table. Production decreases with either a decrease or an increase in the optimum bank height in accordance with the following table. Optimum bank height occurs when the bank is level with the shovel's dipper-stick pivot shaft.

Adjustments to Shovel Production for Bank Height[13]

	Factor
Optimum height	1.00
80% of optimum	0.98
60% of optimum	0.91
40% of optimum	0.80
120% of optimum	0.97
140% of optimum	0.91
160% of optimum	0.85

Rubber-tired, articulated, front-end loader production will be the equivalent of the production of a power shovel that has a bucket two-thirds the size of the front-end loader. This is a rough approximation, for, since loading capacity varies with distance, the loader must travel between the bank and the haulage unit.

Dragline production varies with the type of dragline. Compared to shovel production, cycle time increases more with an increase in equipment size. The production of draglines with a bucket size of 1 cu yd or smaller will be approximately 80 percent of a shovel's production, with an equivalent size bucket. This relationship reduces to 75 percent for $2^1/_2$-cu-yd draglines, to 65 percent for $3^1/_2$-cu-yd draglines, and to 58 percent for draglines of $4^1/_2$-cu-yd size and larger. When draglines are excavating material from below a water surface, production rates should be reduced another 30 to 50 percent.

Towed-belt loaders operating in a suitable long borrow pit can load 1,000 bank cu yd per hr if they are the older type equipped with 54-in. belts; if they are the newer type, equipped with 72-in. belts, they can load 3,000 bank cu yd per hr. This capacity is reduced when the borrow pit is shorter and more equipment turning is required.

A 60-in. portable-belt loader loading large bottom-dump trucks with free-running material from a sidehill deposit, with 4 D-9s pushing the material into the belt loader, has a maximum loading rate of 1,800 bank cu yd per hr.[14] Any change from optimum loading conditions will rapidly decrease this loading rate.

Determination of the Number of Haul Units

The number of operating trucks or scrapers required on the haul from the borrow areas to the embankment is established in the following manner. After the average turnaround cycle for the haulage unit is determined, the cycle time divided into a 50-min hour gives the number of cycles per unit per hour. The number of cycles per hour times the haulage-unit capacity gives the haulage units' hourly production capacity. This number divided into the scheduled hourly production gives the required number of operating units.

The computation of the average cycle time becomes quite complex since hauls lengthen, grades increase, and cycle times increase as the embankment is raised in elevation. Computers are helpful in establishing cycle time. The number of cycles per hour should be computed on a 50-min hour to reflect job efficiency. Detailed explanations of the computations required to determine cycle time are given in Chap. 3.

If scrapers are used on the excavation, the balancing of the scraper fleet to the pushers can be done by dividing the pusher cycle into the scraper cycle to find the desired number of scrapers for each pusher. Fractions should be disregarded and not rounded off.

A typical pushing-cycle computation is as follows:

Scraper loading time	0.90 min
Pusher return	0.45 min
Allowance for stopping and turning	0.25 min
Total pusher cycle	1.60 min

If the scraper cycle is 10.5 min, then this divided by 1.60 is 6.6. Therefore, one pusher will be required for every six scrapers.

If conveyor belts are used for transporting material, their capacities can be secured from published handbooks.

If trains are used for fill haulage, their capacity is dependent on haul lengths and grades, engine horsepower, number of cars per train, capacity of each car, method of loading, and car dumping method.

Determination of Amount of Embankment Equipment

The determination of the number of equipment units required for spreading the fill, for obtaining the proper moisture content, and for compacting the fill is computed from the rate of fill placement, the type of fill material, the lift heights, the number of compaction passes required, the type of compactor, and the speed and width of the compactor.

These are all straight-line computations and should present few problems. Most embankment spreads are organized with one spread of equipment to handle the impervious core and another spread to handle the remaining zones.

Total Equipment Requirements

The foregoing calculations will establish the number of equipment units that must be in operation to maintain the scheduled rate of production. Total production equipment requirements will exceed this number by one-fourth or one-third since equipment units will only be able to work from 70 to 80 percent of the time. This is termed *equipment availability.* Equipment units will be unavailable between 20 and 30 percent of the time because they must be pulled out of production for servicing and for repair.

After the fleet of production equipment has been determined, equipment must be added to this production fleet for road building, for keeping the space clean around excavating units, for road maintenance (i.e., water wagons and graders), for equipment servicing units (i.e., grease and fuel trucks), for tire changing, and for transporting labor and supervisory personnel (i.e., pickups and man-haul trucks). Provisions for water distribution, job lighting, repair shops, tire shops, a rigging loft, and craft-tool storage facilities must also be made. If some of the servicing equipment or plant facilities have been furnished with other construction systems, duplication is often unnecessary.

REFERENCES

1. W. Irvine Low, M. ASCE, *Portage Mountain Dam Conveyor System*, Journal of the Construction Division, Proceedings of the American Society of Civil Engineers, September, 1967.
2. At Tarbella Dam—Computer Controls 15 Processing Plants, *Contractors & Engineers Magazine*, November, 1969, page 47.
3. Keeping the Big Fleet Rolling, *Western Construction*, September, 1969, page 43.
4. Earthmover Train Hauls 308 Tons per Trip to Highway Grade, *Construction Methods and Equipment*, September, 1969.
5. Big Wheel Replaces Dragline, *Engineering News-Record*, May 14, 1964, page 26.
6. Its Push-button Earth Moving at Oroville, *Engineering News-Record*, October 8, 1964, page 74.
7. *Conveyor Belt System for Transporting Zone 1 Material, Trinity Dam*, U.S. Department of the Interior, Bureau of Reclamation, Trinity River Division CVP Construction Office, not dated.
8. Three Scrapers in Tandem Make Fast Work of Moving Earth at Building Site, *Construction Methods and Equipment*, September, 1969, page 66.
9. New Developments in Earth Moving Part 2: Wheel Excavators, a reprint from *Construction Methods and Equipment*, February 1965, copyright 1965, McGraw-Hill, Inc.
10. Excavator Feeding Two Lines of Bottom Dumps Never Stops in Work on 4-Million-yd Job, by David C. Etheridge, Western Editor, *Construction Methods and Equipment*, April, 1970, page 61.
11. A. W. Johnson and J. R. Sallberg, *Factors That Influence Field Compaction of Soils, Compaction Characteristics of Field Equipment*, Highway Research Board Bulle-

tin 272, National Academy of Sciences, National Research Council, Publication 810, 1960, Washington, D.C.
12. *Handbook of Compactionology*, copyright 1968, American Hoist and Derrick Company.
13. *Caterpillar Earthmover Performance Handbook*, Caterpillar Tractor Company.
14. Modified Equipment Moves 45,000 Yards per Shift, *Roads & Streets*, April, 1969, page 48.

CHAPTER SEVEN

Rock-fill Embankment Construction

This chapter describes how the construction of rock-fill embankments is planned, scheduled, and performed. The many differences in the construction of these embankments result from the use of two basic types of water barriers.

When impervious fill is located near the damsite, the embankment is designed to have an impervious core for a water barrier. Embankments designed in this manner contain a main rock-fill zone for dam stability; a sloping, impervious core placed upstream of the main rock fill, which is inclined to the natural slope of the rock and permits dam stability to be obtained by a minimum of rock fill; a narrow rock-fill zone on the upstream side of the impervious zone, which serves as protection and support for the impervious zone when the water in the reservoir is lowered; and narrow zones of graded granular material, which function as fill transitions between the impervious core and the rock fill.

When rock-fill embankments are constructed in high, mountainous areas, impervious fill is often impossible to obtain economically. In this case, rock-fill embankments are designed with a thin, concrete slab functioning as the water barrier. This type of embankment contains one main rock-fill zone with a selectively placed narrow rock facing on its upstream slope. This rock facing supports the concrete slab that forms the water barrier.

Some of the work required for the construction of a rock-fill embankment has been described in other chapters in this book. The planning and the construction of any required diversion facilities are included in Chap. 2. The planning and scheduling of the excavation and preparation of the dam's foundation is adequately covered in Chap. 3. The scheduling of embankment placement should be correlated with the sequence of performing the foundation excavation, as also described in Chap. 3. The construction of any associated concrete structure is covered in Chaps. 4 and 5. The construction of the impervious zone of a rock-fill dam is similar to the construction of the impervious zone of an earth-fill dam, as is described in Chap. 6.

The descriptive information still lacking for rock-fill embankment construction is presented in this chapter and covers the development and operation of rock quarries, the production of granular materials for the transition fill zones, the construction of haul roads, the loading and transportation of fill materials from the quarries and borrow pits to the embankment, and the placement of the fill in the embankment.

CONSTRUCTION-SYSTEM SELECTION

The selection of the most economical system for constructing rock-fill embankments is accomplished by preparing takeoffs of embankment fill quantities, planning the development and operation of the rock quarry, planning the production of filter materials, selecting the method of transporting the fill materials to the embankment, laying out the haul roads, planning the placement of fill in the embankment, scheduling the construction of the embankment, and determining the type and number of construction equipment and plant units required to maintain the scheduled production.

Since the rock quarry is the source of the rock fill that forms the major portion of a rock embankment, the quarry's development and operation is the controlling component of the embankment-construction system.

QUANTITY TAKEOFFS

Quantity takeoffs are made to check the bid quantities; to establish work quantities that are not listed in the bid schedule; and to furnish a breakdown of quantities to use for construction scheduling, for establishing hourly production rates, and for selecting the number and size of the equipment units. Quantities should be tabulated so that the quantities of fill between any two embankment elevations are securable for the entire embankment and for any planned sectional construction of the embankment. An excellent method of tabulating this information is to prepare embankment fill curves with the embankment elevations as one ordinate and fill yardages as the other.

Before the quantity takeoffs are started, the relationship between embankment yardage, loose yardage, and bank yardage should be established. These relationships are used in determining the yardages that must be excavated in quarries and borrow pits and in computing the required number and sizes of the drilling, loading, and hauling equipment units. The only true method of establishing these relationships is by field testing. Approximations of these relationships for impervious and granular material were given in Chap. 6. To assist in initial planning, the following table lists approximations of these relationships for various types of rock:

Estimated Relationship between Bank Yards, Broken Yards, and Embankment Yards

	Bank		Broken		Embankment	
	yd	lb/yd	yd	lb/yd	yd	lb/yd
Granite	1	4,500	1.5–1.8	3,000–2,500	1.35–1.65	3,250–2,750
Limestone ..	1	4,200	1.65–1.75	2,550–2,400	1.5–1.6	2,800–2,630
Sandstone ...	1	4,140	1.4–1.6	2,950–2,600	1.2–1.4	3,450–2,950
Shale	1	3,000	1.33	2,250	1.2	2,500
Slate	1	4,590–4,860	1.3	3,500–3,750	1.2	3,800–4,050
Trap rock ...	1	5,075	1.7	3,000	1.5	3,400

ROCK-FILL SOURCE

The source of rock for the embankment fill may be a rock quarry or a large rock cut required for the dam's spillway or power-intake channel. When the rock is secured from a quarry, the quarry may be a coyote quarry or one of the three types of blast-hole quarries. When the rock fill is secured from a large cut, the development of the cut is quite similar to the development of a blast-hole quarry except that the cost of excavating the rock exceeds what it would cost to secure it from a quarry. This increased cost is caused by more restrictive face development and haul-road construction and by the necessity of presplitting to develop the nearly vertical faces in the rock cut. This extra cost is justifiable since the cut will serve as a permanent facility of the dam as well as a rock quarry.

QUARRY PLANNING

Prior to planning the development of a quarry, all available data on the proposed quarry should be assembled and analyzed. These data should include specification requirements, amount of stripping required, type of rock, rock's boundaries, location and quantities of any unsuitable rock, location of any major shear planes or fault zone, whether rock slippage along shear or fault planes will be a problem, amount of jointing or stratification in the rock, weight of the rock, rock's swell factor, embankment-shrinkage factor, estimated drilling speeds, powder factor (pounds of explosives required to break one bank cu yd of rock), and how the rock will break (size of pieces).

Planning a quarry is started by determining whether the rock deposit should be broken by explosives placed in *coyote* drifts (4 by 6 ft drifts excavated in the quarry face) or by explosives placed in drilled blast holes. Coyote quarry planning consists of developing a suitable quarry face, quarry floor, and a coyote-drift system. Blast-hole quarry planning entails developing the yards of stripping and determining the *burden* (distance the blast holes are back from the quarry face), drill-hole spacing (distance between drill holes measured parallel to the quarry face), drill-hole diameters, drilling speeds, powder factor, bench heights, and number of quarry faces.

Specification Requirements

The specifications should be analyzed to determine whether rock screening, selective quarrying, or wasting part of the quarry's production will be required. The word "analyzed" is used because specification writers seldom state that any of these procedures are necessary. They place the prebid responsibility for this determination on the contractor by specifying permissible rock sizes or by restricting the fines that can be included in the rock fill. Often, in producing fill from local rock deposits that complies with such requirements, it is necessary to perform selective quarrying or rock screening. Also, specifications may include gradations for filter materials so strict that to produce materials that will meet requirements it becomes necessary to separate the material by sizes, stockpile each size, and then reblend the different sizes.

If the rock embankment is the type that has a concrete slab on its upstream slope as a water barrier, the specification for the selectively placed rock facing may be so restrictive that a large amount of work may be required to select the rock.

The specifications for the main fill zone may require that large rocks be placed in its downstream portion and that the smaller rocks be placed in the upstream portion. Often to comply with this stipulation, selective loading in the quarry is required or the rock must be separated on a grizzly.

Exploratory Work Performed by the Owner

The contractor should assemble all the quarry data available from the records of the owner's quarry investigation. This data may be included in the specifications, shown on the plans, or distributed to bidders as supplementary information. Drill-hole cores are often available for inspection at or near the quarry site. Tunnel drifts or surface excavations are usually open for inspection during the contractor's visit to the damsite.

Contractors can seldom secure adequate data for planning the quarry's development from the results of the owner's investigation of the quarry, since these investigations are often limited to drilling a few core holes. A quarry can only be adequately explored by opening up one or more quarry faces with a test blast and by recording drilling speeds, powder consumption, and rock-breakage characteristics during the performance of this exploration. In addition, sufficient core holes should be drilled to prove that the quarry contains the volume of rock required for the embankment's construction. The owner can readily justify the expense of adequately exploring a quarry since contractors will submit lower bids if they are furnished this information.

Often the owner so inadequately explores a quarry that after the contract is awarded, the quarry is found unsuitable or lacking in the required quantities of rock. When this occurs, the contractor must develop a new source of rock, and this will often increase the construction cost of the project. This increased cost is always much more than if the owner had adequately explored the proposed quarry prior to bid advertising.

Contractors' Prebid Investigation of the Quarry

When the owner has inadequately investigated a proposed quarry, then contractors are forced to perform their own quarry investigation. Contractors seldom have the time between bid advertising and bid submittal to properly investigate quarries, and since they are often not the low bidder on the job, it is difficult for them to justify the expense of adequate quarry investigations.

Therefore, a contractor's investigation of a proposed quarry is usually limited to surface investigation and the performance of a few drilling tests. If contractors do not test-drill the quarry, they may send rock samples to drill manufacturers who will then perform laboratory tests on the drillability of the rock. In addition, contractors often employ consulting structural geologists to report on the suitability of the quarry.

DEVELOPMENT OF COYOTE QUARRIES

Coyote quarrying is the removal of large quantities of rock from the quarry face by explosives placed in a drift driven parallel to the face and called a coyote. Such quarrying is economical when the proposed quarry is of the sidehill type, has a 75-ft or higher face, has a quarry floor sufficient to receive the broken rock (which necessitates a floor width of twice the height of the quarry face), and contains rock that will be well-fragmented when the only breaking force is rock displacement. For example, sound, unfractured rock, i.e., a granite monolith, should not be quarried by coyote blasting because the rock would break into such large pieces that an excessive amount of secondary drilling and shooting would be required to reduce them to manageable sizes. Coyote blasting is such a specialized field that it should only be planned and performed by experienced personnel—otherwise results may be undesirable.

Coyote quarry development consists of providing a haul road to the bottom of the quarry face, providing a quarry floor to receive the broken rock, excavating the coyote drifts, and exploding the charge in these drifts. Coyote drifts are excavated by driving an adit into and perpendicular to the quarry face for a distance equivalent to 60 to 75 percent of quarry-face height. The adit is driven a shorter distance into the face than the face height so that the relief path from an explosion will be shorter to the quarry face than it is upward, resulting in proper shearing at grade. At right angles to this adit and parallel to the quarry face, cross drifts are then driven in both directions along the length of the quarry face.

Typical dimensions for the adits and cross drifts are 4 ft wide by 6 ft high. They are short and of such small cross section that expenditures for tunneling equipment are held to a minimum using air legs for drilling and a slusher scraper for muck removal. Typical crew size is three men per shift.

Low-cost explosives are placed in the cross drifts, such as ANFO (ammonia nitrate prills mixed with fuel oil). The amount of explosive used per cu yd varies with rock type but often is between 0.75 and 1.25 lb per cu yd of burden. The farther the adit is driven into the quarry face in relation to face height, the greater must be the powder factor. The explosives are piled in the cross drifts with piles on 25-ft centers and are detonated with primacord, using two interconnected lines to insure detonation. The primacord is protected from damage by hanging the cord from the roof of the drift or by placing it on the floor and covering it with sand bags. To provide blowout protection, the pile of explosive in each cross drift that is adjacent to the main adit is covered with a timber bulkhead, and the remaining length of cross drift between the bulkhead and the adit is stemmed with tunnel muck. The minimum distance for this stemming should be from 12 to 15 ft. As previously mentioned, the explosion will move the rock burden away from the face and deposit it on the quarry floor, which extends out from the quarry face a distance equal to twice the quarry face height.

The volume of rock moved by a coyote blast is bounded in its cross-section area by the rectangle whose height is the quarry-bench height and whose width is the distance from the quarry face to the back of the cross drift. The length of this rock rectangle is the same as the lengths of the cross drifts. A coyote blast will produce such a large volume of broken rock that only a few blasts are required.

The foregoing description is of a simple coyote layout. Coyote layouts should be modified to fit local conditions. Modified layouts may consist of two or more cross drifts extended from one adit, several adits with short cross drifts, a supplementary high coyote level, or a special layout to suit an irregular quarry face.

DEVELOPMENT OF BLAST-HOLE QUARRIES

There are three types of blast-hole quarries: exposed rock sidehill quarries, hillside quarries covered with overburden, and pit-type quarries. All three types of blast-hole quarries are similar in that their development is controlled by quarry bench heights, blast-hole diameter, and blast-hole spacing. These controlling factors do not vary with the type of quarry but rather are determined by the characteristics of the rock contained in the quarries.

The remaining quarry development work, i.e., quarry stripping, bench developing, and constructing haul and access roads to the quarry faces, varies with the type of quarry.

Bench Heights, Blast-hole Spacing, and Hole Diameter

The primary objective in blast-hole quarrying is to economically produce a pile of broken rock of the proper size and bank height for optimum shovel loading. This is accomplished by selecting a bench height, a hole pattern (a combination of burden depth and drill-hole spacing), and a blast-hole diameter that will result in the production of broken rock of the proper size and height for the lowest possible drilling cost. To achieve low drilling cost, a wide drill-hole pattern should be used. To produce small sized broken rocks, a close drill-hole pattern is necessary. Therefore, quarry planning requires that the proper quarry-shovel size be selected and that the widest drill-hole pattern be used that will produce broken rock that the quarry shovel can handle.

There are no criteria except past records and experience that can be used to establish the drill-hole pattern. However, in theory, broken-rock size in a quarry is dependent on rock breakability. One measure of rock breakability is its hardness. The harder the rock, the harder it is to break. The harder the rock is to break, the closer should be the individual explosive charges in the rock. The closer the individual charges, the closer the drill-hole pattern. Therefore, rock hardness can be used as a gauge for establishing the drill-hole pattern. One measure of rock hardness is its Mohs' scale, as given in the following table:

Mohs' Hardness Scale

	Origin	Mohs' hardness scale	Lb/yd in the bank	Specific gravity
Andesite	Igneous	7.2	4,660	2.4–2.9
Basalt	Igneous	7.0	5,080	2.8–3.0
Bauxite	Mineral	2.0	4,290	2.4–2.6
Chalk	Sediment	1.0	3,700	2.4–2.6
Chert	Sediment	6.5	4,320	2.5
Clays	Sediment	1.0	2,700	2.4–2.6
Diabase, dalerite	Igneous	7.8	4,730	2.8
Diorite	Igneous	6.5	5,000	2.8
Dolomite	Sediment	3.7	4,860	2.7
Felsite	Igneous	6.5	4,590	2.65
Flint	Sediment	7.0	4,320	5.5–7.0
Gabbro	Igneous	5.4	4,860	2.8–3.0
Gneiss	Metamorphic	5.2	4,860	2.6–2.9
Granite	Igneous	4.2	4,590	2.6–2.9
Limestone	Sediment	3.3	4,400	2.4–2.9
Marble	Metamorphic	3.0	4,320	2.1–2.9
Marl	Sediment	3.0	3,780	2.2–2.4
Mudstone	Sediment	2.0	2,970	2.4
Porphyry	Igneous	5.5	4,290	2.8
Quartz	Igneous	7.0	4,460	2.65
Quartzite	Metamorphic	7.0	4,320	2.0–2.8
Sandstone	Sediment	3.8	3,920	2.0–2.8
Schists	Metamorphic	5.0	4,660	2.8
Serpentine	Metamorphic	5.0	4,670	2.8
Shale	Sediment	4.0	4,320	2.4–2.8
Slate	Metamorphic	3.0	6,830	4.0
Talc	Sediment	1.0	4,540	2.5–2.8
Trap rock	Igneous	7.0	4,870	2.6–3.0

The drill-hole pattern that is selected establishes the blast-hole diameter, drill type, and bench height. The drill-hole pattern establishes blast-hole diameter since the pounds of explosives required to break a cubic yard of rock does not radically change with rock types. Therefore, as the drill-hole pattern decreases in its dimensions, the diameter of the blast-holes decreases; or, as the drill-hole pattern increases in its dimensions, the diameter of the blast-holes must increase.

Since the drill-hole pattern establishes the diameter of the blast holes, it also controls the type of quarry drill since the diameter of the blast hole controls the selection of quarry drills. And, since the drill-hole pattern establishes drill type, it also limits the maximum height of bench, inasmuch as the different types of quarry drills can economically drill different depths of hole. Safety requirements, however, prescribe maximum bench heights of 60 ft.

After the drill-hole diameter has been established for a given quarry, the following rules of thumb can be used to check the depth of burden, drill-hole spacing, bench heights, depth of subdrilling (hole depth below the quarry floor), and length of hole stemming (top of drill-hole length that is packed with inert backfill).

The burden depth in feet should be from 2 to 3 times the drill-hole diameter in inches. A typical burden is 2.5 times the drill-hole diameter.

The drill-hole spacing, in feet, should be from 1 to 2 times the burden, in feet. A typical drill-hole spacing is 1.25 times the burden.

The bench height, in feet, should be from 1.5 to 4 times the burden, in feet. A typical bench height is 2.6 times the burden.

The depth of subdrilling, in feet, should be 0.2 to 0.5 times the burden, in feet. A typical subdrilling depth is 0.3 times the burden.

The length of hole stemming, in feet, should be 0.5 to 1.3 times the burden, in feet. A typical length of stemming is 0.7 times the burden.

Quarry Drills

There are three main types of quarry drills: percussion drills mounted on air-powered tracks, down-the-hole drills, and rotary drills.

Crawler-mounted percussion drills are used to drill blast holes up to 5 in. in diameter. Drills are sized by the drill's piston diameter and are available with piston diameters of from $4^1/_2$ to 6 in. They can be secured with or without independent bit rotation. A drill of this type with a 6-in. piston diameter is shown in Photograph 35. These drills are used in hard-rock quarries when the rock is to be loaded with relatively small-sized shovels. A close pattern of holes will result in good fragmentation, and secondary drilling and shooting will be minimized. The use of small holes will give better rock fragmentation near the top of the quarry face since less hole stemming is required and the explosive charge can be brought up higher in the hole. Bench heights often used with this type of drill vary between 25 and 35 ft. When it is necessary to break rock into sizes that can be handled by small shovels, a close hole pattern is mandatory. The drilling and explosive-loading cost for this large number of holes will be relatively high, but because secondary drilling and shooting cost will be at a minimum, total drilling and shooting cost will be less than if a wider pattern of holes was used.

Down-the-hole percussion drills are used in hard-rock quarries when the hole pattern can be enlarged, which necessitates the use of blast holes between 6 and 9 in. in diameter. Since these drills are larger and more efficient than the crawler-mounted percussion drills,

Photo 35 Crawler-mounted 6-in.-bore percussion drill. (Joy Manufacturing Company.)

they can drill to greater depths, allowing the use of bench heights of from 35 to 45 ft.

Down-the-hole drills follow the bit down the hole, securing their control and support from drill steel suspended from a drill carriage located on the surface of the excavation. The drill steel (hollow steel rods connecting the drill to the drill carriage) furnishes support for and conveys air to the drill and rotates the drill and bit, receiving this rotation force from a rotary table mounted on the drill carriage. Since there is a minimum length of steel between the drill and the bit, all the drill's striking force is exerted on the bit. Down-the-hole drills are of simpler construction than drifter drills since they do not have to furnish a force for the bit's rotation.

Rotary drills transmit a rotary force to the drill bit in contrast to percussion and down-the-hole drills which transmit a primary striking force and only a secondary rotating force to the drill bit. Rotary drills are used in quarries that contain soft rock that is easily broken, allowing a wide pattern of drill holes and thus requiring drill holes of 9 in. or more in diameter. Since rotary drills can economically drill deep holes, quarry-bench heights can be as high as desired. Photograph 36 is of this type of drill.

Generally, rotary drilling is used in the softer rock types, and percussion drilling is used in the harder rock types.

An example of quarry planning that illustrates the application of the foregoing discussion is shown in the performance of estimating task XII in Chap. 11. This quarry planning was done to select the quarry equipment for the example estimate of cost for the construction of a concrete dam.

Inclined Drilling

There is another explosive distribution procedure that is important in quarry operations. For each blast hole, the explosive should be distributed through the hole in accordance with the work to be done. A quarry face normally forms a slope from the toe of the face to the top of the bench. Therefore, vertical blast holes drilled from the top of the bench have a greater thickness of rock to break at the bottom of the hole than at the top. To take care of this change in burden, the greatest concentration of explosives should be at the bottom of the hole and should decrease proportionately up the hole. This gradation of explosive concentrations throughout the hole can be readily accomplished when certain types of slurry are used.

Another method of solving this problem is to incline the blast holes so the bottom of the hole is the same or at a lesser distance from the quarry face as the top of the hole. An additional advantage secured by inclined drilling is that, because of the inclination of the hole, the explosives in the bottom of the hole will have more explosive force reacting toward the quarry face and less wasted in a downward reaction into the quarry floor.

Quarry Stripping, Bench Development, and Access Roads

The best method of determining the amount of stripping, bench development, and access road construction required to develop a blast-hole quarry is to prepare a quarry layout. A layout is helpful in developing the required amount of stripping; the number and extent of the quarry benches, quarry-bench heights, the tonnage of rock available in each bench; and the amount of work necessary to provide haul and access roads. If a quarry layout is not prepared, there will be a tendency to underestimate the cost of the quarry's development.

The amount of work required to develop a quarry varies with the type of quarry. Exposed rock sidehill quarries require the least amount of quarry-development work since stripping is not required.

Photo 36 Rotary drill. (Joy Manufacturing Company.)

Multiple bench development is often required because, as each quarry bench is excavated back into the hillside, the bench height increases, necessitating the starting of an additional bench. As each additional bench is started, a haul road must be constructed to the bottom of the bench and an access road to the top of the bench.

When hillside rock deposits are located underneath overburden, large areas must be stripped to provide space for the required number of quarry benches. The quarry layout should reflect the tons or yards of stripping required for each ton or yard of quarried rock. Similar to other types of quarries, these benches must be developed and haul and access roads must be provided to each bench.

Pit quarries do not require stripping when the rock is exposed, but more frequently the rock is located beneath overburden which necessitates a large amount of stripping. Face development for these quarries is costly because the face excavation must start at the surface and must be ramped down until economical face height is achieved. Side drilling is also required to trim up the sides of this pit-type excavation. Generally, only one quarry bench is developed, but if more benches are required, then each additional bench must be developed in a similar manner and haul and access roads must be provided to each additional bench.

Quarry Blasting

Economical quarry blasting is achieved by the use of ammonium-nitrate fuel-oil mixture (ANFO) or a slurry.

ANFO is suitable for use in coyote blasting and in dry blast holes. It can be used in wet holes when it is purchased in plastic bags or is bagged by the contractor at the jobsite. Explosives trucks can be used to bulk-load ANFO into dry holes. When bagged ANFO is used, the advantage of bulk loading is sacrificed.

Job delivery of bulk ANFO can be made with large bulk carriers that discharge the ammonium nitrate into elevated storage tanks. From these storage tanks, it can be gravity-loaded into the bulk-loaded trucks. These trucks can then be driven to the drilled area and from one location they can load a large number of holes.

Bulk explosives loading trucks are of two types. One type uses pneumatic pressure to discharge the ANFO through a 3-in. hose into the blast holes. The other type elevates the ANFO to the top of a boom by use of a screw. From the boom top the ANFO flows by gravity through a loading hose into the blast holes.[1,2] The trucks can mix the ammonium nitrate with the fuel oil as they are loading the holes, or the mixing can be done when the ammonium nitrate prills are unloaded into the storage silos. Many contractors do the latter as they prefer using ANFO after it has aged in its mixed state.

Slurry is used in preference to ANFO when a stronger blasting agent is required or when wet holes are encountered. Slurry has a density of 1.0 to 1.35, while ANFO has a density of 0.85 if it is gravity loaded and 0.90 if it is pneumatically loaded. This higher density of slurry makes it a stronger blasting agent and gives it sufficient weight so it will replace water in a blast hole. Slurry is a mixture of water, from 65 to 75 percent ammonium nitrate, a sensitizer, and other ingredients.[3] The ingredients used in slurries vary between powder companies, and each company adjusts their ingredients for changes in rock types. Slurry mixing is done by the bulk-loading truck as it is loading the holes. One supplier of slurry uses a patented mixing process by which the density of the slurry is controlled by the inclusion of small nitrogen bubbles in the mix. With this type of mixing, the slurry has a greater density at the bottom of the hole, which is desirable, than at the top, because the weight of the column of explosive compresses the nitrogen bubbles in proportion to hole depth. These nitrogen bubbles also increase the explosive power of the mixture.

Slurry hole loading requires a tank-type mixing truck equipped with pumping facilities, since the slurry is mixed during the hole-loading operation. The mixed slurry is pumped through hoses into the blast holes. If jobsite storage of slurry is desired, each ingredient should be stored separately.

Both ANFO and slurry blasting agents require the use of an explosive for detonation. ANFO can be detonated by using an electric firing circuit to detonate blasting caps, which detonate an explosive, which in turn detonates the ANFO. A preferred method is to use primacord to detonate the explosive. Slurry is detonated by using primacord and a solid, self-contained slurry detonator that is manufactured for use in this mixture.

The use of bulk-loading trucks permits fast and efficient blast-hole loading with a minimum crew. Using bulk-loading trucks, blast holes can be loaded with ANFO at rates from 200 to 600 lb of mix per minute and with slurry at rates from 400 to 750 lb per minute. Often bulk-loading trucks will be furnished by the explosives dealer. Some contractors prefer to use their own trucks, for they believe this puts them in a better bargaining position to purchase explosives.

The number of blast holes that are loaded and exploded at one time varies with the diameter of the holes. Multiple rows of small holes, drilled in a staggered pattern, are often exploded at one time with different delays used for each row. Large-diameter blast holes are commonly exploded one row at a time.

Secondary Breakage of Quarried Rock

Except in very soft rock quarries, secondary drilling and shooting must be done to break large rocks into sizes small enough for the quarry shovel to handle. This is done by drilling small holes into the rock, loading these holes with an explosive, and then exploding these charges. This procedure is called *block holing*.

Another method of performing secondary breakage, *mudcapping*, is to place explosive charges on the surface of the large rocks, cover the charges with mud, and then explode these charges. This method of performing secondary breakage eliminates the need for jackhammers, steel, bits, and compressed air. Packaged charges of explosives for mudcapping are available.

The required amount of secondary drilling and shooting varies with the quarrying method, with the hardness of the rock, and with the drill-hole spacing. For well-planned quarry work, the cost of secondary drilling and blasting seldom exceeds 20 percent of the primary drilling and blasting cost. If it is neglected, then the production achieved by the quarry shovel will decrease since it must then spend valuable loading time to maneuver its bucket to pick up the large rocks. The speed of the shovel's swing must then be reduced, since the large rocks must be balanced on the bucket.

SOURCE OF FILTER MATERIAL

If the rock-fill embankment is designed with an impervious zone for water cutoff, it will also contain filter zones which protect and separate the impervious material from the rock fill. Two sets of two adjacent filter zones are used: fine granular filter zones are located on each side of the impervious zone, and coarser granular filter zones are located between the fine filter zones and the rock fill. Specifications often state that gravel from borrow pits or quarry fines can be used in these zones. They also are often misleading in that they imply that this material can be hauled directly from the source without the necessity of processing. This is seldom the case. Often specification restraints are so strict that borrow-pit material must be processed by removing oversized material with a grizzly, by separating the grizzlied material into sizes, and then by reblending the sized material to produce the two types of filters.

If the filter materials are to be produced from gravel pits, the gradation of the pit materials should be examined and flow sheets and processing plant layouts should be prepared showing the required processing equipment. If the filter material is to be produced in a quarry, it may be necessary to provide crushing facilities as well as screening and blending facilities.

PLACING EMBANKMENT FILL

One of two methods is commonly used for placing rock fill into the embankment. The method used influences the layout of the haul roads and the schedules of fill placement.

Placing Rock Fill by the End-dumping Method

When the available rock fill is sound, of large size, and with a minimum amount of fines, then rock embankments are often designed for the use of the end-dumping method of placement of the main rock-fill zone. Specifications designate the percentage of large rock required and the maximum amount of minus 4-in. material that will be acceptable. When the end-dumping method is used, the main rock-fill zone is often placed in two lifts with the bottom lift higher than the top lift. Compaction is achieved by the rock cascading down the lift slope after it has been end-dumped over the edge of the fill. As the rock is dumped, it is struck with high-pressure water jets, which wash all the fines into the previously placed rock fill, permitting all large rocks to have three points of contact. Photograph 37 shows monitors being used for this purpose.

Photo 37 Rock washing at Wishon Dam, California. (John W. Stang Corporation.)

Loads of rock are distributed on the fill so that the larger rocks are placed in the downstream portion of the fill and the smaller ones in the upstream portion. One theory for this is that if there is any leakage through the rock fill, the water will flow through expanding water channels and this will eliminate erosion of the embankment. Water pressure and the volume of water that must be used are defined by the specifications. A typical specification is that the water pressure must be 100 psi and the volume of water must be twice that of the rock fill.

The angle of repose of the rock fill establishes the upstream and downstream fill slopes. The design often includes a downstream setback at the top of the first lift.

End dumping is the most economical method of placing rock fill in the embankment since only minor amounts of embankment equipment are required. A bulldozer is required to maintain a roadbed on the surface of the fill, monitors are required for water jetting, and pumps and water distribution lines are required to furnish water to the monitors.

Placing Rock Fill in Compacted Lifts

The rock fill in an embankment must be placed in shallow compacted lifts when the available rock breaks down into small pieces, when it contains an excess of fines, or when it contains common material. Also, when a rock embankment is designed with an impervious core and the main fill of a rock embankment is placed by the end-dumping method, it is still necessary to place the upstream rock zone and the impervious and filter zones in shallow, compacted layers. In addition, it may be a requirement that a narrow zone of compacted rock be placed on the upstream side of the main rock fill, adjacent to the filter and impervious zones.

Compacted rock fill is placed in lifts that vary between 3 and 10 ft in height. Typical compaction requirements state that each lift must be compacted with four passes of the crawler tracks of a 50,000-lb minimum-weight tractor.

Selective Rock Placement

If the embankment is designed with a concrete face slab on the upstream slope for a water barrier, then often the specifications will require that the construction of a foundation for this face slab be composed of the selective placement of a narrow layer of rock on the upstream face of the main rock fill. A typical requirement for selectively placed rock is that each rock must be rectangular-shaped, from $^1/_2$ to $2^1/_2$ cu yd in size, and individually selected from quarried material. Chinking of the large rocks may be allowed after they are placed in position.

Placement procedure is to dump these rocks near the upstream edge of the main rock fill. They are then picked up and lowered into position by a crane operating on the surface of the fill. Crane handling can be expedited if each rock is drilled so that it can be handled with a bridle. The drilled holes should be angled toward each other and should be approximately 16 in. deep. The bridle consists of a center lifting ring, two 12-in.-long iron pins, and cables connecting eyelets on the pins to the center lifting ring. The pins are inserted into the angled holes and as the crane lifts the bridle, the pull on the cables exerts enough side stress on the pins that they remain in position while the rock is lifted and positioned.

The placement of all other zones in the dams must be keyed to the placement of the selectively placed rock fill. After the first layer of the main rock fill has been placed, then the placement of the rock facing must commence, since this placement equipment must be supported by the surface of the main rock fill.

Placement of the selectively placed rock is time-consuming and its placement cost is often more than contractors estimate. Equipment requirements consist of a bulldozer in the quarry to select the rocks, jackhammers for drilling the lifting holes, shovel or crane time in the quarry for loading the rocks, haul trucks, and cranes on the embankment for placing the rocks.

Placement of Impervious Core and Filter Materials

Irrespective of the method used for rock-zone placement, when rock fill embankments contain impervious cores, the impervious core and the protective-filter zones must be placed in lifts between 6 and 12 in. in height; the moisture content of the impervious material must be brought to close to optimum; and each lift must be compacted in the specified manner. This necessitates that truck access be provided to every embankment elevation.

If the rock-fill embankment has a vertical impervious core, there will be no particular fill-placement scheduling problems. Most rock-fill embankments, however, are designed with sloping impervious cores so that the amount of rock fill required for embankment stability can be kept to a minimum. When the embankment has a sloping impervious core, this core and the filter zones are partially supported by the main rock-fill zone. This necessitates that fill placement start with placement of rock in the main rock-fill zone and that the impervious core and filter-zone placement not start until part of the supporting main rock fill has been placed. Also, since the upstream rock zone is partially supported by the impervious fill, it must follow its placement.

As discussed in Chap. 6, rain and freezing weather will restrict the placement of impervious fill. When impervious borrow pits are too wet, the impervious fill can be dried by draining, blading, harrowing, or by passing the fill material through a pug mill. In extreme cases, it may be necessary to dry the impervious material in mechanical dryers. When the impervious borrow pits are too dry, the water content of the impervious fill can be raised by sprinkling the borrow pits. In some cases, several months of sprinkling may be required.

Placement of the filter zones is relatively simple. The only problem is the dumping and spreading of the material in the narrow-zone limits. To prevent materials from mixing, the filter zones are maintained at a slightly different elevation from the impervious zone.

As described in Chap. 6, the impervious- and filter-zone placement equipment will consist of blades or bulldozers for spreading the fill, water wagons, and compactors.

Concrete-slab Placement

When the embankment has a concrete slab as a water cutoff, this concrete is not placed until the rock-fill facing has reached sufficient height to permit the concrete slab to be placed in long, continuous strips up the slope of the embankment. The production, hauling, and placement of concrete are described in Chaps. 4 and 5. Placement of this slab is often done using slip forms. Concrete can be delivered to these forms by belts, skips, or by pumping. Photograph 38 shows one method of placing a concrete slab on a rock embankment.

HAUL-ROAD LAYOUT

The preparation of a haul-road layout is a necessity for successful embankment-construction cost estimating. This layout can be used to compute the quantities of excavation, fill, surfacing, and culverts required for the roads' construction. The layout will establish haul distances and haul grades to use in computing truck turnaround cycles. Truck turnaround cycles are used to establish equipment requirements and to estimate the direct cost of fill haulage.

The haul roads should be designed to permit maximum truck speeds and the continuous flow of traffic to and from the embankment. The road layout should show the initial haul roads and how they must be relocated as the embankment is raised in elevation. The number of road-access points required for a rock-fill embankment varies with the method of fill placement and the design of the

Photo 38 Placing a concrete face slab on a rock-fill dam. (Morgen Manufacturing Co.)

embankment. This subject was discussed in the preceding section. Often, truck access to the embankment between two different haul-road elevations is assured by constructing short lengths of road within the embankment. These haul roads can be graded out when they are no longer needed.

Haul roads should be wide enough to allow haul trucks to overtake and pass each other, so that the speed of all the haul trucks is not controlled by that of the slowest unit. Preferably, the haul-road grades should be under 3 percent, but they should never exceed 7 percent. The roads should be surfaced, maintained, and watered so that haul trucks can operate at maximum speeds with a minimum of tire wear.

EMBANKMENT SCHEDULING

A schedule for a rock-fill embankment construction must allow time for access-road construction, for mobilization of plant and equipment, for the construction of diversion facilities, and for excavation of the embankment's foundation. The scheduling of the construction of diversion facilities was discussed in Chap. 2. The scheduling of the excavation of the dam's foundation was discussed in Chap. 3.

To properly schedule fill placement, the embankment control dates should be established. These dates are the date that embankment placement can start, the date that maximum placement of fill material can be achieved, the date that maximum fill placement must be reduced because working areas will be restricted as the embankment is topped out, and the date that the embankment must be completed.

Maximum monthly fill-placement requirements can be determined from work quantities and the foregoing embankment control dates and can be reduced to maximum hourly requirements by selecting the number of days that will be worked per month, the number of shifts that will be worked per day, and the number of hours that will be worked per shift. These choices were discussed under earth-fill embankment scheduling in Chap. 6.

The maximum hourly rate of fill placement divided by the production achieved by each equipment unit establishes the number of operating construction-equipment units that will be required on the project. The number of operating construction-equipment units also are needed for computing the direct cost of constructing the embankment.

The date that fill placement can be started in the embankment is dependent on the topography of the damsite, the design of the embankment, and on specification requirements.

The topography of the damsite controls the start of embankment placement since rock-fill embankments are normally constructed at sites where the river channel is narrow and abutments are steep. At these sites, foundation excavation can be completed from the top of the abutment down to water level before the stream is diverted. This is often the easiest part of the dam's foundation to excavate, and it often can be done in less time than is required to excavate the section of the foundation below streambed level. The portion of the dam's foundation that is beneath the stream often cannot be efficiently excavated until the stream is diverted. In most cases, excavation below the stream must be carried to greater depths than is required on the abutments. Considerable time must be scheduled for completing the excavation below the original stream level, since working space will be restricted and pumping will be required.

When a rock embankment is designed for end-dumping of the

main rock-fill zone, then it is possible to start the placement of rock fill on one abutment before the center of the dam is excavated. If the damsite has steep abutments, only a small amount of rock fill can be placed on this abutment, but a truck-dumping area can be developed. This truck-dumping area will expedite the maximum placement of the dumped rock fill, upon completion of the excavation of the dam's foundation.

When the rock fill is to be placed in compacted layers in the dam embankment, in most cases fill placement cannot be started until the excavation of the entire foundation is completed. Exceptions occur when specifications permit and when the slope of one abutment is so flat that a longitudinal section of fill can be hung on the abutment.

Impervious-zone fill-placement scheduling is dependent on the method of placing rock fill in the dam and whether a vertical or sloping impervious core is to be placed. This subject was discussed when impervious-fill placement was described in this chapter. The scheduling of impervious fill must be correlated with rock-fill placement. If the embankment is located in an area that has a long rainy season, its rate of placement may control the scheduling of the main rock fill. Impervious-core placement will be slow when it is started because hand compaction will be required and the working area will be restricted. When the impervious zone lengthens out, placement can be done at maximum capacity until the embankment approaches its maximum height. Then production will be again retarded because of lack of working room. Impervious-core placement must be suspended during rainy or freezing weather.

Selective rock placement must also be scheduled to coordinate with the placement of the main rock fill. When embankments contain selectively placed rock fill, its placement often controls the placement of the main rock fill. Also, after the rock-fill layer has been selectively placed, time must be allowed for concrete-slab placement.

EQUIPMENT SELECTION

Upon completion of the construction schedule, the maximum hourly rate of fill placement is used in the computation required to determine the number of operating productive plant and equipment units that are required to maintain the schedule. To eliminate the chance of making errors in these computations, it is preferable to convert embankment yardage to bank yardage and base all computations on these units.

Determining the Number of Drills

The required number of operating drills is determined by multiplying the scheduled production of rock in bank yards per working hour by the lineal feet of hole that must be drilled to break one bank yard of rock and by dividing this product by the lineal feet of hole produced per hour per drill. Many studies have been performed for the purpose of establishing a set of criteria for determining drill-penetration rates for different types of rock. The results of all these studies is that drill-speeds and rock-breakage characteristics can only be determined by drilling and breaking each rock deposit. Therefore, drill-hole diameters, drill-hole spacing, and drilling speeds should be established for each rock quarry from past experience, by performing drilling tests, or by making a laboratory drilling analysis. Field drilling tests or laboratory tests will establish drill penetration rates.

Drill penetration rates must be reduced to establish hourly drill production rates. This allows for delays attributable to the operator for time required to move drills, for steel and bit changing, for hole cleaning, for freeing stuck steel, etc. Drill penetration rates were discussed in detail when drilling was described in Chap. 3. Indications of proper drilling speeds for different rocks were also given in Chap. 3.

Mohs' scale of hardness (in this chapter) of the different types of rock will give a rough comparison between the drilling speeds probably required and the breaking characteristics of different rock, but its use is not accurate enough to establish drilling speeds for a cost estimate. For instance, andesite with a Mohs' hardness of 7.2 is easier to drill than basalt which has a Mohs' hardness of 7.0.

Quarry Loading Equipment

After the rock is broken in a quarry, it must be picked up and loaded into haulage vehicles for transportation to the dam embankment or, if required, to a rock separation plant. The number of operating loading units that will be required is determined by dividing their production capacity into the total production capacity required.

In most hard-rock quarries, large shovels are used to load out the rock. A typical large quarry shovel is shown in Photograph 39. Large shovels must be transported in sections to the job's location and then erected. At the job's completion, they must be dismantled and then transported in sections to their next location. Photograph 40 shows how the undercarriage of an 8-yd shovel was transported to an isolated location.

For estimating purposes, the number of bank yards that can be

Photo 39 A 15-cu-yd quarry shovel. (Marion Power Shovel Company, Inc.)

loaded by quarry shovels should be established by evaluating their production capacity on previous work. Since these records may not be available to all estimators, the following table shows the production rates that should be obtainable with average loading conditions. These production rates should be adjusted for loading conditions that are peculiar to each project.

Production Rates for Quarry Shovels; Average Conditions, 70% Equipment Efficiency

Shovel-bucket size, cu yd	Well-broken rock, bank cu yd/hr	Poorly broken rock, bank cu yd/hr
$3^1/_2$	220	150
$4^1/_2$	300	200
6	370	260
8	490	320
10	600	400
14	800	520

The production rates shown in the foregoing table are for optimum bank height. For other than optimum bank heights, these production rates should be reduced in accordance with the table included in Chap. 6.

Photo 40 Transporting the undercarriage of an 8-cu-yd shovel to a mountainous location.

Front-end loaders are used to load out finely broken rock in some soft-rock quarries. Track-mounted front-end loaders can load broken rock more efficiently than rubber-tired front-end loaders, but the largest track-mounted loader produced is one with a 5-cu-yd bucket. Rubber-tired front-end loaders are available with a bucket capacity of 15 cu yd. When they are used in quarries, chains are often placed over their tires to increase traction and reduce tire wear. The production rates achievable with front-end loaders are discussed in Chaps. 3 and 6.

Other Quarry Construction Equipment

A powder house, a cap house, and a pickup are always necessities. Depending on the equipment supplied by the powder company and on the type of explosive used, a truck for bulk loading the explosive into the blast holes may be required. It may also be necessary to supply facilities for bulk storage of explosives.

Secondary drilling and shooting require that the quarry be supplied with jackhammers and portable compressors.

If the primary drills do not contain their own source of compressed air, then compressors are required. Since electric-driven compressors are more economical to operate than diesel-driven, they should be used and installed at a central point with pipe lines distributing air throughout the quarry. The air required for percussion drills was listed in Chap. 3. The air required for other drills is securable from the manufacturers of the drills.

If the quarried rock must be separated into sizes, a separation plant will be required. This was discussed under quarry development.

Bulldozers will be required for building haul and access roads, for cleaning up around each quarry shovel, and for selective quarrying. If several shovels are used, a rubber-tired bulldozer has the speed to clean up around several shovels.

Quarried-rock Haulage Equipment

Rear-dump off-highway trucks are used for hauling quarried rock. The trucks should be sized so that they will be loaded by three shovel buckets. Although rear-dump trucks as large as 200 tons have been used by the mining industry, the largest size used by the construction industry to date is one that has a capacity of 110 tons. Photograph 41 shows a truck of this capacity.

The number of operating trucks that is required to maintain any production rate can be found by dividing the required production rate by the trucks' hourly production capacity. This hourly truck-production capacity is determined by multiplying the truck capacity times the number of cycles the truck will make in an hour. Cycle time is composed of loading time, haul time, dumping time, return time, and delay time. An example of the computations required to determine the truck cycle time is given in Chap. 3. When rock-fill embankments are placed in two lifts, then only two truck cycles need be computed. When shallow lifts are used, then truck cycles are continually changing and a large number of truck cycles must be computed to arrive at an average cycle. In this second case, a computer can be utilized to determine the average cycle time and its use will eliminate many routine calculations.

Photo 41 A shovel loading a 110-ton rear-dump truck. (K. W. Dart Truck Co.)

Impervious- and Filter-zone Equipment

The equipment required to load and haul the impervious-zone material was discussed in Chap. 6.

Screening, stockpiling, and reblending equipment may be required to produce the filter-zone material. This material can be loaded by front-end loaders into haulage equipment. The daily requirements for filter material are so low that any available equipment can be used for this haul. If equipment must be purchased, then light, highway-type rear-dump trucks will provide sufficient haulage capacity.

Embankment Equipment

If dumped fill is used in the embankment, the installation of a monitor system may be quite complex. At many sites, the stream flow is not large enough to supply sufficient water during the dry season. Thus it may be necessary to construct downstream storage dams so that the water used on the fill can be collected, stored, and reused. Pumping requirements are controlled by maximum rock-placement rates. Pumps may be capable of lifting the water to the top of the embankment and still furnish 100 psi water at this elevation. Banks of vertical high head pumps are required. Water distribution lines on the embankment must be continually moved and extended. Often a main line is carried up one abutment. Monitors are often mounted on tractors or on air-powered crawler tracks to provide flexibility of movement.

Bulldozers will be needed on the embankment to spread and compact rock in the compacted fill zones and to maintain the road surface on top of any dumped-rock zone. Graders will be required to spread the impervious- and filter-zone material. Water wagons will be required for controlling the moisture content of the impervious fill. Compactors will be needed for compacting the impervious fill and filter zones.

Total Embankment Plant and Equipment Requirements

The total plant and equipment requirements are established by increasing the required number of operating units by 25 to 30 percent to compensate for equipment availability. This provides for the equipment units that must be pulled out of operation for servicing or for repairs.

To this fleet of equipment must be added the bulldozers, graders, and water wagons needed for haul-road construction and maintenance and the servicing equipment and facilities. Servicing equipment includes grease trucks, fuel trucks, tire repair trucks, pickups, etc. Servicing facilities include the warehouse, shops, tire shop, rigging lofts, craft buildings, etc. When the amount of servicing equipment and facilities required for rock-embankment construction are determined, the overall project demands should be reviewed so that there will not be duplication of these facilities in any two construction systems.

REFERENCES

1. When Bulk ANFO is Best, *Construction Equipment,* December, 1969, pages 50–51.
2. Newest Trend in Blasting, *Roads & Streets*, January, 1969, page 34.
3. Blast Drilling: New Materials, New Techniques, New Equipment, a reprint from *Construction Methods and Equipment*, June, 1962, and August, 1962, copyright 1962, McGraw-Hill Publishing Co., Inc.

CHAPTER EIGHT

Powerhouse Construction

This chapter describes the planning, scheduling, and selecting of construction equipment for the construction of powerhouses. A dam contractor must be capable of estimating the cost and performing the construction of powerhouses since many dam-construction contracts include either the first-stage construction or the complete construction of a powerhouse. Often, because of inadequate construction planning, the successful bidder of a project that contains both a dam and a powerhouse will make a profit on the construction of the dam but will incur a loss in constructing the powerhouse.

Few dams are constructed without power-generation facilities, since power generation is the primary justification for the construction of many dams. Even dams constructed to store water or control floods are often equipped with power-generating facilities to help justify their construction. When the construction of both a dam and a powerhouse is included in the same construction contract, the powerhouse may be located as a separate structure near the downstream edge of the dam, it may be located below one of the dam's abutments, or it may form an integral part of the dam. On large rivers, powerhouses and their intake structures have often been designed to function as longitudinal sections of low-head, spillway-type dams. Dams and powerhouses of this type have been constructed on the lower Columbia and Snake Rivers. A recent new design concept features the powerhouse structure functioning as a spillway for a low-head dam. This is known as a hydrocombine; Wells Dam on the upper Columbia River is an example of this type.

The main difference in the interiors of powerhouses results from the installation of different types of turbines, and the head of water available for power generation controls turbine selection. Kaplan turbines are used for waterheads up to 75 ft in height; Francis turbines are used for heads of between 75 and 900 ft; impulse turbines, also called Pelton turbines, are used for heads exceeding 900 ft. (This practice differs in Europe where Francis turbines are used for heads up to approximately 3,000 ft.)

STAGE CONSTRUCTION OF POWERHOUSES

Most powerhouses are designed so that they can be constructed in two stages, as shown on Drawing 5. The first-stage construction is often the only powerhouse work included with a dam contract. First-stage powerhouse construction will vary in scope, but it generally consists of the erection of the basic powerhouse structure, including the main permanent crane and the draft-tube gates. First-stage construction does not include the concrete that encases the embedded turbine parts. Thus concrete placement is not delayed by the installation of turbine parts, and first-stage construction is expedited.

Second-stage construction consists of the fabrication, installation, and concrete encasement of the draft-tube liners, the scroll cases, and the turbine's embedded parts; the placement of the remainder of the second-stage concrete; the installation of the turbines; the assembly and installation of the generators; the installation of the transformers; and the installation and wiring of all major powerhouse equipment. This second-stage construction may be included with the main dam contract but more often it is advertised as a separate powerhouse completion contract.

Some powerhouses are designed so that all concrete is placed at one time. When this type of powerhouse is constructed, the draft-tube, pit, and scroll-case liners, and the turbine's embedded parts must be fabricated and installed prior to the placement of the surrounding concrete. This design eliminates the formwork required between the first- and second-stage concrete and also eliminates

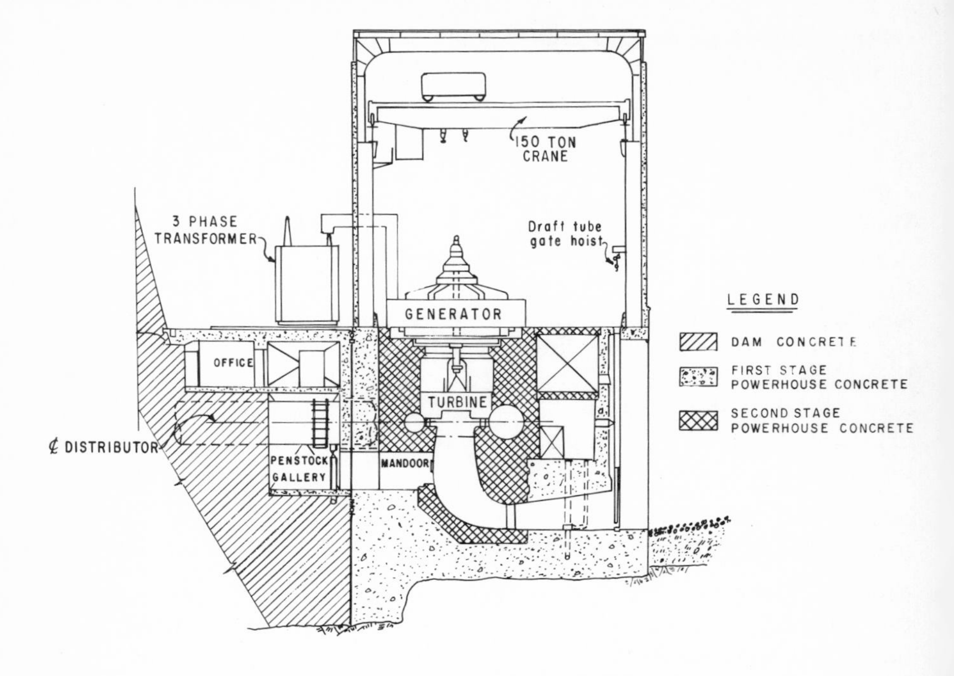

Drawing 5 First- and second-stage concrete for the generating bays of a hydro power plant.

minor amounts of concrete. Concrete placement is more difficult, however, since it must be suspended while the embedded liners, scroll cases, and embedded turbine parts are being installed.

Often powerhouses are constructed with extra generating bays that provide space for future installations of power-generating facilities. When this is done, only first-stage construction is completed in these extra, or "skeleton," bays. When additional power-generating units are needed, the draft-tube liners, scroll cases, and turbine embedded parts are placed in the skeleton bays and encased with second-stage concrete.

The foregoing paragraphs describe the power-generating bays of a powerhouse. In addition to generating bays, powerhouses contain two other bays or equivalent areas along the length of the powerhouse. One bay or area is required for assembling and repairing generators and other powerhouse equipment. The other bay or area contains the powerhouse control equipment and provides office space.

POWERHOUSE EXCAVATION

Foundation excavation for powerhouses and dams is described in Chap. 3. Special attention should be given to the amount of presplitting that may be required to maintain the sides of the powerhouse excavation. Many designers prepare drawings that depict excavation pay lines for the finished foundation surface of the powerhouse as a series of humps, valleys, and trenches. When this is done, the cost of excavation and foundation preparation is greatly increased, and the quantity takeoff should reflect this situation. When the drawings establish excavation and concrete pay lines, the specifications frequently require that the contractor place backfill concrete in any area excavated beyond these pay lines, without receiving reimbursement for this work. Since it is economically impossible to excavate any foundation exactly to prescribed lines, the takeoff quantities should show the estimated amount of overbreak so that the estimator may weigh the cost of performing this additional excavation and the cost of placing the overbreak concrete against the appropriate pay quantities.

PENSTOCK CONSTRUCTION

Penstocks are required to convey water from the reservoir to the turbines when powerhouses are located adjacent to the downstream toe of the dam or are located a distance from the dam. When the powerhouse is located at the downstream toe of the dam, the penstock is embedded in the dam concrete. When the powerhouse location is at a distance from the dam, open penstock runs are installed between the dam and the powerhouse. Penstocks of this type are shown in Photograph 42.

Penstock construction consists of aligning, fitting, and welding curved plates into lengths of penstocks at an assembly yard at the jobsite; transporting these assembled sections to their job locations; placing these sections on their foundations; joining these penstock lengths by the use of Dresser couplings or by welding; anchoring the penstocks to the foundations by ring guides bolted to concrete foundations or by concrete encasement; and sandblasting and painting the exposed surfaces.

A penstock assembly yard may be outfitted with concrete footings to furnish a level support while the curved plates are aligned and fabricated. Assembly-yard equipment consists of cranes, plate-alignment tools, welding machines, weld-testing equipment, and heat-treatment equipment. *Spiders* (interior supports) are installed in each section of penstock to maintain its shape. These spiders are

Photo 42 Constructing Guri Dam Powerhouse, Venezuela.

left in the penstock sections until the sections have been joined together in their final location and anchored to their foundation or encased in concrete.

Heavy-duty transport trucks or truck tractors pulling special trailers are required to transport penstock sections from the assembly yard to the placing cranes. If the project is equipped with cableways of a capacity sufficient to handle the penstock sections and if the cableways can supply hook coverage of the penstock area, they can be utilized to place the penstock sections on their foundation. Otherwise, guyed derricks or other types of cranes are required.

When penstock sections are placed in position, their ends must be in alignment and spaced the correct distance apart for welding or for Dresser coupling installation. This final alignment is often done by welding bolt anchors on the inside and close to the edge of each penstock section. Each penstock section is then pulled into alignment by adjusting bolts that span the openings between penstock sections and between adjacent bolt anchors. Kickplates are also often welded to the edge of each penstock section to force the penstock's edges into alignment. After the sections have been aligned and joined together by welding or with Dresser couplings, they are anchored into place, the spiders are removed, and the penstocks are sandblasted and painted. Where the penstock joins the scroll case, a blocked-out concrete section is left. This blocked-out concrete is poured after the connection is made.

POWERHOUSE CONCRETE PLACEMENT

When powerhouse construction is limited to first-stage construction, the placement of this first-stage concrete presents no special problems. When the contract work includes the placement of second-stage concrete or when the powerhouse design does not provide for first- and second-stage construction, special concrete-placement procedures are necessary to prevent misalignment, distortion, and uplift of the embedded liners, scroll cases, and embedded turbine parts. These special placement procedures include: placing concrete in shallow lifts and maintaining a minimum height differential between adjacent pours to reduce the fluid pressure of the freshly poured concrete; dividing encasing lifts into a group of pours and controlling the time interval between pours to reduce concrete shrinkage stresses on the embedded steel; eliminating the bracing or attachment of concrete forms to the embedded parts; preventing the contact of concrete buckets or vibrators with embedded parts or supporting members; prohibiting the use of large vibrators and the dumping or chuting of concrete from high distances into the pour; restricting the rate of concrete placement; and using cold-water sprays on all internal surfaces of embedded parts during the concrete pouring and curing period to dissipate the heat of hydration that is produced in the concrete.

STORAGE OF POWERHOUSE EQUIPMENT

If the contract work includes the complete construction of the powerhouse, the powerhouse equipment must be received and stored in suitable building and storage areas.

Specifications may delineate storage requirements for the powerhouse equipment, but storage often remains as the contractor's responsibility. Specifications may require that all critical items be stored in buildings where the temperature and humidity are accurately controlled. If storage requirements are not covered by the specifications, the requirements recommended by the equipment manufacturer may be used as a guide to determine the extent and type of the required facilities.

All powerhouse equipment should be protected from weather damage, dirt, corrosion, and distortion. Whether buildings are required, whether plastic covering is sufficient, what extent of heating is required, etc., depend upon the protection provided by the manufacturer when the equipment was crated and upon the site's weather conditions.

Electrical equipment and delicate equipment should be stored in a dry building. Governing equipment should be stored in a clean, dry, well-protected place. In cold or damp weather, space heaters should be used to heat the storage areas. If equipment susceptible to damage by dust or moisture is installed for a considerable period of time before operation, it should be enclosed with clean tarpaulins or plastic to prevent ingress of dirt, dust, and moisture during construction. Space heaters and moisture absorbents may be required within the enclosure.

Turbine parts that control the turbine's final alignment and running clearance are machined to a fine tolerance. During storage, these parts should be well-blocked to prevent distortion of the machined surfaces. Runners, head covers, bottom rings, wearing rings, and the like should be stored with their axes vertical to maintain roundness. Shafts, runners, and other attached equipment should be supported so that their weight is uniformly distributed to prevent any deformation. Supports should not be placed under bearing surfaces. The shaft should be stored so that it has no direct contact with wood supports. The shaft should be rotated 180° periodically and lifted so that additional rust-preventive material can be applied to prevent pitting at the points of support.

INSTALLATION OF POWERHOUSE EQUIPMENT

First-stage construction of a powerhouse does not present any construction problems in regard to the installation of permanent materials and equipment that have not already been discussed under concrete-dam construction. If the project work involves the construction of the second-stage work in the powerhouse or the construction of a powerhouse without first- and second-stage construction, specific problems are encountered during the installation and testing of powerhouse equipment and the fabrication, alignment, and encasement of the embedded parts.

The installation of powerhouse equipment and the embedded plate-steel liners is such specialized work that often the prime contractor will prefer to subcontract this work. To prevent scope of work disputes by subcontractors, it is preferable to subcontract this work to contractors who perform both mechanical and electrical work and who specialize in installing powerhouse equipment. If the prime contractor undertakes the installation of powerhouse equipment, he should make maximum use of manufacturers' field representatives and his construction superintendent should be experienced in this field. There should be close cooperation between the manufacturers' representatives and the contractor's supervisors, and they should agree on erection procedures before work is started. Adequate space must be available for turbine subassemblies and for the assembly of the generators and other equipment. Since space for this work is restricted, the equipment for one generating bay is completely assembled and is moved into position before the assembly of the equipment for the next generating bay is started.

Following are descriptions of the procedures required to install the main embedded parts and equipment for one generating bay of a powerhouse. The first description is of a generating bay where the first-stage concrete has already been placed and a Kaplan-type turbine is to be installed. (The main parts of a Kaplan turbine are shown in Drawing 6.)

There are work variances from this description when concrete is not placed in two stages or when other types of turbines are installed. The order of work and work procedures is an additional variable dependent upon the installers. The installations of Kaplan and Francis turbines are quite similar. If the powerhouse contains an impulse wheel, however, its installations will be different. Therefore, a brief description of the installation of a Pelton turbine will follow that of the Kaplan turbine.

These installations are so complex that cost estimates for this work should only be performed by engineers who have had experience in this field. The purpose of these descriptions is only to provide the background required to schedule powerhouse construction, based on the assumption that the prime contractor will subcontract the equipment installations. Therefore, the following descriptions should not be used as a basis for estimating installation costs.

Installation of Kaplan-type Turbine Equipment

The embedded parts for a turbine must be set with fine accuracy since the final leveling and alignment of the turbine's operating parts depend on their correct elevation, concentricity, and trueness. Before any placement of embedded parts is undertaken, the proper grade and center lines for each powerhouse unit should be accurately determined and referenced.

The machined surfaces of all parts of the powerhouse equipment must be cleaned of rust, paint, or protective coating, and burrs and nicks must be removed with a fine emery cloth before assembly. This often requires months of man labor.

The first embedded part to be installed is the draft-tube pier

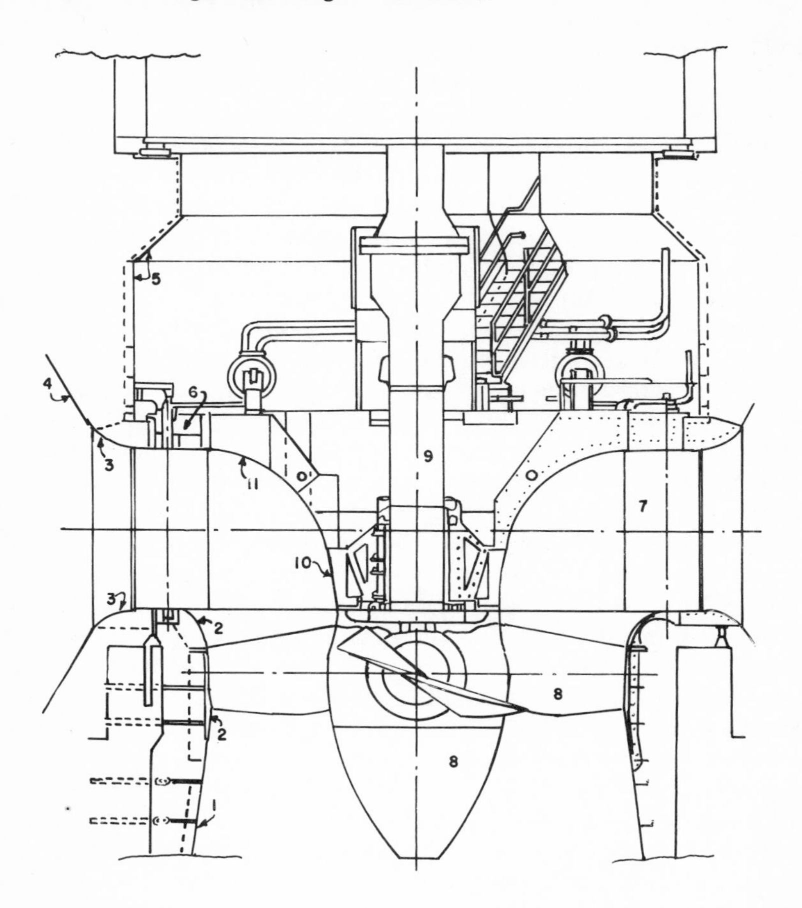

LEGEND:

1. DRAFT-TUBE LINER
2. DISCHARGE RING
3. STAY RING
4. SCROLL-CASE LINER
5. PIT LINER
6. OUTER HEAD COVER
7. WICKET GATES
8. RUNNER
9. TURBINE SHAFT
10. INNER HEAD COVER
11. INTERMEDIATE HEAD COVER

Drawing 6 Parts of a Kaplan turbine.

nose. When the draft-tube liner extends past the dividing pier in the draft tubes, it forms a pants leg, and the pier nose is a part of the pants leg. When this liner stops short of the pier nose, draft-tube pier noses are required. The installation of these pier noses does not present any particular problem. Pier noses must be received at an early date, however, so that their installation will not delay concrete placement.

The next embedded part is the draft-tube liner. The draft-tube liner is located below and connected to the discharge ring. Its length varies with the powerhouse design. If the turbines are small or if the draft-tube liners are short, the liners can be installed in one piece. Otherwise, they are installed in sections; the sections are then aligned and welded together. One installation method is to place spiders in the prefabricated liner sections; they are then lifted and set on adjustable jacks mounted on concrete pedestals poured with the first-stage concrete. The liner sections are connected to steel rods fastened to anchors placed in the first-stage concrete and equipped with turnbuckles. The liner sections can then be positioned and aligned by adjusting the jacks and tightening the turnbuckles on the anchored rods. The rods prevent the fresh concrete from floating and moving the draft-tube liner. Welding is then done; the welds are tested, and depending on specification requirements, heat treatment may be performed. The concrete encasement can then be poured as described under concrete placement, and after it has set, the spiders can be removed. After the draft-tube liner has been partly encased in concrete, the discharge ring can be installed.

Another name for the discharge ring is the throat ring. The bottom of the discharge ring is attached to the draft-tube liner and it extends to the stay ring. It contains the bushings for the lower stems of the wicket gates. It may come in two sections, depending on the particular design. If it is in two sections, the section next to the stay ring, which contains the stem bushings for the wicket gates, is called the bottom ring, and the remainder is called the discharge ring. Also, the discharge ring may have a removable section. Installation of the discharge ring is done by using jacks and turnbuckle-equipped anchored rods to bring it into alignment. It must be positioned accurately because the final leveling and alignment of the turbine depend on an exact setting of the discharge ring. After alignment, it is connected to the draft-tube liner and the concrete that encases it can be poured.

After the discharge ring has been partly encased in concrete, the stay ring, also known as the speed ring, is installed with the inner surface of its bottom ring flush with the top of the discharge ring. The stay ring consists of a top and bottom ring separated by wicket vanes. Its purpose is to guide water into the wicket-gate openings. The bottom ring spans the distance from the discharge ring to the scroll case. The top ring fits between the scroll case and the pit liner. The top ring is also connected to the outer turbine cover (a ring). The stay ring can be assembled on the powerhouse floor and then set in place or it can be assembled in place. Like the other embedded parts, it is positioned on jacks and aligned with rods equipped with turnbuckles. It must be set to exact elevation and alignment since it also controls the turbine's position. The stay ring is bolted to the discharge ring and their water surfaces must match. If there are any surface discrepancies, these must be removed by grinding.

The scroll-case liner, also known as the spiral-case liner, is next in the order of installation. This is a steel-plate liner for the water passage that partly encircles the turbine pit and supplies water to the turbine. A penstock section connects the scroll case to the turbine shutoff valve. The scroll cases may also be connected to a discharge pressure-relief valve located so that it can discharge out the downstream face of the powerhouse. For Kaplan and Francis turbines, the scroll cases are of welded construction. For impulse turbines, scroll cases are bolted construction to prevent distortion and consequent misalignment of the nozzles. Scroll-case liners are assembled, fabricated, and encased in concrete with the same type of procedure as that used for the draft-tube liners. Jacks and anchor rods equipped with turnbuckles are used in alignment. After they are aligned, welding, weld testing, and heat treatment can be performed and their concrete encasement can be poured. The exterior surfaces of some scroll-case liners may be covered with cork before the encasing concrete is poured.

The pit liner is then erected above and fastened to the upper ring of the stay ring. The pit liner is a steel plate that lines the pit between the scroll case and the turbine gallery. It is the last major embedded part and is placed with the same procedure as that used for the draft-tube liner. It is set back along one wall to form a shelf for support of the servomotors. When a Pelton turbine is installed, the liner comparable to the draft-tube liner is commonly called a pit liner, though its proper name is a wheel pit liner.

The remaining pieces of major equipment are nonembedded. The first nonembedded item to be set is the outer head cover, which is a ring that fastens to the upper ring of the stay ring and contains bushings for the top stems of the wicket gates. It also is equipped with removable cover plates which allow removal of the wicket gates. Its other purpose is to provide the upper support for the wicket gates and to support the intermediate turbine cover, which in turn supports the inner head cover, the runner, and the turbine shaft.

If the wicket gates have not been previously installed, they should be positioned at this time. They are installed between the discharge or bottom ring and the outer head cover. Their purpose is to control the water flow past the turbine runner.

The turbine assembly, consisting of the runner, shaft, inner head cover, and intermediate head cover, is next installed. These parts can be assembled before being lifted into place. Photograph 43 shows a bridge crane setting a Kaplan wheel and an inner head cover at the McNary Dam powerhouse. The intermediate head cover rests on the outer head cover and supports the inner head cover. The inner head cover contains the shaft bearing and seals and its bottom edge is equipped with a renewable bronze wearing ring to take the upward thrust of the runner. Final positioning of the turbine moving parts is not done until the turbine shaft has been coupled to the generator shaft. Lubricating, cooling, and control piping can be connected at this stage of erection.

The gate-operating ring can then be mounted on top of the intermediate head cover. It is connected to and controls the openings of

Photo 43 Bridge crane moving a Kaplan wheel and inner head cover, McNary Dam powerhouse. (General Electric Company.)

the wicket gates. In turn, it is connected to the servomotors located in a shelf provided in the turbine pit.

In the meantime, second-stage concrete placement can be completed and the installation of the generator can commence. Drawing 7 shows the generator parts whose installation is described in this chapter. Anchor bolts can be installed and plates can be set on the concrete and grouted to support the stator, the bottom-generator frame, and the rotor brakes. The stator on large generators comes in sections. These sections are erected and assembled on the sole plates; the stator plates are stacked and all coils are installed. A minimum time of one month is required for this work. The lower bracket arms for the generator and the generator shaft also can be erected in place. The lower bracket arms support the generator shaft and contain the shaft bearings. The rotor is assembled and the stator sections are connected on the rotor-erection stand in the assembly bay. All windings are connected. After the stator work is completed, the rotor can be placed on and connected to the shaft. Photograph 44 shows a bridge crane moving a generator rotor at the McNary Dam powerhouse. Next, the generator shaft is connected to the turbine shaft. This is usually done by reaming and grinding all holes to fit the bolts. After this connection is made, final adjustments are made to the generator and turbine bearings to be sure that they are in perfect alignment. Ladders, platforms, stairways, oil piping, cooling

Photo 44 Bridge cranes moving a generator's rotor, McNary Dam powerhouse. (General Electric Company.)

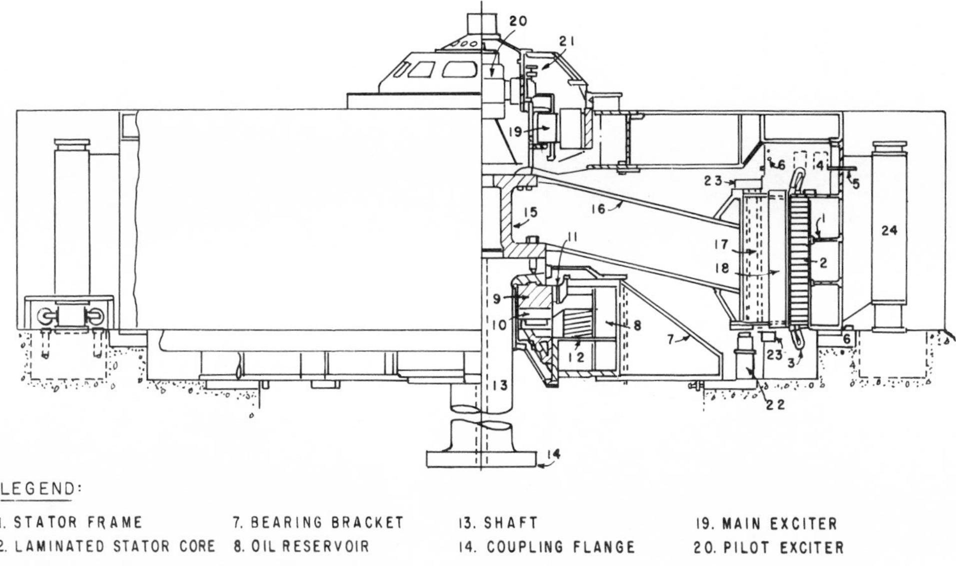

Drawing 7 Parts of an electric generator.

Photo 45 Generator floor, McNary Dam powerhouse. (General Electric Company.)

piping, fire-protection piping, and electrical connections can be installed. The upper generator bracket, the exciters, the air housing, and the generator coolers can be placed, controls connected, and painting completed.

The governor should be installed last, because it should not be installed and then be exposed to other construction. The governor is mounted on the floor adjacent to the generator. Piping and control wiring can then be connected. Photograph 45 shows the generator floor of the McNary Dam powerhouse.

Transformer installation consists of unloading and positioning the transformers; installing their wheels, heat exchangers, and bushings; hooking up the cooling and oil piping; filling them with oil; and connecting them to the generator and to the lines leading to the switchyard.

Installation of Pelton-type Turbine Equipment

The installation of a Pelton-type turbine is quite different from the installation of a Kaplan turbine. For a Pelton turbine, the major work is the placement, alignment, anchoring, and bolting of the scroll case; the installation of the needle valves and nozzles; and the turbine encasement. Bolted construction is used to prevent misalignment of the turbine nozzles. After these parts are bolted and aligned, they are encased in concrete. Passageways to the needle valves and chambers behind the needle valves must be constructed to permit removal of the needle valve and nozzle when repairs are necessary.

The installation of these turbine parts is illustrated by three progress photographs taken during the construction of Middle Fork powerhouse on the American River, California. Photograph 46, taken during the installation of the bolted scroll cases, shows the nozzles and the pit liner. Photograph 47 was taken after the scroll cases and pit liners had been installed in concrete and while concrete formwork and reinforcing steel were being placed for the generator supports. The penstock and the penstock bifurcation are visible at the left of the photograph. Photograph 48 was taken while the generators were being installed.

TESTING POWERHOUSE EQUIPMENT

Testing, starting up the units and bringing them on the line are done under the direction of the equipment suppliers' representatives. Since the owner will operate the completed powerhouse, it is to the owner's advantage to supply the testing crews and have them trained

Photo 46 Installation of the bolted scroll case, nozzles, and pit liner for a Pelton-type turbine, Middle Fork Powerhouse, American River, California.

Photo 47 Installation of concrete formwork and reinforcing steel for the support of a generator, Middle Fork Powerhouse, American River, California.

by the equipment representatives during the testing period. Since this work is handled by the equipment suppliers and by the owner, its substance will not be discussed. For scheduling purposes, testing and start-up take approximately 2 months per unit, of which 1 month is required for drying out the generator during slow-speed operation.

Photo 48 Installation of generator, Middle Fork Powerhouse, American River, California.

In special cases, the contractor must provide operating crews to the equipment suppliers' representatives. When this is the case, the cost of supplying these crews must be included in the project cost.

POWERHOUSE-CONSTRUCTION SYSTEM

The powerhouse-construction system is established by determining the work quantities, scheduling the construction of the powerhouse, and selecting the required construction equipment. Powerhouse construction is so rigidly controlled by the specified requirements for equipment installation that construction systems vary only in the use of different equipment units for concrete placement and the handling of the powerhouse equipment and embedded materials.

QUANTITY TAKEOFFS

Similar to the procedures required for planning the construction of all other types of heavy construction, quantity takeoffs of the work required for powerhouse construction furnish the basic data for planning and estimating its construction. If the powerhouse contains multiple power units, the work should be scheduled and the quantity takeoffs based on constructing the units in a stepped manner. Stepped construction means that each unit is started, built, and finished with approximately a 2-month interval between units. This allows better utilization of trained crews and maximum reuse of forms, since forms can be progressively used on each power-generating unit.

Decisions must be made on the extent to which powerhouse work will be subcontracted before quantity takeoffs are started. Most prime contractors subcontract the electrical, piping, heating, plumbing, and air conditioning items; the installation of reinforcing steel, turbines, generators, transformers, governing and associated equipment; the erection of gates, stop logs, and much of the architectural work. If this work is to be subcontracted, the quantity takeoffs need only cover excavation, backfill, concrete forms, concrete pay quantities, overbreak concrete, cement grout, floor finishes, wall finishes, structural steel framing, and miscellaneous embedded metalwork such as embedded frames, guides, anchor bolts, etc. This miscellaneous embedded metal is installed by the prime contractor since it is closely tied to concrete placement.

Special attention should be given to the forms for the galleries and passageways, block-out forms, draft-tube forms, and the scroll-case forms for low-head powerhouses where steel-lined scroll cases are not used. Form cost increases when large amounts of embedded metal project through the forms. The forms for the draft tubes are expanding curved forms that must be set to strict alignment tolerances. Resultingly, they are costly to build and require a long erection time. When powerhouses contain multiple units and when the powerhouse pours are designed to permit their use, steel forms are used in preference to job-built wooden forms. When concrete-finished scroll cases are used, this forming is also expensive and often difficult. In most cases steel forms cannot be used for forming scroll cases, since, after stripping, the forms must be dismantled into small enough pieces to go through the wicket-gate openings. This requirement also makes the reuse of wooden draft-tube forms impractical. If the forms for all water passageways in a powerhouse are not constructed with smooth surfaces, a large amount of hand grinding and finishing will be required.

As the quantity of subcontract work is reduced, the quantity of takeoff work is increased. Piping, liquid storage, liquid pumping, and electrical installations are quite extensive in a powerhouse and detailed takeoffs must be prepared for estimating this work. As an example of work complexity in a powerhouse, separate piping systems must be installed for raw-water supply, potable-water supply and treatment, sewage, drainage, cooling water, compressed air, piezometer installations, the turbine shaft seal, the bearing thermometer, governor oil supply, fuel supply to the standby generator, CO_2 fire protection, generator and transformer fire protection, lubrication oil supply, and transformer oil supply.

Takeoffs alone cannot indicate the extent of work involved in the installation of embedded turbine parts and the erection of turbines, generators, transformers, governing equipment, gates, etc. However, takeoffs can indicate the weights to be lifted and the amount of welding required. Estimating the cost of installing these items and determining the time required for their installation should be based upon past job records or on estimating experience.

SCHEDULING POWERHOUSE CONSTRUCTION

The scheduling of first-stage construction work for a powerhouse consists of determining the time required to make each concrete pour, the time required for erection of any powerhouse structure, and the time required to erect and test any cranes. The total time required is dependent on work complexity and size of the powerhouse. The first powerhouse unit often can be completed in 8 months from the start of concrete placement. Reuse of forms and utilization of construction equipment and trained crews are accomplished more efficiently if powerhouse units are constructed in a stepped manner. For stepped construction, an additional 2 months of construction time are required for each additional generating bay in the powerhouse.

Powerhouse scheduling is more complex when a completed powerhouse is to be constructed. If every pour is made on schedule, approximately 12 to 14 months from the start of concrete placement is required to place all the concrete for the first powerhouse generating bay, erect the crane rails, and complete the crane erection. Two more months are required to complete the concrete pour in each additional powerhouse generating bay. After the powerhouse permanent cranes are in operation, approximately 4 additional months are required to set the first turbine, the first generator, and the first governor and transformer and to complete the piping and electrical work. Again there will be a 2-month lag from unit to unit when multiple-unit powerhouses are constructed.

To determine accurately the time required for the placement of powerhouse concrete, a detailed pour schedule must be prepared. The powerhouse generating bays will control job completion since concrete placement will not be delayed by major equipment installation in the control and assembly bays. The pouring schedule should show the number of pours, the pouring sequence, and the periods when pouring must be suspended for the erection of the draft-tube forms and the installation of the embedded steel liners and turbine parts. Powerhouse scheduling requires that time be allowed on the schedule for placing special forms and for installing pipes, electrical conduit, and embedded turbine parts. Concrete placement will also be controlled by the dates that embedded materials will be received at the jobsite. The first embedded item to be installed is the draft-tube pier noses; therefore, the pier noses must be delivered to the job at an early date. Following this, the draft-tube liner, scroll-case liner, stay and discharge rings, and pit liner must be received in time to make the pour schedule.

It is preferable to have the same individual who made the concrete takeoff prepare the powerhouse-construction schedule since he will have knowledge of the amount and complexity of the required forms.

CONSTRUCTION-EQUIPMENT SELECTION

If the powerhouse is an integral part of the dam, the powerhouse concrete is often placed in the same manner and with the same equipment as the mass concrete. When the powerhouse is located as a separate unit at the downstream toe of the dam, a powerhouse concrete-placing method and placing equipment are selected that will not interfere with dam concrete placement since the placement of mass concrete is of prime importance on any dam project.

When the dam concrete is placed with a cableway, this separation of powerhouse concrete-placement equipment is of special importance. A large cableway has such a high hourly operation cost that low concrete-placement cost can only be obtained when the cableway is being used to its maximum capacity. The cableway's maximum capacity can only be achieved during the placement of mass concrete. Placing powerhouse concrete with a cableway is slow because most of the powerhouse pours are small and placement is delayed because accurate bucket spotting and bucket stability are required to allow concrete to be deposited in narrow form openings. The pendulum action or drift of a bucket suspended from a cableway and the jump that occurs when the bucket is discharged do not particularly delay the placement of mass concrete, but on powerhouse pours these cableway characteristics result in many pouring delays because the concrete must be discharged from the bucket into restricted areas and it is necessary to protect protruding reinforcement steel and other embedded items from bucket damage. If cableway-tower runways must be extended to provide cableway coverage to the powerhouse, this will also increase the cost of using a cableway to pour the concrete in a powerhouse.

In consideration of these factors, the cableways used for massive concrete placement are seldom used for powerhouse pours; instead, powerhouse concrete is usually placed with separate equipment. Occasionally, a smaller-capacity cableway is used. This was done at Hungry Horse Dam where powerhouse concrete was placed with a 4-cu-yd-capacity cableway with its traveling tower located on a separate runway from the three 8-cu-yd-capacity cableways used for placement of the dam concrete. In other instances, the main dam cableways are only used to transport the concrete buckets to the powerhouse site, and final placement of the concrete is done with other equipment. This does not tie up a main cableway exclusively for powerhouse pouring and permits more accurate concrete-bucket handling for powerhouse pours. More often concrete buckets are transported from the mix plant to the powerhouse site by trucks and then handled with other equipment. Since the powerhouse is often located at the bottom of the canyon, requiring steep road grades between the mix plant and the powerhouse, trucks are used instead of rail equipment for hauling concrete.

If the dam concrete is placed with a trestle-crane placement system, often powerhouse-concrete placement then can be done with a similar type of crane added to the system for this purpose. If the placement of the dam concrete necessitates the use of a trestle, it is sometimes possible to locate this trestle so that it can service both the dam and the powerhouse, or it may be possible to provide trestle service to the powerhouse by constructing a branch trestle curving off the main trestle.

When the concrete in the powerhouse is to be placed with a separate system, the selected system should be one that will result in low concrete-placement cost and will also provide hook service to handle most of the lifts required in its construction. If only first-stage powerhouse construction is required, then hook coverage is only needed for form moving, for handling reinforcing steel, for placing other embedded materials, for erection of any structural-steel framework, and for crane erection. If the work consists of constructing a completed powerhouse, then a large amount of additional hook time is required. Hook time is needed for handling the draft-tube liners, discharge rings, stay rings, scroll-case liners, pit liners, wicket gates, stop logs, water-control gates, pumps, control motors, transformers, turbines, and generators. Some of these lifts will be beyond the capacity of one crane. When the powerhouse is equipped with a permanent crane, this crane can be used to handle the heavy lifts required for setting the turbines and generators and to handle the transformers.

When cranes are provided for powerhouse-concrete placement and for hook service, it is necessary to check their required boom length, lifting capacity, and gantry height before their size can be determined. This information should be plotted on a drawing containing the plan and sections of the powerhouse. This drawing can also be used to determine whether a trestle will be required. It is usually necessary to try cranes with different boom lengths and gantry heights. If a trestle is required, the drawing will show its location and elevation. It is desirable to keep the trestle outside the powerhouse area, preferably upstream. For some powerhouses, however, hook coverage cannot be provided except by placing a trestle on the center line of the powerhouse.

Gantry cranes provide a fast and efficient method of pouring concrete and give good hook coverage for powerhouse construction. Gantry cranes provide bucket stability and can accurately spot and maintain the bucket's position. They have sufficient boom and hook capacity to handle all but the heaviest of lifts. The crane tracks can be located on the ground or on a trestle. The choice between the two depends on the results of the previously described hook-coverage study.

Self-propelled crawler or truck cranes also provide concrete coverage and hook capacity. Whether they should operate from ground level or from a trestle is also determinable by the hook-coverage study. As crane capacities keep increasing, the need for trestles keeps decreasing. Because the undercarriage of crawler cranes is not supported in as rigid a manner as that of rail-mounted gantry cranes, crawler cranes will not provide as much concrete-bucket stability. The advantages to the use of crawler cranes when compared with gantry cranes are purchase price and erection cost are less; a higher percentage of the purchase price will be recovered when they are sold because of their greater versatility; and they are more mobile, which allows them to be used on other project work. The boom of one of these crawler cranes is shown on the left side of Photograph 42 of the Guri Dam powerhouse. This photograph is deceptive since the steep excavated rock surface that is shown in the left of the photograph stops short of the generating bays. Excavation in front of the generating bays permitted water to flow freely from the draft tubes. Two crawler-type cranes were operating on this excavated surface when the construction of the powerhouse was started; later they were operated from a timber trestle placed on this excavated area. These cranes handled the concrete buckets for most of the powerhouse-concrete pours.

Tower cranes can be used for placing concrete in small powerhouses. They have a limited hook capacity, so that they can only handle small concrete buckets and make light lifts. For heavy lifts, other types of cranes such as guyed derricks must be provided. The guyed derrick shown in Photograph 42 of the Guri Dam powerhouse was used to handle the heavy lifts during the construction of the powerhouse structure.

Stiffleg cranes have been used for placing concrete in powerhouses. They are so slow that it is hard to keep the pour "alive" while it is being placed. Because they are anchored to one position, many stifflegs are required to furnish complete hook coverage for a powerhouse. This is an older type of construction equipment and its use has been largely replaced by modern self-propelled cranes. The two stiffleg cranes shown in Photograph 42 of the Guri Dam powerhouse were so slow in operation that they were mostly used for moving forms and setting embedded parts and materials.

The use of conveyor belts for making powerhouse pours should increase as specifications are revised to permit their use. They can only be used for concrete placement, however, and hook coverage of the powerhouse must be provided by material-handling cranes. This is a fast and economical method of making the small pours in a powerhouse structure.

If the powerhouse must be completely enclosed before second-stage concrete is started, then special equipment will be required for second-stage concrete placement. When specifications permit or when its use is made mandatory, pumped concrete or intrusion-prepacked concrete is used for second-stage powerhouse concrete. Second-stage concrete placement can be accurately and evenly controlled by either method. The intrusion-prepacked method will reduce the lifting force against the scroll-case liners. Permanent powerhouse cranes may be available for placing this concrete, but because they are so slow, they are unsuitable for this work.

CHAPTER NINE

Descriptions of Estimating Procedures

Successful dam estimating is accomplished by planning the dam's construction and by estimating the cost of performing this construction. Descriptions of the work required to plan and estimate the cost of constructing dams located within the United States are presented in this chapter.

To follow these descriptions properly, a knowledge of the information presented in Chaps. 1 to 8 is a prerequisite. The overall concept of dam estimating was presented in Chap. 1. Diversion methods were discussed in Chap. 2. In Chaps. 3 to 8, descriptions were given of the work required for the construction of dams, the methods and equipment used to perform this work, the production capacities of this construction equipment, and the selection of economical systems for their construction.

The work involved in preparing a cost estimate for dam construction is shown in the work-flow diagram in Figure 6. This diagram divides the work into estimating tasks and arranges these tasks into three phases: collection of necessary estimating data, selection of the construction system, and costing the estimate.

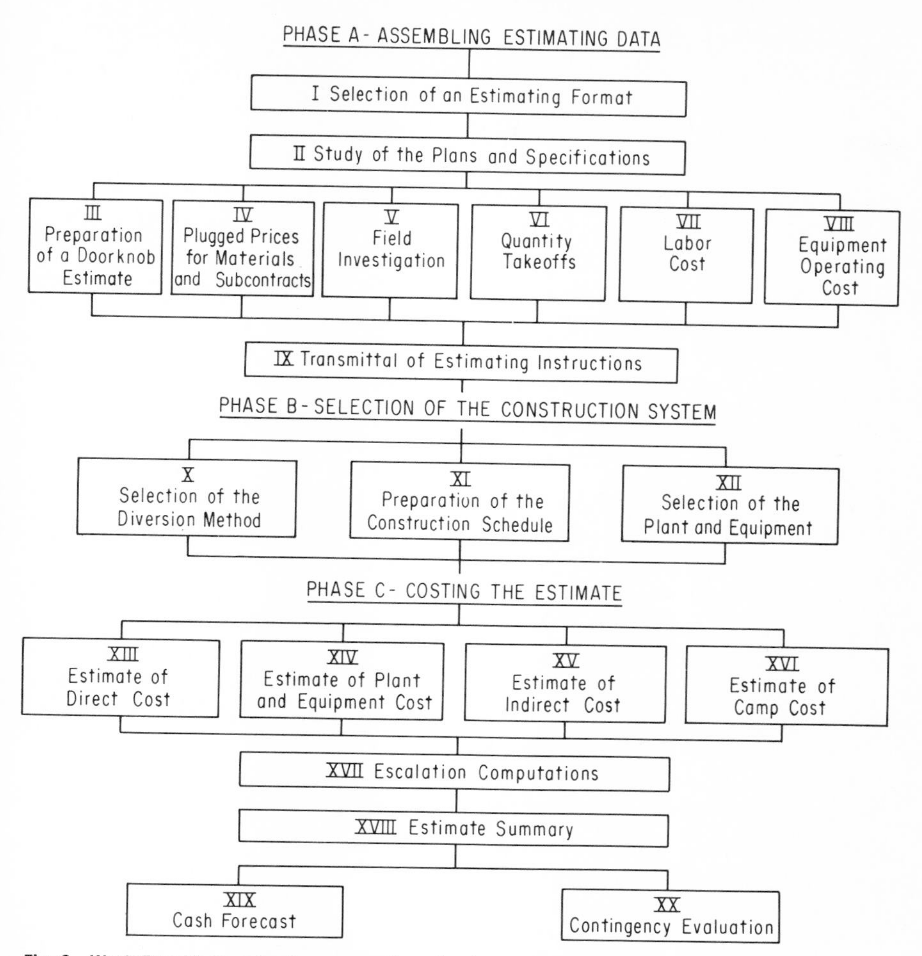

Fig. 6 Work-flow diagram for the preparation of a dam estimate.

In Figure 6 each estimating task has been titled and identified with a Roman numeral. This was done to simplify cross-referencing within this volume and does not indicate the sequence of work performance. The descriptions of the estimating tasks are presented in the remainder of this chapter and identified by the same titles and Roman numerals as shown on the work-flow diagram.

The work-flow diagram, Figure 6, shows that many estimating tasks can be performed simultaneously and that the work done for one estimating task flows into the succeeding task. (This is covered in more detail in the descriptions of the estimating task.) Because many tasks can be performed simultaneously, an estimating team can be used to prepare a dam estimate; this is desirable because of the large amount of work required to prepare it properly and because the time between bid advertising and bid submittal is very limited.

During the preparation of the estimate, as each task is completed, those tasks previously completed should be reviewed to determine whether they should be changed. This is necessary because new estimating concepts may develop. As an example, when direct cost is planned and estimated, the development of construction details may make it necessary to change the construction schedule or change the construction equipment. On occasion, after an estimate has been completed, more economical construction systems are developed, making it necessary to rework the estimate completely.

I. SELECTION OF AN ESTIMATING FORMAT

A suitable estimating format should be selected prior to starting to prepare an estimate. Some estimators use printed forms to expedite estimate preparation and to prevent them from neglecting any item of cost. Other estimators adjust their forms to fit each project. The standardization of forms is becoming more common because this expedites estimate comparisons at joint-venture prebid meetings.

A suitable heavy-construction estimating format should be easy to use, have a minimum of cost divisions, allow the estimate to be prepared in a logical progression of steps, and provide arithmetical checks of additions and extensions. Cost proration should be minimized since the more cost is prorated, the less accurate will be the estimated unit cost. The format should produce unit labor cost for each bid item so that, after construction is started, comparisons can be made between the estimated labor cost and the actual labor cost. The format should provide the total project labor cost since this is required for escalation computations and for markup determinations.

The format should follow sound accounting practices in separating direct cost, capital expense, and indirect cost. It should be arranged so that the cost engineers and cost accountants can use it as the guide for establishing the project's cost report, since, if both reports use the same form, estimated cost can be rapidly compared to actual cost, and the cost report can supply unit cost in the proper form for future estimating.

The format should tabulate cost into natural divisions that have similar exposure to cost overruns. It should arrange cost details so that they will be available for quick reference at joint-venture prebid meetings.

An estimating format that complies with all the foregoing

requirements is used in Chap. 11 for the preparation of the sample estimate. The forms used for tabulating the estimated cost are shown in the performance of Estimating Tasks XIII to XVIII in Chap. 11. The estimated cost items are separated into five divisions and each division forms a separate part of the estimate.

Each cost division is subdivided into cost components with labor cost a component common to each cost division. The total project labor cost is obtained by adding the labor component of each cost division.

The five cost divisions are:

1. *Direct Cost.* This is the cost of labor and supplies expended for the performance of the bid item work, the cost of purchasing permanent materials and equipment that form a part of the completed work, and the cost of subcontracting any portion of the bid item work. Direct labor cost is the cost of all on-site labor, except that classified as indirect labor and the cost of labor required for the erection of the construction equipment, for the construction of the construction plant, and for the construction and operation of the construction camp. Supply cost is the cost of purchasing consumable job supplies, including fuel, oil, grease, repair parts, powder, drill steel, bits, form materials, etc.

2. *Plant and Equipment Cost.* This is the cost charged to the project for the use of the construction equipment and plant facilities required for the construction of the project. It is composed generally of the acquisition, erection, and construction costs of the construction equipment and plant facilities less the net salvage value, or revenues received for their sale at project completion, plus the cost of renting any nonoperated, nonmaintained construction equipment. If maintained and operated construction equipment is rented, this constitutes a subcontract and the applicable rental cost should be listed as a subcontract under direct cost.

One method used in the heavy-construction industry to distinguish between equipment and plant cost is to classify as equipment the construction facilities that have a resale value at project completion, e.g., tractors, trucks, rock crushers, concrete mixers, etc. Plant items are classified as those construction facilities that do not have resale value at project completion. Plant cost also includes the cost of assembling or erecting equipment, since this work does not increase the salvage value of the equipment. Typical plant cost items are assembling and erecting equipment, constructing access roads, grading the plant areas, building plant foundations and structures, erecting construction buildings, etc.

The net salvage value of the plant and equipment is the revenue received from the sale of the equipment after it is no longer required on the project, less the cost of dismantling and moving the equipment off the project, project cleanup cost, equipment storage cost, and any selling expense. Equipment storage cost may continue for several years after project completion, since specialized construction equipment can only be sold when a new project is to be constructed that requires this type of equipment.

3. *Indirect Cost.* This is the cost of supervising and controlling the construction project. It includes the cost of providing job management, job supervising, job engineering, surveying, accounting, purchasing and warehousing, nurses and safety engineers, a personnel staff, and watchmen. Insurance premiums, bond premiums, property and special taxes are part of this cost. Sales tax is not charged to indirect cost but rather is added to the acquisition cost of plant and equipment, permanent materials, and job supplies.

4. *Camp Cost.* This is the cost of constructing and operating any required construction camps.

5. *Escalation.* Most estimates are prepared using wage rates and material and supply prices that are in effect at the time the estimate is prepared. To arrive at the total construction cost, it is necessary to estimate the amount that these costs will increase during the construction of the project. The cost increases are termed *escalation.* The amount that labor costs increase is dependent upon increases in wage rates and fringe benefits that have already been negotiated and on increases that must be negotiated prior to project completion. Since insurance rates and payroll taxes are computed as a percentage of labor cost, these costs will increase in the same proportions.

The amount that materials and supplies will increase is dependent upon whether fixed-price quotations are received for these items. Since these quotations are not received until shortly before bid submittal, the amount of material escalation can only be determined at that time.

II. STUDY OF PLANS AND SPECIFICATIONS

The study of plans and specifications should be initiated as soon as the bidding documents are received. A review should be made of the work constraints; work performance requirements; completion times; and working conditions set forth by the owner on the plans, in the specifications, and in the construction contract. When these documents are reviewed, consideration should be given to the fact that they are the owner's, that they are written for the benefit of the owner, and that they will be interpreted and enforced by the owner's representative. As the review progresses, the information controlling job scheduling should be plotted on a construction schedule. This scheduling procedure is explained in detail later in the chapter.

Some estimators brief the specifications, make photostats of general drawings, bind them together, and distribute this booklet to company management and to the engineering and supervisory personnel who are or will be connected with the construction project. Such a brief is included in Chap. 11.

Specifications and bidding information should be checked to determine whether it is necessary for potential bidders to prequalify. Prequalification is required by certain government agencies and private owners. If it is required, efforts to prequalify should be expedited. Prequalification requirements vary from proof of financial responsibility to the listing of similar projects constructed by the contractor.

Contractor's licensing requirements should also be reviewed. This point is not mentioned in the specifications but is covered by local laws. Some states classify bids as nonresponsive if joint-venture bids are not presented in numbered bid forms checked out to the joint venture, and will rule out bid forms checked out to individual members of the joint venture.

The general and special clauses of the specifications should be carefully examined. Special attention should be paid to whether they furnish any financial relief to the contractor and whether they allow extensions of contract time in the event of unforeseen conditions: acts of God, strikes, war, weather, unexpected floods, etc. Contract completion dates and liquidated damage provisions should be understood, and the bond and insurance provisions should be reviewed.

Payment provisions and the bid items should be reviewed to determine whether they contain stipulations that will enable the contractor to receive mobilization payments or plant prepayments, whether a percentage will be withheld from each progress payment until job completion (known as *contract retention*), and how often and when progress payments will be made.

The source and the extent of the funds available for contract payments should be checked. In certain cases the contracting agency may have limited funds for the present fiscal year, which will then restrict the speed with which the dam can be constructed.

A review should be made of construction work that is included in the contract; construction work that will be done by others before the start of the contract, during the contract period, or after the contract is completed; the location of the work; any requirements or descriptions of job access; the names of the owner, contracting agency, and inspecting agency.

The bid date, the length of time until contract award, any intermediate control dates, the date of completion, bidding procedures, and the information required for presentation with the bid should all be carefully noted. Additionally, insurance requirements and time-extension provisions should be checked.

Work restrictions should be reviewed. These often encompass diversion methods or requirements, locations or restrictions on the contractor's construction plant areas, access-road location or restrictions, and construction constraints. Concrete-pouring constraints often include control of placing temperatures, placing methods, mixing-plant equipment, bucket size, height of pours, time limit between pours, and the time that must elapse before form stripping. Excavation constraints may include drilling and blasting controls, the required use of spoil areas, the required operation of borrow pits or quarries, control of the maximum thickness of embankment lifts and their methods of compaction, etc.

The labor rules and regulations should be reviewed to see that they do not conflict with existing labor contracts in the area. The major quantities should be examined and used in the preparation of the initial construction schedule.

Before bidding, the contractor should bring to the owner's attention any discrepancy between the plans and specifications, mistakes in the specifications, or any instance where the owner's intention needs clarification. Written clarification should be requested since the engineer's verbal clarification of the plans and specifications does not hold the owner to these opinions or interpretations. In all cases, the specifications place the responsibility on the contractor for their proper interpretation; therefore, it is important that they be thoroughly understood by the contractor's estimator.

III. PREPARATION OF A "DOORKNOB" ESTIMATE

The preparation of a *"doorknob" estimate* is an optional task, since its purpose is to provide an early indication of the total-contract volume and to highlight the items of work that form the major portion of the total cost. Highlighting major cost items indicates what work should receive the estimator's major work efforts.

If a doorknob estimate is desired, it should be prepared as soon as the plans and specifications are received. Only 2 to 3 hr should be spent on this work, since a longer period of time would not increase the accuracy of the estimate. The doorknob estimate is begun by listing the bid-item numbers, descriptions, and quantities. Unit bid prices for similar work items secured from past bids for dam construction are used to extend these quantities to arrive at the bid price for each work item. The total of these extensions gives an approximation of the contract value.

IV. "PLUGGED" PRICES FOR MATERIALS AND SUBCONTRACTS

To accomplish this task, the estimator must first determine the amount of work to be subcontracted. Since subcontractors supply the permanent materials for the work that they subcontract, this also fixes the amount of permanent materials that the prime contractor must purchase. As soon as the extent of the subcontracted work has been established, the estimator should request quotations from suppliers and subcontractors for subcontract and permanent material prices.

The subcontractors and the suppliers will not submit their quotations until shortly before bid submission. When these quotations are received, the contractor makes adjustments to his bid that equal the difference between the prices used in the estimate and the lowest quotation received. The method of making these adjustments to the bid price is described in Chap. 11.

Since subcontract prices and permanent material prices are adjusted in the estimate prior to bid submittal, these prices do not have to be accurately determined until just prior to bid submittal. For this reason, *plugged* or *bogie* prices are used in the estimate. These are prices that the estimator approximates for the subcontract work and for the permanent materials. They are used initially in the estimate to permit its completion and to allow the establishment of the approximate total estimated cost of the project. This approximated total is used to establish capital requirements, interest expense, and markup.

Very little estimating time should be spent on establishing plugged prices since their evaluation does not affect the final bid; estimating time can be more profitably expended on more important estimating tasks. The quickest method of establishing plugged prices is to use unit costs from previous projects or adjusted bid prices from past projects. A typical tabulation of these plugged prices is shown in the accomplishment of this estimating task for the Denny Dam estimate in Chap. 11.

When a joint venture is formed to bid a construction project, the sponsor's estimator often sends the list of plugged prices to each participating partner. This simplifies the joint venture's prebid estimators' meeting and decreases the amount of work that must be done by each partner's estimator. All partners will then have the same plugged prices for permanent materials and subcontracts in their estimates. Thus, it is not necessary to waste meeting time on the comparison of these prices, and more time is available for discussion of estimated cost items that vary due to differences in estimating concepts and labor productivity.

The amount of permanent materials that the prime contractor must purchase varies depending on the type of dam project, the amount of work subcontracted, and whether the owner supplies the permanent materials. Permanent materials that are often typically part of the prime contractor's construction cost are cement, concrete admixtures, water stops, joint filler, structural steel, and miscellaneous metal. If the prime contractor plans to use his own forces to construct most of the bid items, it may be necessary to secure material quotations on reinforcing steel, penstocks, gates, bulkheads, cranes, turbines, generators, transformers, and similar types of materials.

Many prime contractors contract all specialty work. Typical specialty items include the furnishing and installation of reinforcing steel, structural steel, ventilation facilities, architectural work, piping, plumbing, heating, and electrical work; the installation of turbines, generators, and transformers; the performance of roadwork; the manufacture of aggregates; and the construction of access roads.

V. FIELD INVESTIGATION

As soon as the plans and specifications have been reviewed, a field investigation should be made of the proposed damsite. Occasionally dam projects are located at such high elevations that it is impractical to visit the damsite during the winter season. This creates a problem when the dam-construction project is advertised for bidding during the winter months. To solve this problem, contracting agencies may issue preliminary project information and schedule field-inspection trips during the preceding summer months.

The proposed dam project must be investigated since climatic conditions, work access, job logistics, camp requirements, and the availability of power and water influence the construction cost. The site conditions control the diversion method, work sequence, work performance, and the selection of construction equipment and plants. A preliminary review of these factors can be accomplished by office research, but their final determination requires a field investigation.

During the field-investigation trip, emphasis should be placed on establishing the total work concept and on solving any special problems resulting from specification review or preliminary work planning. Specific areas to cover during the field-investigation trip are:

1. *Labor.* The type and availability of labor must be determined. Experienced labor is essential to economical construction. The availability of experienced construction labor in an area is dependent on the labor force and the amount of heavy construction that is being performed in the area. The chance of securing experienced dam-construction labor will be good if other dams have been constructed in the same area. If heavy-construction projects are new to the area, considerable training of labor may be necessary, and labor cost will be high until sufficient crews have been trained. The number of days that must be worked per week on the project is often dependent upon the number of days worked per week on adjacent construction projects. If other projects are working 6 days a week, it will be necessary to plan on the same work week in order to secure competent workmen.

Local labor agreements should be secured. These agreements will list for each area: craft classifications, the existing wage rates, any future wage rates that have been negotiated, hours of work, overtime rates, fringe benefits, subsistence rates, subsistence areas, travel-time regulations, and working rules.

2. *Camp Requirements.* The area surrounding the project should be examined to determine whether there are sufficient living facilities for the construction force in the surrounding towns and adjacent areas. If existing facilities are inadequate, a single-status camp may be required. Most employees, however, prefer to live in established family quarters and drive long distances to work rather than to live in single-status camps. It should be determined whether a married-quarters camp is desirable. The available schools should be reviewed and the location of the nearest hospital determined.

3. *Climatic Conditions and Work Elevation.* Weather conditions and the elevation of the project should be reviewed to determine their effect on construction. If the work is located in heavy snowfall areas, the construction equipment must be winterized and allowance made for the removal of snow from access roads and plant areas where required. The weather might be so extreme that work must be shut down completely during the winter months or perhaps

only severe enough to cause a drop in work efficiency. In any case, the weather's effect on construction must be determined.

High altitudes will result in lower production from workmen and a loss in production from equipment unless the equipment has been designed for high-altitude operation.

4. *Surrounding Areas.* The nature of the area surrounding the project may influence project cost. In forested areas, work may have to be shut down during dry, hot periods when the risk of forest fires is high. The contractor may be required to maintain fire-fighting equipment and assist in the control of any fire. The construction of access roads may be expensive in forested areas where side casting of excavated material is not permitted, where access roads are constructed across swampy areas requiring deep fills, or where access roads are constructed up steep canyons, requiring excessive grading.

If the work is located in a state park, a national park, or a national forest, waste disposal may present problems since tree preservation and stream pollution are rigidly controlled by the governing agencies. Work in military installations may cause additional costs due to security requirements. Such costs may include the cost of security clearances, special transportation of men because of private-car restrictions, and additional travel-time payments.

5. *Remote Areas.* The work may be located in such a remote area that special off-site cost is required for labor recruitment, purchasing, and expediting. At remote work locations, large stocks of spare parts, materials, and supplies are a necessity because of the difficulty of obtaining more. Initial job-personnel transportation from hiring hall to jobsite may be required, as well as daily transportation. Payment for travel time to and from the job may be a requirement.

The work may be located in a subsistence area. Labor turnover, material-handling cost, and haulage cost increase as the job's remoteness increases. It may be desirable to provide helicopter or airplane service to the job area, including landing areas and servicing facilities at the jobsite.

6. *Access, Communication, and Utilities.* Access to the job should be investigated to ascertain what difficulties will be encountered in transporting equipment, supplies, and materials to the project site and to determine whether access roads need improvement or whether new access roads must be constructed. The location of the nearest highway, rail connection, and airport should be determined.

Communication facilities should be reviewed to determine whether new telephone facilities must be installed or whether radio communication should be used. The cost of any required modernization of the communication system should be secured.

The availability of power and water should be examined so that cost connected with any expansion, improvement, or installation of these facilities can be determined.

7. *Damsite Inspection.* The topography of the damsite and the river width and size should be reviewed to determine the applicability of different construction systems. As background information for this preliminary determination, concrete construction and earth- and rock-fill placement construction systems have been described in the preceding chapters. The abutments should be examined to determine the most economical method of performing foundation excavation. The type and the desired location of diversion facilities must be correlated with the type of site and topography of the abutments. The damsite should be examined to determine favorable locations for haul roads, construction equipment, construction plants, storage and shop areas, warehouses, and offices.

The streambed at the damsite should be examined and its width, depth of water, low- and high-water marks should be determined. Stream-flow records should be examined and correlated with the existing stream channel. A study of the flow records will furnish information on minimum-flow periods, maximum-flow periods, flood flows, and the frequency of flood-flow occurrence. This information will provide data for design of the diversion method and for cofferdam design.

The overburden and rock should be examined to determine their characteristics. If the owner has explored the foundation by drilling, the location of these holes should be determined with respect to the dam's location and used to establish the depth of the required excavation. Drill cores should be examined to determine rock types and rock hardness. Geological reports should be studied, including any published by the owner, the state, or the government. This information will provide the basis for selecting excavation methods, estimating drilling rates, and establishing powder consumption. A structural geologist is often retained to assist in this investigation.

8. *Disposal Areas, Aggregate Pits, Borrow Areas.* If the owner has provided areas for spoil disposal, these areas should be examined to determine if clearing is needed, whether drainage facilities will be necessary, and what haul roads will be required. If the owner has not provided disposal areas, the estimator must make complete provision for the cost of spoil disposal.

If natural aggregates are to be used, the aggregate pit should be examined and the length of the required haul determined. If manufactured aggregates are to be used, the quarry site should be examined and haul distance determined. If the work includes the construction of a rock embankment, the proposed quarry site should be inspected, any drill-hole or other exploratory work examined, and quarry development and haul-road construction visualized. Sources of filter material and impervious-core material should be examined. If the work includes earth-fill embankment construction, the borrow pits that are to be the source of the fill should be investigated.

9. *Property Taxes.* The county tax collector should be contacted to obtain the property tax rates and to secure the basis for determining the assessed valuation of the contractor's plant and equipment.

VI. QUANTITY TAKEOFFS

The quantity takeoffs should be started as soon as the plans and specifications have been received. As soon as any portion of the quantity survey is completed, this information should be distributed to the estimators who are preparing the construction schedule and those who are estimating the direct cost items. All quantity takeoffs must be completed before direct-cost estimating can be finished.

Prior to discussion of quantity takeoffs, it is necessary to properly define three terms: *takeoff quantities*, *bid quantities*, and *final pay quantities*. Each of these quantities may differ from the others for any item of work.

Takeoff quantities are quantities taken off the bidding plans and specifications by the estimator and represent to the contractor the most accurate estimated quantities of work to be performed.

Bid quantities are quantities printed on the bid comparison sheets. These quantities are furnished by the owner's engineer and are used for comparison of bids. The sum of the extensions of all bid quantities, each multiplied by the unit bid price, will equal the total bid submitted by each contractor.

Final pay quantities are the pay quantities of materials that are handled during construction of the project. The sum of the extension of all final pay quantities, each multiplied by the appropriate unit bid price, will equal the total payment to the contractor.

Accurate quantity takeoffs of all work quantities are required for the preparation of a dam estimate for the following reasons:

1. *To establish work quantities for lump-sum bid items.* These quantities must be determined so that they can be used for construction cost estimating. The specification writers often show bid quantities as a lump sum when the work required to construct the item is of a composite nature requiring the performance of many different tasks. Or, they may show bid quantities as a lump sum for work items that are of such a nature that work quantities are difficult to es-

tablish. Use of the lump-sum bid quantity, in this latter case, is to relieve the specification writers of responsibility and to force the contractor to accept the responsibility.

2. *To determine work quantities that form part of the contractor's cost but that are not listed in the bid comparison sheets.* The preparation of these quantity takeoffs has been discussed in Chaps. 2 to 8. Depending upon the extent of the work items included in the bid schedule, quantity takeoffs are required to determine: the pounds of powder consumed per cubic yard of excavation, the nonpay cubic yardage of excavation in road construction, the estimated yards of overbreak excavation, the square footage of concrete forms, the form ratio, the square footage of concrete cleanup and finish, the tonnage of aggregates, the lineal footage of welding, etc.

3. *To check bid quantities.* If takeoff quantities vary greatly from the bid quantities, this will influence the method of spreading the cost of the plant, equipment, indirect cost, escalation, and markup to the bid items to arrive at unit bid prices. The spread of these costs to bid items whose bid quantities are greater than takeoff quantities should be held to a minimum to permit the contractor to recover this cost. This is explained in greater detail in Chap. 12.

4. *To determine unit cost more accurately.* An estimator may estimate the total direct cost of performing some bid item work and then establish unit cost by dividing this total cost by work quantities. If takeoff quantities are used as the divisor, the resulting unit cost will be the most accurate cost obtainable, since takeoff quantities are considered by the contractor to be more accurate than bid quantities.

Detailed descriptions of the quantity takeoffs required for each type of dam construction have been presented in Chaps. 2 to 8.

VII. LABOR COST

The work on this item of cost should be started as soon as the area labor agreements are secured and the insurance quotations are received. This work must be completed prior to sending out the instructions to the partners and before any direct cost is estimated. This tabulation will ensure that each member of the joint-venture partners' estimating team uses correct and identical labor cost. The sponsoring joint-venture partner may send these computations to all partners in the joint venture to relieve them of a clerical estimating task and to ensure that all estimates are prepared using identical wage rates. Typical computations required for this tabulation are shown under Estimating Task VII for the Denny Dam sample estimate in Chap. 11.

This tabulation should show the contractor's cost for one shift of work or for one hour of work for every classification of labor. The data required for its preparation are the hourly wage rates, hours of straight-time work, overtime requirements, overtime rates, amount of fringe benefits, subsistence areas, subsistence rates, travel-pay regulations, and the amount of travel pay for each classification of labor. All of this information is found in the labor agreements applicable to the working area.

The workmen's compensation insurance rates, the public liability and property damage rates, and the state and federal taxes for social security and unemployment insurance that form the payroll burden can be secured from insurance brokers.

The hours worked per shift, the number of shifts that will be worked per day, and the number of days that will be worked per week establish the percentage of overtime payments. The chief estimator must make these decisions, basing his selection on the production that must be achieved to meet the job schedule and whether overtime pay premiums are necessary to secure labor. The other condition that determines the percentage of overtime is set by labor agreements, which establish the overtime rates and the standard straight-time work week for the particular area.

To illustrate how the percentage of overtime is computed in an area where the standard work week is 40 hours, overtime for the first six days in the week is time and one-half, and for shift work, eight hours are paid for seven hours' work, the following percentages of overtime have been computed for different working conditions. If the job works three shifts per day, five days per week, there will be no overtime payments, but the workmen will only work seven hours a shift or 21 hours a day while receiving straight-time pay for eight hours a shift or 24 hours a day. If the job runs three shifts a day, six days a week, the workmen receive the equivalent pay per week of 52 straight-time hours for 48 shift hours, so the percentage of overtime is $8\ ^1/_3$. The hours worked per day remain the same at seven hours a shift or 21 hours per day. If two 10-hour shifts are worked five days per week, the workmen receive the equivalent pay of 11 straight-time hours a day or 55 hours a week for 50 hours of shift time, so the percentage of overtime is 10 percent. However, they only work 9 hours a day or 45 hours a week. If two 10-hour shifts are worked six days a week, the workmen receive the equivalent of 70 hours per week of straight time pay for 60 hours of shift time, or a percentage of overtime of $16\ ^2/_3$ percent. The workmen will work only 9 hours per day or 54 hours per week.

VIII. EQUIPMENT OPERATING COST

A tabulation of equipment operating cost should be started for a dam project as soon as the wage rates applicable to the project have been established, but it cannot be completed until all equipment selections have been made. Most estimating teams will designate one member to be responsible for the preparation of this schedule. The estimator can prepare a partial tabulation covering basic construction equipment, e.g.: front-end loaders, rear-dump trucks, drills, air compressors. This partial tabulation is distributed to each member of the estimating team; then, as more types of construction equipment are selected, they are added to the tabulation, and these revisions are distributed. Continued revisions must be made until all direct cost items have been estimated and all plant and equipment units have been selected.

Equipment operating cost constitutes another basic cost tabulation that should be established and used by each member of the dam estimating team so that all work will have the same cost basis. Tabulation of equipment operating cost consists of listing each construction plant and equipment unit, and estimating each unit's average operating cost per shift, excluding depreciation. The average operating cost includes the operating labor; maintenance labor; the consumption of fuel, oil, and grease; the cost of repair parts; and the cost of tire repair and replacements. These costs are estimated by adjusting past equipment operating-cost records for changes in wage rates and for changes in supplies and tire prices. Some contractors exclude operating labor and others include it. This inclusion or exclusion of operating labor results from the use of different cost-accounting procedures.

The estimators use this tabulation in the same manner as accountants use clearing accounts. In estimate preparation, the direct cost-estimating details will show the number of shifts that each plant and equipment unit is required to work in constructing each work item. To determine the cost of these operating shifts, they are multiplied by the equipment average operating cost secured from this tabulation. Labor and supplies are extended separately so that their cost will be included in the total labor and supply project cost. If labor and supplies are extended separately by the estimators and cleared separately by the accountants from clearing accounts, then operating labor can be included in the equipment operating cost. If equipment use is extended as one total by the estimators and cleared as one sum by the accountants, then it is preferable to exclude operating labor from the tabulation, since it will be lost and could not be readily included in the total project labor. The disadvantage of this second method of estimate extension and the clearing of clearing accounts is that maintenance labor will be lost and cannot be readily included in the total project labor.

One of the great advantages of the use of these tabulated costs is the assurance they provide that the service and maintenance labor required to keep the equipment in operation is included in the estimated cost.

IX. TRANSMITTAL OF ESTIMATING INSTRUCTIONS

In the heavy-construction industry, when the work is to be bid by a joint venture, it is the practice for the sponsoring company's chief estimator to inform all partners of the date, time, and location of the prebid estimators' and principals' meeting, and to send them estimating instructions and estimating data. Estimating instructions and data are distributed to simplify the comparison of estimates at the prebid estimators' meeting, which will free meeting time for comparisons of cost differences resulting from the use of different construction systems and evaluations of labor productivity.

These instructions should be sent to each partner as soon as the applicable wage rates and payroll burdens are available and after plugged prices have been established. Distribution should be made of most or all of the following data:

1. *Estimate Comparison Forms.* For typical estimate comparison forms, refer to Chap. 12.

2. *Descriptions of the Cost Divisions Shown on the Estimate Comparison Forms.* These descriptions should be similar to those presented in the first portion of this chapter under Estimating Task I.

3. *Wage Rates, Fringe Benefits, and Payroll Burdens.* An easy method of transmitting this information is to distribute the wage rate tabulations described under Estimating Task VII.

4. *Insurance Rates.* The rates for insuring the contractor's construction equipment and plant facilities, for providing builders' risk insurance, and for securing any special insurance that is required during construction of the project should be included in the estimating instructions.

5. *Tax and Assessment Rates.* The tax and assessment rates for personal property taxes should be determined, as should the percentage of sales tax and of any business tax. This information should all be included in the estimating instructions.

6. *Bond Rates.* These rates are given when Estimating Task XV is discussed.

7. *Power Cost and the Up-and-Down Charges.* The power cost and the up-and-down charges for the installation of any required substation must be included. This information may have been established during the field-investigation trip or it can be secured from the power company that services the area.

8. *Plugged or Bogie Prices.* Estimating instructions should tabulate the plugged or bogie prices for permanent materials and subcontracted work. This has been described under Estimating Task IV.

9. *List of Available Equipment.* A descriptive list and the acquisition cost of any suitable, available construction equipment that is owned by the sponsoring partner should accompany estimating instructions.

10. *Request for Equipment Lists.* All partners should be requested to distribute a similar listing and pricing of any suitable construction equipment that they have available for use on the project.

11. *Geological or Materials Engineering Reports.* Any geological or materials engineering reports on the project site, quarry site, or borrow areas should accompany the instructions. If a consultant has been retained to examine any of these areas, a copy of his report should be distributed to each partner.

X. SELECTION OF THE DIVERSION METHOD

When the plans and specifications define and design the method of diversion, the contractor must plan the dam's construction around this method. When the selection of the diversion method is the responsibility of the contractor, he should coordinate that selection with the preparation of the construction schedule (Estimating Task XI), and the selection of the plant and equipment (Estimating Task XII).

Prerequisite to the accomplishment of this construction planning is the review of the required work, the time available for construction, the damsite's topography, the stream-flow characteristics, and the work constraints, as shown on the plans, as stipulated in the specifications, or as developed during the field-investigation trip. It is not a necessity that the quantity takeoffs be completed before this planning is done, but the quantities that are available will be of use in systems selection and in construction scheduling.

The selection of the diversion method is a complex task and requires that the estimator have a working knowledge of diversion methods, the risks involved with the use of each method, the effect that different diversion methods will have on the construction system, and the diversion methods that have economical advantages to their use at different types of damsites. All this information has been well-presented in Chap. 2 and its inclusion here is not warranted.

XI. PREPARATION OF THE CONSTRUCTION SCHEDULE

In Chap. 1 it was described how the construction schedule is used in the systems approach to dam construction and how it can assist the estimator in coordinating all work and in selecting the type and quantity of construction equipment. A discussion was also presented on the advantages of the use of a precedence diagram for work scheduling.

The preparation of the construction schedule should be started simultaneously with the review of the plans and specifications. During this review, the bid date, anticipated award date, start of construction, completion date, climatic conditions, river-flow information, periods when concrete placement is restricted because of climatic conditions, dates when different diversion stages can be started, and dates when owner-supplied permanent equipment and material will be received should be plotted on the schedule.

The scheduling of a dam's construction must be correlated with the selection of the types of construction equipment and plant units and with the selection of diversion facilities. Detailed descriptions of the scheduling of all work required for the construction of the different types of dams have been presented in Chaps. 1 to 8.

The construction schedule must develop the quantities of work that must be completed in each time period in order that the work can be finished prior to the specified completion date. These scheduled work quantities are used in performing Estimating Task XII, to establish the required number of construction equipment units, and for preparing the direct cost estimate, Estimating Task XIII.

Scheduling the construction of a dam is a continuing operation since it must be changed as new or additional work concepts are developed. For example, during direct-cost estimating (Estimating Task XIII) it may develop that changes or additions must be made to the construction equipment, which may necessitate changes in the construction schedule. Or it may be found that cost can be reduced if the schedule is rearranged. In this case it may be necessary to completely reschedule the work. For these reasons, construction scheduling is never finalized prior to estimate completion.

A typical bar-chart construction schedule is shown under Estimating Task XI for the example estimate in Chap. 11.

XII. SELECTION OF THE PLANT AND EQUIPMENT

The types and number of units of construction equipment and plants should not be selected until the plans and specifications have been received, the quantity takeoffs completed, and the field investigations made. Then, this selection should be done simultaneously with diversion-method selection and work scheduling. The performance of these three estimating tasks establishes the dam's construction system. Proper selection of the diversion method and the types of plant and equipment and proper scheduling of construction are primary requirements for successful dam estimating, since the low bidder for a competitive bid project is usually the bidder who plans on constructing the dam with the most economical construction system.

While the plans and specifications are being reviewed and during the field investigations, the estimator should start to visualize the best construction system for performing the dam's construction and begin to establish in his mind the construction plant and equipment units that will be needed to complete the construction system. During the preparation of the estimate, the construction plant and equipment should be constantly reviewed to determine whether a new item of construction equipment or new applications of existing equipment might be economical to use when constructing the project.

The originally selected diversion method, the construction plant and equipment, and the initially prepared construction schedule may have to be changed if there are changes in construction concepts developed during the costing of the estimate. Often estimates have been completed and then more economical construction systems have been conceived, resulting in the necessity of a total revision of the complete estimate prior to bid submittal. One of the failings of some estimators is that they are so rigid in their thinking that they resist these last-minute revisions. This rigidity may result in an unsuccessful bid. One of the greatest advantages of joint-venture partnerships is that the partners can adjust the agreed estimate to take advantage of any new approach to construction brought to the meeting by any partner.

To select the proper construction equipment and plant units, the estimator must have a broad knowledge of this subject. He must know what types of construction plants and equipment have economical advantages to their use at each type of damsite. At some damsites, the solutions to these problems may be self-evident, but at other damsites it may be necessary to prepare comparative estimates to arrive at the right solution.

The selection of construction plant and equipment for each

phase of the dam's construction has been described in the preceding chapters. In Chap. 1, construction systems were defined, the selection of construction systems was discussed, and the controls for subsystem selection were described. In Chaps. 2 to 8, descriptions were given of the types of construction systems and their plant and equipment components that have been developed for performing foundation excavation, concrete construction, earth- and rock-fill embankment placement, and powerhouse construction. In these chapters it was explained how plant and equipment requirements are computed by dividing the work quantities by the time period given on the construction schedule for performing this work. The resultant of this division is then divided by the production rates that will be achieved by each equipment and plant unit, which yields the required number of operating equipment or plant units. This number of operating units, divided by their percentage of availability, will give the number of units that must be supplied to the project. The service vehicles and facilities required for the project must be added to this list. The completed plant and equipment list should be reviewed to determine whether all systems have been completed and to make sure there is not a duplication of any equipment or facility. An equipment-use schedule is helpful in making this check. This schedule will also indicate if it is desirable to adjust the construction schedule by rearranging the time that some noncritical items are being constructed in order to accomplish *equipment leveling.* Equipment leveling is the adjustment of the schedule so that equipment demand will be flattened and any short demand peaks will be leveled and slack periods reduced.

It is necessary to prepare construction-facility drawings to establish the quantities of work that are required to construct these facilities. These quantities are secured from layouts of major plant facilities, i.e., cableway, trestle, mixing plant, aggregate plant, and cooling plant; from construction road layouts; from construction area drawings; and from layouts of the construction camps. Such quantities include: common excavation, rock excavation, grading, surfacing, concrete, reinforcing steel, lumber, structural steel, and building construction quantities. A review of these drawings will also establish whether proper plant and equipment coverage is furnished for the dam's construction.

XIII. ESTIMATE OF DIRECT COST

The preceding estimating tasks were performed to collect estimating data for estimate costing and to plan the dam's construction so that these planned activities could be costed. Therefore, direct-cost estimating cannot be completed until all the foregoing estimating tasks are completed.

Since the majority of time spent on costing an estimate is spent on direct-cost estimating, this work should be started as soon as possible. For instance, the direct cost of excavating the dam's foundation can be estimated as soon as the excavation takeoff has been completed and the excavation sequence and equipment have been selected. As more takeoff quantities become available and as the construction schedule and the construction system become more finalized, more direct cost items can be priced.

Direct cost is composed of the cost of direct labor, construction supplies, permanent materials, and subcontracts. As explained under Estimating Task IV, plugged permanent material and subcontract prices are used in the estimate. This leaves direct labor and construction supplies as the direct-cost components whose cost must be estimated. These are the most difficult components of an estimate to cost since they vary with the use of different construction equipment and plant units, with changes in labor rates, with differences in labor productivity, and are dependent on proper work supervision. For dam construction, the direct labor cost is the largest cost item that can be controlled by the prime contractor; therefore, its accurate determination is next in importance to the selection of the most economical type of construction system for successful dam estimating.

The direct cost of work items that are constructed by the application of large construction equipment or plant units are estimated using the heavy-construction estimating method. To save a large amount of detailed work, the direct cost of work items that are constructed by hand labor can be estimated using the unit-cost estimating method. Any individual bid item may contain work that has been estimated by the heavy-construction method and work that has been estimated by the unit-price method. The reasons for the use and the different applications of these two estimating methods were discussed in detail in Chap. 1.

The heavy-construction method of direct-cost estimating is used to estimate the direct cost of machine excavation, truck haulage, embankment construction, aggregate production, concrete cooling, concrete mixing, concrete placement, and all other work items that are constructed by the use of large equipment or plant units. It consists of listing the man-hours of each type of labor and the operating hours of each construction equipment and plant unit that are required to construct each phase of the work. The consumption of construction supplies is also listed. These listed items are then costed, using labor cost from the wage rate tabulations prepared under Estimating Task VII, equipment and plant hourly operating cost from the tabulations prepared under Estimating Task VIII, and by pricing miscellaneous supplies from past records or by using recent quotations. Data required for determining man-hours and equipment and plant operating hours are presented in Chaps. 2 to 8. Briefly, this presentation consisted of describing the required quantity takeoffs; presenting data on the productivity of labor, equipment, and plants; explaining how truck and scraper cycles are determined; describing construction procedures; and comparing the differences in measurement yardages. Typical examples of how direct cost is estimated by the use of the heavy-construction estimating method are shown in Chap. 11.

While estimating direct cost, new concepts in work productivity and equipment usage may be developed. If this occurs, then the construction schedule and the plant and equipment list should be adjusted to reflect these changes.

To save a large amount of detailed work, the unit-cost estimating method can be used to estimate the cost of hand excavation, foundation preparation, form construction, form setting, concrete cleanup, curing and finishing, electrical work, welding, plumbing, piping, architectural work, and work of similar nature performed by hand labor. The preferred manner of applying this method is to estimate unit direct labor cost for any work item by securing the man-hours of labor spent on a unit of this work on past projects and multiplying these man-hours by the applicable wage rate. Unit supply cost is established from the supply costs recorded on previous projects, adjusted for changes in purchase price. These units can then be used to extend work quantities.

If man-hours per unit of work are not available from past records, then recorded unit direct cost of past work must be used. These unit costs should be adjusted for changes in labor rates and supply costs before they are used.

When any work item is constructed by the use of operated and maintained rented equipment, then estimated direct cost consists of determining the production that will be achieved by the use of this equipment. The rent paid for the equipment is similar to the payment of a cost-plus contract and it should be tabulated with the cost of other subcontracted work. Since it is not a firm subcontract, it is the responsibility of the prime contractor's estimator to estimate how long its use will be required. This estimate, times the rental rate, will total the subcontract price.

XIV. ESTIMATE OF PLANT AND EQUIPMENT COST

The gross plant and equipment cost is the cost of providing the construction equipment and erecting the plant facilities that are used for constructing the project. The cost chargeable to each dam project is the net plant and equipment cost, or the depreciation, which is the gross cost less the net salvage value. The net salvage value is the estimated sales price of the plant and equipment at the time of project completion, less dismantling, shipping, storage, and selling expense.

The distinction made between construction plant and construction equipment was given in this chapter. Since these definitions are extremely broad and since different definitions are used by many construction companies, plant and equipment costs are often added together and the sum used for the purpose of comparisons at joint-venture meetings.

The performance of Estimating Task XII, the selection of the construction plant and equipment, will establish a list of the major construction equipment and plant units required for the project and a list of all the quantities of work required to erect the plant facilities, construct the roads and parking areas, and build the service facilities. The work required to perform this estimating task and Estimating Task XIV flows together and is often handled by the same estimator.

Costing the construction plant and equipment can begin as soon as the first equipment is selected. Prices should be solicited from equipment manufacturers (so that cost will be based on the latest prices). If possible, it should be determined if the quotations received from the suppliers were for estimating purposes, and whether they can be reduced by a percentage to arrive at actual *buy out* prices or prices paid at the time of equipment purchase.

The plant and equipment cost should be tabulated on a form that includes identifying descriptions, the required number of units, weights, sales tax, freight charges from factory to delivery point, transportation cost from receiving point to jobsite, labor cost for assembling equipment units and erecting plant facilities, invoice consisting of plant materials and consumable supplies, used equipment purchase price, new equipment purchase price, total cost, net salvage value, and net job cost. A typical tabulation of this kind is shown by the performance of this estimating task for the Denny Dam example in Chap. 11.

Plant and equipment estimating cannot be completed until all direct-cost items have been estimated, since, during the performance of this work, it may develop that different types of equipment or additional equipment units and plant facilities may be required, or a change in the construction system may be more economical than the original plan. Also, direct-cost estimating may indicate a need for a change in construction scheduling, which in turn may change plant and equipment requirements.

Depending on each construction company's accounting procedure, the purchase of sedans, pickups, office equipment, surveying equipment, and first-aid equipment is either charged to plant and equipment cost or to indirect expense. Accounting practice also varies in regard to how a rock quarry's development cost is handled. This cost can be carried as part of the construction plant and equipment cost, or it can be placed in a quarry clearing account and cleared out of this account at an applicable rate per cubic yard of rock produced.

Both the number of years that the equipment will be in service and the number of hours it will work should be considered when setting equipment salvage value. The maximum equipment life of construction equipment similar to trucks, tractors, and scrapers is often set at five years, or 10,000 hours of service. The usable life of large equipment units such as mixing plants, aggregate equipment, cableways, gantry cranes, and large shovels is comparatively much longer, but it is dependent on whether there will be construction projects that require this type of equipment at the time of project completion.

Several publications are available to aid in determining the salvage value of secondhand construction equipment. One of these is the *Green Guide*, published by Equipment Guide Book Company in Palo Alto, California. If allowances are not made in the salvage value for the cost of dismantling the equipment, moving it out, general job cleanup, and projected cost of storage until resale, a separate item containing this cost should be added to the estimate.

If part of the construction work is done with nonoperated, nonmaintained rental equipment, the cost for renting this equipment should be tabulated on a separate schedule and shown as a separate sum in the estimate summary. Some estimators include this equipment rental in direct cost, but this has the disadvantage of distorting direct unit cost so that it will be unsuitable for use in future estimating. Other estimators mix this rental cost with plant and equipment cost, but this does not follow good accounting practice since it is not a capital expense, and if handled in this manner, it will distort the relationship of plant and equipment acquisition cost to salvage value. This may also confuse company management when they review this relationship during markup determination.

The net cost of plant and equipment in any construction cost estimate is based on the use of a selected construction system. If the work is constructed using the same system and equipment that formed the basis of the estimate and if scheduled production is maintained, then plant and equipment cost will not deviate from that included in the original estimate. However, if the job falls behind schedule, it is often necessary to increase the amount of equipment and plant facilities, with a corresponding increase in cost.

XV. ESTIMATE OF INDIRECT COST

The estimate of indirect cost must be based on the total construction time, the length of time that each subsystem will be operated (e.g., excavation subsystem), the complexity of the project, the number of workers that will be employed, the number of shifts that will be worked, the number of days worked per week, the extent that the work will be subcontracted, the contract volume, the application of computers, the efficiency of the engineering and office staff, and the complexity of the cost records and reports that are required to be submitted to the owner and to the contractor's main office. The majority of the foregoing information can be obtained from the construction schedule and other sections of the estimate. How to compute the prime contractor's work force is explained when the estimate of camp cost is discussed (Estimating Task XVI).

The indirect cost estimate can be started as soon as the construction schedule is established, but it cannot be completed until the contract volume is available for bond rate calculating, and the value of the plant and equipment is available for computations of insurance rates.

The following are listings and brief descriptions of the construction cost items that belong in indirect cost. These costs are separated by accounts to provide a check list so that items of expense will not be neglected. Many construction companies use printed estimate forms listing all these items, which serve as a checklist.

1. *Job Management.* This includes the salaries and payroll burdens of the project manager, his staff, and secretaries. Whether the wages and payroll burdens of the project engineer and office manager are included in this account or in the engineering labor and office labor accounts depends on the practice followed by each construction company.

2. *Job Supervision.* The wages, salaries, and payroll burdens of the general, shift, and major craft superintendents, and of their staffs

and secretaries are placed in this account. The personnel charged to this account should be held to a minimum. It is preferred that as much supervisory labor as possible be charged to direct cost.

3. *Engineering Labor.* The wages, salaries, and payroll burdens for the project engineer, office engineers, draftsmen, field engineers, and surveyors should be charged to this account.

4. *Office Labor.* This category includes the wages, salaries, and payroll burdens of the administrative manager, accountants, payroll clerks, timekeepers, personnel manager, and other clerical labor.

5. *Purchasing and Warehouse Labor.* The salaries, wages, and payroll burdens of the purchasing agent, chief warehouseman, warehousemen, and if a service truck is used, its driver's wages should all be charged to this account.

6. *Safety and First-aid Labor.* The salaries and payroll burden of the safety engineer and any required nurses are placed in this account. The specification requirements determine if it is possible for the safety engineer to have other duties. This will determine whether all or only part of his time is chargeable to this account.

7. *Janitorial and Security Labor.* The wages and payroll burdens of janitors and job watchmen are chargeable to this account.

8. *Operation of Pickups and Sedans.* This account includes the cost of operation and maintenance of automobiles and pickups that are required on the project.

9. *Office Supplies.* The cost of postage, stationery, office forms, and all other types of office supplies are charged to this account.

10. *Computer Cost.* This account covers the cost of renting computer time from the home office or from an outside agency.

11. *Telephone, Telegraph, and Cable.* This account covers telephone and telegraph charges. Most estimators do not allow sufficient funds to cover this item of cost.

12. *Office Rent.* This account is only used when applicable.

13. *Heat, Light, and Power for the Office.*

14. *Engineering Supplies.* The cost of engineering supplies for the office and field, including surveying supplies, comprises this account.

15. *Office Equipment.* The cost of office equipment may be charged either to this account or to plant and equipment. This should be the net cost to the job of the office equipment provided. The net cost is first price less salvage, or the rental cost of this equipment.

16. *Engineering Equipment.* This is another cost item which may be carried either in indirect cost or in plant and equipment. It is the net cost to the job of engineering equipment.

17. *First-aid and Safety Supplies.* This account should receive the cost of first-aid and safety supplies. On some jobs, the industrial compensation insurance company may provide first-aid services, and as a result, it is not a separate item of expense to the contractor but rather is included in the workmen's compensation rates.

18. *Entertainment and Travel Expense.* The cost of entertaining the client and other visitors should be included in this account. This cost also covers the expense connected with any traveling done by supervisory personnel.

19. *Blueprints, Photostats, and Photographs.* This account covers the cost of making prints and photostats and taking job pictures.

20. *Outside Consultants.* This account provides funds for hiring outside legal or engineering help to assist on the job or to help present claims.

21. *Legal Cost.* This covers the cost of any legal help supplied by the home office or employed on the jobsite.

22. *Audit Cost.* This covers the cost of yearly audits, performed by outside auditing firms.

23. *Testing-laboratory Expense.* This cost account is used, when applicable, to accumulate charges for the testing of concrete or other items.

24. *Job Fencing.* Many work areas at damsites are fenced in preference to hiring watchmen and guards.

25. *Mobilization of Personnel.* This account covers the cost of moving key personnel and their dependents and possessions to and from the jobsite.

26. *Licenses and Fees.* This account covers the cost of licenses and fees required by the project.

27. *Sign Cost.* This is intended to cover the cost of erecting both the owner's and the contractor's signs. For example, the Army Engineers Department always requires the contractor to erect a sign designating that it is an Army Engineers' job.

28. *Drinking Water.* The cost of distributing iced drinking water to the construction personnel should be charged to this account.

29. *Chemical Toilets.* The rent of portable toilets is charged to this account.

30. *Miscellaneous Cost.* Any item of cost not defined above.

31. *Insurance.* This account covers the cost of insuring the plant and equipment, builders' risk insurance, vehicle insurance, individual bonding, and any special insurance required. A special insurance cost occurs when construction is carried near railroad tracks.

In this case, the railroad often requires bonds or insurance to be furnished it by the contractor.

32. *Taxes.* The cost incurred by the assessment of county and city property taxes on the contractor's plant and equipment is covered by this account. Some states have an additional business tax and other special taxes. Sales tax should not be included in indirect cost, but should be added to the cost of plant and equipment and to the direct cost of materials and supplies.

33. *Bond Premium.* In the majority of jobs, the contractor is required to post bonds to guarantee the successful completion of the job. Typical bond premium rates are as follows:

Contract price	Bond premium	
	Towner group of bonding companies	Non-towner group of bonding companies
First $ 100,000	$10.00/1,000	$7.50/1,000
Next 2,400,000	6.50/1,000	5.25/1,000
Next 2,500,000	5.25/1,000	4.50/1,000
Next 2,500,000	5.00/1,000	4.00/1,000
Over 7,500,000	4.70/1,000	4.00/1,000

The foregoing bond premiums are based on a contract period of under 24 months. If the contract covers a longer period of time, a charge by the bonding companies of 1 percent per month of the total premium is added to the above rates. Bond premiums are computed on the total contract volume and not on the amount of the bond.

XVI. ESTIMATE OF CAMP COST

There are two types of construction camps, married camps and single-status camps, which are used for different purposes. Married quarters are provided to house the contractor's key supervisory, office, and engineering personnel. Single-status camps are provided to house the remainder of the working force. At very remote areas, both types of facilities are required. Since contractors often lose money on camp operations, single-status camps are only provided when their use is a necessity. On the other hand, contractors often provide married-quarter camps on nonremote locations in order to attract desirable personnel and to enable supervisory personnel to live near the jobsite and be available when any construction problems arise.

The simplest method of providing married quarters is to purchase house trailers. These are subsequently rented at nominal rates to the key personnel. Camp size will vary from one unit for the project superintendent to enough units to house every key man. In addition to the trailer cost, the cost of constructing the camp includes clearing the site, grading, providing water-supply facilities, providing a sewage-disposal system, and furnishing and distributing electricity. At remote areas, either school facilities and instruction or the furnishing of daily transportation from campsites to schools for the children of the camp occupants may be required.

The facilities provided for a single-status camp consist of sleeping quarters, washrooms, mess hall, and may include a clubhouse. Local regulations should be checked to determine the number of men that are permitted to occupy one room and the square footage of room area that is required for each man. Locked storage for personal articles and clothes is a necessity. It is now possible to secure quotations on prefabricated single-status camps, completely erected, including mess hall, clubhouse, sewage disposal, water supply, heating, and electrical distribution. The prime contractor must furnish the graded camp location to the camp subcontractor. The advantages in using this type of camp are that it has good salvage value at the end of the project and the cost of construction can be definitely determined before bid submission.

Single-status camp operation is often subcontracted to a caterer who will contract for a price per man-day to supply meals and run the camp. Prices for catering service can be secured before bidding and this establishes the operating loss or profit before bid submission. When a caterer is used, the camp operating cost or profit will be the difference between the caterer's per diem charges and the per diem revenue received from the men for staying in the camp. This per diem revenue, by union agreement, is limited to the amount each craft receives per day for subsistence.

The field investigation will determine whether a married-quarters camp and/or a single-status camp are required. If camps are required, then quotations should be requested for house trailers, for the furnishing and erection of single-status trailer camps, and for the operation of these camps. Firm quotations cannot be received for furnishing and operating single-status camps until the estimated occupancy can be determined. Since the prime contractor often provides camp space for subcontractors' personnel, their demands on the camp facilities must also be established. In addition, some project specifications require that the prime contractor provide camp facilities for the owner's engineers.

The total number of employees and subcontractors' personnel cannot be estimated until the direct cost, the plant and equipment cost, and the indirect cost have been estimated. When these tasks are completed, the number of construction men for each time period can be determined by using the construction schedule and the estimated labor cost to make a labor-time distribution. This labor distribution is then divided by the average man-day cost to determine the prime contractor's work force. Each subcontractor's labor cost must be estimated and his labor cost distributed so that his total work force can be determined. This total work force can be reduced by the number of men that will be housed in married quarters. A judgment must be made to determine what percentage of this remaining work force will wish to stay in a single-status camp. This is a difficult number to establish, since construction personnel often prefer to drive long distances to work rather than live in single-status camps. To this total, then, must be added any owner's engineering personnel that must be housed.

As soon as this estimated camp occupancy has been established, it should be transmitted to the subcontractor who is quoting on the camp construction and to the caterers, so that they can *firm up* their quotations. The caterers' firm quote will be a cost per man-day that varies with the camp occupancy. The prime contractor's estimator must base his estimate of camp operation on an estimated occupancy and use the applicable man-day quote from the caterer. This can never be accurately estimated, so for any project there will always be differences between estimated cost and actual cost for camp operation.

XVII. ESCALATION COMPUTATIONS

Escalation computations cannot begin until all of the preceding estimating tasks have been completed, since these computations are based on the total project labor cost and permanent material cost.

Dam estimates are based on wage rates and material prices applicable at the time of bidding. To determine the true cost of constructing the project, the increase in cost that occurs during its construction must be determined. These cost increases are called *escalation*. The basic data required for escalation computations are: the time of year that labor rates will increase; the amount of labor expenditures that will be made in each yearly time period, starting at the time when labor rates increase; and the percentages that labor cost will increase for every yearly period. Each year's escalation cost is computed by multiplying the yearly expenditures for labor times the yearly increase. The sum of these yearly increases will equal the total labor escalation. A typical labor escalation computation is shown in the performance of this estimating task for the Denny Dam example estimate in Chap. 11.

The yearly percentages of labor cost increases are established by reviewing labor contracts to determine when the yearly labor increase will occur and through what date future yearly wage increases have been negotiated. For the yearly periods that are not covered by existing labor contracts, it is necessary to estimate the anticipated labor increase. The trend of labor increases in the area for each craft should be reviewed, and future wage increase predictions should reflect this past information, as well as be influenced by overall wage increases throughout the country. The labor increase that will be applicable to any yearly period is the total of all preceding yearly increases. The total project labor can be distributed over yearly periods by using the construction schedule to determine what work will be performed in each period.

Material and supply escalation cost is often very minor since many quotations will be firm for the life of the project. It cannot, however, be determined if the material and supply quotations will be firm, prior to their receipt. These final quotations are not received until shortly before bid submittal time. Therefore, material and supply escalation costs must be determined at this late date. These computations are performed using the same procedure as that used for labor escalation. The estimated yearly increases are based on records of past increases.

XVIII. ESTIMATE SUMMARY

After all construction costs have been estimated, they should be summarized in a form that is consistent with company practices and one that is readily understood by the construction company's principals. Labor, permanent materials, subcontracts, and plant and equipment purchases should be listed separately to simplify markup determination. A recommended tabulation of this type is shown for the performance of this estimating task for the Denny Dam estimate in Chap. 11.

XIX. CASH FORECAST

The construction company's management cannot properly evaluate the markup that should be placed on the estimated cost to arrive at the bid price unless they know the amount of capital that must be invested in the project, the length of time this capital will be required, and the interest applicable to this invested capital. Therefore, after the estimate is completed, a cash forecast should be prepared to establish capital requirements and interest expense.

The contractor must invest capital in a dam project to provide funds for the purchase of construction plants and equipment; to buy bonds, insurance, etc.; to pay for several months of indirect expenditures; and to finance one month's expenditures for the construction of the dam, since the contractor will not receive any large progress payments from the owner at the start of the job's construction.

When the specifications provide for the prepayment of the contractor's expenditures for plant and equipment or when they include an item in the bid schedule for payment of the contractor's mobilization cost, it is possible for the contractor to receive prompt recovery of some of this expense and thus reduce his capital investment.

The method used for determining the amount of capital that must be invested in a construction project, and the amount of interest on this capital, is to prepare a cash forecast showing the cash expenditures and receipts during the project's construction.

A preliminary cash forecast can be made by totaling the following items and assuming that cash will be required for 50 percent of the contract time.

Preliminary Cash Forecast

Plant and equipment outlay necessary to start the work	________
Cost of payment and performance bond	________
One year's insurance	________
Inventory	________
Camp construction cost	________
Cost of any bid item whose payment is to be deferred for any reason	________
Overhead required to start job before any bid item revenue is received	________
Working capital	________
Total outlay	________
Less any plant and equipment prepayment or mobilization payment	________
Total cash requirement	________

A more detailed and more accurate cash forecast is prepared by estimating and listing the amount of cash revenue that will be received in each period and the cash expenditures in each period for direct cost, plant and equipment acquisition, camp construction, and indirect cost. Time increments of months, quarters, and halves of a year are used, depending upon the accuracy required and the total length of the construction period. A cash forecast cannot be accurately prepared until the estimated cost and the completed construction schedule are available. To arrive at the total revenue, the estimator must assume a percentage of markup which, when added to the total estimated cost, will give the total revenue. The simplest method of determining the bid revenue per time period is to multiply the direct cost by a factor found by dividing the total direct cost into the estimated total bid price. A more accurate method of determining period revenue would be to spread the plant and equipment to the direct cost items, which often gives a higher factor to use on the excavation items which have early completion, and a lower one to use on the concrete items which are completed later in the job. This type of spreading will reduce the cash requirements.

Receipt of revenue will lag from 30 to 45 days behind the completed construction of pay quantities, because progress payment requests are submitted near the end of each month and because additional time is required by the owner to check the progress payment estimates before making payment.

The Denny Dam example in Chap. 11 shows a typical cash forecast. This cash forecast was prepared using four-month periods, to simplify its presentation in this book.

When a cash forecast is prepared for a job that is already under construction, it is necessary to vary the form to include additional items such as accounts receivable, accounts payable, deferred charges, and miscellaneous revenue in order to tie into the project's balance sheet and profit and loss statement. Engineers should understand enough accounting to use these statements to establish a starting point for a cash forecast. A sample form for this type of cash forecast is as follows:

Cash Forecast Format
for a
Job under Construction

	Last financial statement	Future monthly or quarterly period	Future monthly or quarterly period
Revenue:			
Bid item revenue			
Advances from client			
Subtotal			
Less retention			
Net from client			

	Last financial statement	Future monthly or quarterly period	Future monthly or quarterly period
Camp revenue			
Miscellaneous revenue			
Plant and equipment salvage			
Camp salvage			
Total revenue			
Payables:			
Accounts payable			
Accrued payroll charges			
Accrued other cost			
Total payables			
Revenue plus payables			
Disbursements:			
Direct construction cost			
Indirect cost			
Plant and equipment cost			
Camp construction			
Camp operation and maintenance			
Inventories			
Advances to suppliers			
Accounts receivable			
Deferred cost			
Future labor escalation			
Total disbursements			
Revenue plus payables less disbursements:			
For the period			
Accumulative			
Working capital:			
Capital requirements*			
Financing:			
Advances from partners			
Bank borrowing			
Payments to partners			
Total			

*The sum of the accumulative revenue plus payables less disbursements plus the required working capital.

XX. CONTINGENCY EVALUATION

The remaining information that the construction company's management should have for markup determinations is the contractor's exposure to contingency items. A contingency can be defined as a condition that may or may not occur. If the condition occurs, its extent and duration may be so variable that the cost to the contractor resulting from this condition will vary from a minor to a maximum amount. This cost exposure is often not accurately predictable but the contractor's exposure to the maximum amount can be approximated. This cost exposure should be computed, but not included in the summary of estimated cost; instead, it should be listed separately so that the construction company's principals can evaluate its effect on the project markup.

The contractor's exposure to contingency items will be lessened if the specifications contain force majeure and unforeseen condition clauses, provided specification restrictions do not restrict their application. The contractor can also reduce his exposure to contingency items by taking out builder's risk insurance and by employing a competent structural geologist.

When the contractor is exposed to contingency items, these items are often of one or more of the following types:

1. *Unforeseen Dam Foundation Excavation Conditions.* The extent of the contractor's exposure to added cost is dependent on the amount of foundation exploration performed by the owner. This exploration is done with drill holes, tunnel adits, open-cut trenches, seismographic investigations, and by geological mapping. Unforeseen conditions may result in deeper excavation, which will result in increasing concrete and excavation quantities, necessitating a longer construction period. When insufficient foundation exploration has been done, the owner may require the contractor to remove the excavation in shallow layers and do exploratory cleanup after each layer is removed, which greatly increases the contractor's excavation cost.

Another cost exposure that cannot accurately be predicted is the amount of confined and dental excavation that may have to be performed. Or if the final rock surface is very irregular, the cost of foundation preparation is increased because more hand placement and compaction of the initial dam fill is required. Abutment slides may be encountered, introducing more work quantities and delaying project completion; artesian water may be encountered; large flows of water may come in through the sides of the excavation, etc.

2. *The Possibility That Unexpected Floods Will Occur.* This exposure increases with the size of the river's watershed. This cost exposure was described in detail in Chap. 3.

3. *Unforeseen Inclement Weather.* The exposure to extra cost in placing impervious-core material and earth fill in the embankment is encountered because the working season may be shortened by many days of rain or freezing weather.

4. *Extra Labor Cost.* This may be classified as a contingency

item when it is indeterminable whether competent labor is available, if there is a possibility that future wage increases will be larger than normal, or if there is a possibility that extra overtime must be worked to attract labor. On the other hand, the job may extend for a considerable period of time with a large exposure to raises in wages, changes in fringe benefits, subsistence increases, and changes in working hours and working conditions.

5. *Concurrent Projects.* Equipment or supervision may not be readily available due to many similar jobs being bid or under construction.

6. *Locale.* The job may be in so remote an area that there is difficulty in securing competent workers, maintaining access, and supplying camps and living accommodations.

7. *War, Strikes, Earthquakes, Etc.* The maximum cost exposure resulting from these occurrences is beyond estimating ability. When the contractor is subject to this exposure, it should be brought to the principals' attention so that they can make their own determination of how it may affect the markup.

CHAPTER TEN

Estimating Foreign Dam Construction

The first nine chapters of this book have described how construction cost estimates are prepared for dams located within the United States. The purpose of this chapter is to describe the adjustments that must be made for estimating the construction cost of dams located outside the United States. This chapter is not complete within itself, but serves as a supplement to all the preceding chapters. This supplementary information is presented in the same categories and numerical divisions as were the estimating tasks described in Chap. 9.

These additions to the previously presented data explain how construction cost differs between construction projects located outside the United States and those located within the United States. The cost differences are caused by changes in monetary standards; differences in governmental regulations; reductions in labor rates and in labor productivity; increased cost of equipment operation; decreased equipment productivity; and increased transportation cost, logistic expense, and camp cost. The extent and number of the cost differences that apply to any project vary with the location of the project. For dams constructed in many foreign countries, very few cost differences occur, but when dams are constructed in some of the less developed nations, there are many major cost differences.

I. SELECTION OF AN ESTIMATING FORMAT

Before starting a foreign dam estimate, the currency to be used for preparing the estimate must be selected. Some construction companies always prepare estimates for foreign work in United States dollars. Others prepare foreign estimates using the currency applicable at the work's location. Unless estimators are familiar with foreign currencies, fewer mistakes in judgment and in arithmetic occur when the estimate is prepared in United States dollars.

Before estimating forms can be established, the specifications and contract drawings should be reviewed to determine whether it is a requirement that the construction expenditures in each applicable currency be reported separately. Assuming this is a requirement, the estimating format described in Chap. 9 should be enlarged by adding columns so that American labor cost, third-nation labor cost, local labor cost, United States dollar expenditures, and local currency expenditures can be totaled separately.

II. STUDY OF THE PLANS AND SPECIFICATIONS

In addition to the work described under this estimating task in Chap. 9, the specifications study should include the following:

A review of contractual provisions should be made to establish how the contract will differ from contracts that have become standard for United States construction. If differences occur, a method of evaluation of these differences should be established. The procedures that apply to the settlement of differences of opinions between the contractor and the owner's representative on contract and specification interpretation should be investigated.

The currency to be used for contract payments, the availability of funds, and the procedures used for determining the amount of contract payment should be reviewed. Unless these are clear and well-defined, the risk involved in contract reimbursement may be so great that the submission of a bid is undesirable.

Specifications should be checked to see whether they contain restrictions on the employment of Americans or third-nation labor and whether they contain requirements for the maximum employment of local labor.

It should be determined to what extent the contractor and his employees will be subject to local taxes and whether restrictions or duties will apply to the importation of construction equipment, supplies, materials, food, and commissary items.

The review of contractual provisions should determine whether the contract documents specify that the contractor's construction equipment, construction-plant facilities, and construction camp become the property of the owner at project completion. If this is required, the specifications should be reviewed to determine if there is any way that the contractor will receive special reimbursement for these costs.

The specifications should be reviewed to establish if the contractor is required to construct and/or operate hospitals or schools. Specification study should also establish whether the contractor must furnish living accommodations for the owner's engineering staff and whether he must furnish electricity and/or water and/or sewage disposal to the owner's camp.

III. PREPARATION OF A DOORKNOB ESTIMATE

A doorknob estimate for a foreign project will be less accurate than one for a domestic project, because the estimator will not have access to unit cost that could be applied to each location.

IV. PLUGGED PRICES FOR MATERIALS AND SUBCONTRACTS

On foreign projects, the plugged unit prices for permanent materials and subcontracts often must be increased over their United States counterparts, since their cost may be greater due to increased transporation cost.

V. FIELD INVESTIGATION

The field-investigation team must assemble all the data described under this heading in Chap. 9, plus the additional data described in this section. It is imperative that the field investigation be done in a detailed manner so that the estimators will be well-informed of the cost resulting from local conditions, local regulations, and the use of local labor. This type of investigation will require a team of investigators for a relatively long period of time. Savings in investigation time may be accomplished by assembling all the data available from the specifications and any supplementary information distributed to the proposed bidders by the contracting agency, prior to field investigation. Some pertinent data are securable from the appropriate consular offices. The remainder of the data that must be developed by the estimating team must be secured when it visits the work site. The team should interview port authorities, transportation agencies, labor representatives, local construction-company executives, and governmental agencies to secure the information available from these sources. It may be desirable to retain local construction engineers and a geologist to advise on local conditions. Some of the required data can only be obtained by having the inspection team observe local conditions and local construction projects to form opinions on local labor productivity and camp requirements.

Some of the special subjects that should be investigated are listed and described as follows. The extent to which these items must be investigated is dependent on the nation in which the construction project is located.

1. *Monetary and Tax Investigation.* The conversion rate of the local currency should be investigated, and a determination made of whether the currency is stable or whether it has a record of continuous devaluation. If the work is located in countries with devaluating currency, then it should only be considered when contract reimbursement is made with a stable currency.

The tax laws of the country that are applicable to the construction of the project should be reviewed to determine if the contractor will be subject to any local taxes. To simplify and eliminate any outside profit exposure to local income taxation, the contractor may find it desirable to form a subsidiary company to handle construction. Taxes and charges assessed on the importation of equipment and materials must be investigated. In many cases, such taxes and charges are waived for the particular project but they may still apply when the contract is completed and the contractor wishes to dispose locally of his construction equipment, construction plant, and any excess materials. Under regulations of this kind, many contractors have found that upon job completion it was more economical to ship the construction equipment back to the United States than to pay the local taxes that would have been assessed on local sales.

Income tax regulations must be reviewed both in respect to any tax applicable to the construction company and to any tax applicable to American or third-nation individuals. Some countries have income tax agreements with the United States. These agreements often specify that American employees are subject only to United States income taxes for the first three years they work in the country, but after this time period, they will be subject only to the payment of local income taxes. When American employees must pay high local income taxes, their salaries are increased to take care of any increased tax payments over what they would normally pay in the United States. This becomes a pyramiding type of salary increase and the final monthly salary that must be paid may be ridiculously high. To reduce this pyramiding effect, it is sometimes possible for the construction company to reach an agreement with the local authorities to allow the construction company to pay the difference between the local taxes and what employees would normally pay in United States taxes. If agreements of this type are reached, pyramiding of employees' salary is not required.

2. *Investigation of Transportation Facilities.* It takes longer and costs more to ship construction equipment, plant facilities, camp trailers, materials, and supplies to damsites located in foreign countries than it does to sites located within the United States. This necessitates that a longer time period be allowed for job mobilization; that greater attention be paid during construction to the scheduling of the receipt of construction materials; and that larger inventories of parts, materials, and supplies be kept at the jobsite.

To determine the required transportation time and cost, a review should be made of all transportation facilities and the cost involved in transporting the material from its source to the project site. This extra cost includes added cost to the equipment and material supplier, who must spend additional money for special crates to protect products from shipping damage.

There will be freight charges for shipping the material from the point of origin to the port of departure. Shipments are often delayed at the port of departure while waiting for ships to transport the material to the port of entry. Ocean shipping charges and average shipping times from port of departure to port of entry can be secured from the shipping lines. Ocean freight charges are assessed on long tons or equivalent cubic feet, whichever is greater, with 40 cu ft being the equivalent of a long ton.

At the port of entry it may be necessary to pay port charges and heavy-lift charges. Taxes on the importation of equipment and materials were discussed in the preceding section. Existing interior transportation systems within the country should be investigated in regard to their capacity, the time required to make delivery, adequacy, charges, and the locations where they can make delivery. When high ocean freight charges apply and if interior transportation facilities are inadequate, contractors may elect to use chartered ships, do their own ship unloading, and use their own trucks to provide interior transportation to the jobsite for the initial shipment of construction and camp facilities. Then the receipt of material at the delivery port, transportation to and storage at the jobsite for initial shipments must be scheduled so that the contractor's personnel will be present when delivery is made; equipment to make this transfer and fenced storage areas must be available. This means that shipments must be coordinated with the arrival of construction personnel at the jobsite and that the first equipment received should consist of cranes, trucks, and material-handling equipment.

When all the foregoing cost is totaled, the handling and freight charges on materials and equipment will often run between 10 and 20 percent of their landed cost. The time required for shipment from the factory to the damsite may be as much as three to four months.

3. *Investigate Sources of Construction Equipment.* Whether all construction equipment must be imported or whether some can be secured from local equipment dealers is dependent on import restrictions, import taxation, and the extent to which equipment manufacturers are represented by local dealers. If equipment is locally available, then prices should be secured from the local dealers so that the prices can be compared to the cost of importing similar equipment. The equipment dealers' installations should be inspected to determine if they are set up for equipment servicing and repair and to determine the amount of repair parts that they carry in stock.

If competitive prices on suitable equipment can be secured from local equipment dealers, then local purchase is advantageous. Local equipment servicing and repair will reduce the maintenance work that must be performed by the contractor and the extent that repair parts are stocked by the dealers will reduce the amount of parts that the contractor must carry in inventory.

Import restrictions or purchase price differentials may make it desirable to use construction equipment produced outside the United States. Many products produced abroad are the equal of construction equipment produced within the United States. Examples of excellent foreign-produced equipment are compressors, small drills, rock bits, and drill steel produced in Sweden; French backhoes; and English motors. There is also a large quantity of American equipment manufactured, or assembled by American manufacturers in foreign countries, as well as equipment produced by licensed foreign manufacturers.

The amount of spare parts the contractor must carry in inventory is a function of the time required to ship from the source of supply to the job, whether it is possible to secure experienced equipment operators, and whether local mechanics are experienced in the repair of the equipment.

4. *Materials and Subcontractors.* The specifications and local regulations should be reviewed to determine if there are restrictions or import duties and taxes on the importation of materials and supplies. If local materials and supplies are available, it must be determined if their use is made mandatory by the specifications, if they will pass specification requirements, if they are available in sufficient quantity to meet job requirements, and if their cost is competitive with imports. If construction materials and supplies are imported, their arrival on the job is often delayed by shipping irregularities. This necessitates that the contractor carry a larger than normal stock in inventory and provide warehouse capacity for this larger inventory. This added cost of increased inventories and warehouse facilities is a factor of cost when price comparisons are made between local and imported materials.

In regard to local supplies, it should be determined if the local lumber is suitable to use for concrete forming. The local lumber may be of such poor quality that it will warp and shrink badly when it is fabricated into forms, making it economical to pay a price premium for importing form lumber. It should be determined whether there is an adequate supply of diesel fuel, explosives, detonators, drill steel, bits, etc., or whether they must be imported into the country.

5. *Labor Investigations.* The local regulations should be reviewed to determine if there are limitations on the number of Americans that can be employed on the project, if there are any regulations on the employment of third-nation labor, or if there are any requirements or demands for the employment of local labor. The number of Americans employed on the project should be of a minimum from the contractor's viewpoint because their high wage rates and the high cost of providing them with suitable camp facilities make their employment on foreign projects very costly compared to the employment of other nationalities.

Investigations should be made to determine if there is a sufficient supply of local skilled and unskilled labor. If skilled labor is in short supply, then the supply of skilled labor from a third nation should be investigated. It is preferable to have third-nation labor that speaks the same language as the local labor so that a language barrier will not be encountered. A determination should be made of the productivity of local and third-nation labor compared to the productivity of American labor.

Local wage rates, overtime provisions, taxes, insurance, vacation allowances, severance pay, and other wage benefits should be secured for both local labor and any proposed third-nation labor. In some countries the payroll burden will run as high as the labor rates. Local labor may have a vested interest in their position after a short term of employment. These benefits may increase with prolonged employment and may result in substantial service gratuities when the employee is discharged. The pattern on strikes and other labor disputes must be examined. Often when a large construction job is undertaken, strikes, labor trouble, and increases in labor rates are experienced. A determination should be made whether it is necessary to improve the diet of the local labor so that they can maintain a sustained work effort. The stature of the local labor should be observed since they may be so slight that it might take two laborers to handle some tools.

In addition to determining the amount of local vacations, it should be determined whether construction activities must be shut down at various times because local labor may be accustomed to taking extended holidays. This may occur at Easter time, Christmas time, or whenever holidays occur near weekends. Local practices in regard to furnishing transportation to labor should be ascertained. It should be determined whether transportation of labor from the surrounding communities is a requirement.

Upon completing the labor investigation, if the supply of local skilled labor is inadequate, the field investigation team should make a recommendation on whether local labor should be trained, whether training schools will be required, or whether skilled labor should be imported from a third nation.

6. *Investigations of Engineering, Supervisory, and Office Personnel.* The experience, availability, and required compensation of local personnel as work supervisors, engineers, accountants, stenographers, clerks, and warehousemen should be investigated. Construction cost is greatly reduced when local employees are used to perform the majority of these tasks.

The investigation of this labor force can be performed by reviewing employment on other construction projects and in other types of industry, by contact with other employers, and by having job-application interviews. Typical wage scales and wage burdens are more reliable when they are secured from other employers. Wage burdens, severance pay, and guaranteed employment may often total to a larger amount than salaries. When this investigation is completed, the inspection team should submit its opinion of the competency of the local personnel and determine as accurately as possible what the cost of their employment will be.

7. *Investigation of Camp Requirements.* Camp construction and operation are such large items of cost on foreign construction

projects and their cost is so dependent on local conditions that a detailed study of camp requirements should be made by the inspection team. Prior to this study, the specifications should be reviewed to determine whether they contain any requirements or regulations regarding the construction or operation of construction camps. In some nations, the importation of food and commissary items may be regulated or highly taxed. On some projects, the use of trailers for married-family quarters has been prohibited. The specifications may make it mandatory that the prime contractor furnish the owner's camp with water and electricity and treat the sewage. Or the prime contractor may be required to provide married quarters and/or single quarters for the owner's engineering personnel.

The remoteness of the area and the availability of suitable existing living facilities will determine whether camp facilities and commissaries are required. Local contractors should be contacted to secure their opinions on whether camps for local labor are necessary, whether married-quarter camps should be provided for local labor in order to secure and hold local craftsmen, the type of camp facilities that should be provided for local and third-nation labor, the cost of constructing and operating these facilities, and the amounts that local and third-nation labor can be charged for the use of the camp facilities. If competent local contractors are interested, they should be encouraged to submit quotations for the construction of any required camp.

If other construction work is being performed in the area, these construction projects should be visited to observe the camp facilities that have been provided and to determine what is charged for the use of these facilities.

The inspection team should determine the types of building construction that will suit local climatic conditions and still be economical to construct. Often it is more economical, quicker, and more practical to import trailers and trailer-type single-status camps than to perform on-site building construction. When climatic conditions are severe, this type of construction may be unsuitable.

The sources of food supplies should be checked, and the availability of suitable local cooks should be determined. If the project is so remote that entertainment and retailing facilities are not in existence, it is often necessary for the prime contractor to build and operate commissaries and construct clubhouses, swimming pools, tennis courts, and playing fields.

The need for air conditioning and/or living-facility heating should be determined. Investigations should be made of the existing utilities in the area to determine if they can service the contractor's camps or whether the contractor must develop water sources, generate electricity, and provide sewage-disposal plants.

Finally, the inspection team should make recommendations on the need for camp facilities and the type that should be provided. If local labor camps are required, the inspection team should make estimates of camp construction and operating costs and recommendations on how much employees should be charged for using the camp.

8. *Investigation of Hospitals and Schools.* The specifications should be reviewed to determine if they contain provisions regarding the construction and operation of hospital facilities and/or schools. It should be determined whether existing hospitals, schools, and local doctors can handle the additional load that will be placed on their facilities.

The investigation team should make recommendations on whether a hospital and/or schools are required. If they are recommended, the team should prepare estimates of their construction and operating cost. They should investigate whether local teachers, doctors, and nurses are available to staff these complexes and should determine their salary scale. Many environmental illnesses can be treated better by local doctors than by imported ones. If existing school facilities are adequate but instruction is done in a language other than English, the construction and operation of schools for American children may be a necessity. When American schools are required, employees' wives are often employed as teachers.

9. *Other Investigations.* The area surrounding the dam, the climatic conditions, access, communication, existing utilities, the damsite, disposal areas, aggregate pits, and borrow areas should be investigated as described in Chap. 9. The dam must often be constructed in an area where modern heavy-construction operations have not previously been performed, so previous construction records on the river's characteristics, the geological formation, or local weather conditions will not be available. This necessitates that these field investigations be performed in a very thorough manner. Often less dependence can be placed on past stream-flow and weather records, because they may only cover a short-time period.

VI. QUANTITY TAKEOFFS

Quantity takeoffs are prepared in the same manner as for domestic construction projects, but the takeoff personnel must often think in different units, i.e., long tons, super board feet, kilograms, metric tons, meters, cubic meters, meters per second, etc.

VII. LABOR COST

For foreign projects, emphasis must be placed on the calculation of wage burdens and on determining the special compensation for performing particular work, i.e., wet work, high work, etc., as these often form a large portion of the labor cost. Separate tabulations are required for American labor, local labor, and third-nation labor.

VIII. EQUIPMENT OPERATING COST

When equipment operating cost is being estimated, special attention should be given to the increase in equipment operation cost that often occurs. The cost of repair parts is greater than in the United States since they are subject to higher transportation cost. When Americans are used for operators and mechanics, the hourly labor cost for operating and maintaining the equipment increases since Americans receive additional compensation for working outside the United States; if their pay must be pyramided to take care of the payment of local income taxes, then equipment operating cost will also pyramid.

When local labor has not had experience as operators and mechanics on heavy-construction equipment and it is planned to use them in this capacity, then high hourly operating and maintenance labor cost will be experienced. The high cost results from low labor efficiency, causing a greater increase in cost than any decrease resulting from lower wage rates. Using inexperienced operators will also result in increased equipment downtime causing loss in production and increased maintenance cost.

IX. TRANSMITTAL OF ESTIMATING INSTRUCTIONS

In addition to the description given in Chap. 9 for this estimating task, the partners should be informed of the currency to use for estimate preparation and the manner in which estimate-comparison forms must be enlarged to show the cost of American labor, third-nation labor, local labor, expenditures in United States dollars, and expenditures in other currency. It will also be of value at the estimate-comparison meetings if some of the larger indirect expenses unique to foreign construction are individually listed and thus highlighted on the estimate-comparison forms. Indirect expenses of this type include the mobilization of American personnel, local severance pay, and the cost of training inexperienced local labor.

X. SELECTION OF THE DIVERSION METHOD

The only point to add to the descriptions presented in Chaps. 2 and 9 of the performance of this task is to stress that it should be done with a conservative approach, because the stream- and flood-flow records often have only been recorded over a short-time period.

XI. PREPARATION OF THE CONSTRUCTION SCHEDULE

When a construction schedule is prepared for constructing a dam located outside the United States, ample time must be allowed for the mobilization of personnel, construction equipment, and construction camps. Sufficient time must be scheduled for the mobilization of American personnel to allow for their recruitment; for physical examinations, inoculations and vaccination, for securing passports, for transportation to the site, and for the establishment of living facilities. Time allotted for construction equipment and camp mobilization must be sufficient to allow for shipment to the originating port, holding the shipments at the originating port until scheduled shipping is available, ship transportation time, time required at the receiving port for ship unloading and customs clearance, time required for interior shipment, and the time spent on receipt of materials and on their transportation by truck to the construction project. Upon completion of the project, time must also be allowed for the reverse flow of personnel and construction equipment.

The scheduling of the project's construction must allow for low labor productivity and low equipment productivity in countries where local labor has not had experience in dam construction. Increased time will be required if it is a local practice for labor to take extended holidays. Scheduling should allow for delays resulting from local climatic conditions and from any anticipated high water or flood flows.

XII. SELECTION OF THE PLANT AND EQUIPMENT

Prior to selecting the construction plant and equipment for foreign projects, consideration should be given on how the construction system could be altered to fit local conditions. American construction systems have been developed to perform construction economically using American labor. These systems cannot always be used economically in their entirety at foreign locations. At some locations, due to low wage scales for local labor or due to the familiarity of local labor with the local methods, it may be advantageous to adapt or modify American systems so that greater productivity can be secured from the local labor. However, the use of modern construction equipment is so economical that equipment can seldom be replaced economically by hand labor.

On foreign construction projects, there is often a lack of operators and mechanics who have been trained to operate and repair large equipment units, and the project may be located a long distance from commercial repair facilities and manufacturing plants. For these reasons, foreign projects should not be used as proving grounds for new or larger equipment units. Equipment selection should be made of units that have been used successfully on previous construction projects. Standardization should control equipment selection, including standardization of engine types, equipment makes, and tire sizes. This will reduce the number of spare parts and tires that must be stocked and simplify the training of local mechanics, since they will have only one type of engine to repair.

XIII. ESTIMATE OF DIRECT COST

Direct-cost estimating for dams constructed outside of the United States requires that the estimator adjust his thinking to lower labor productivity and to a lower wage scale. The adjustment for a lower wage scale is readily achieved but it is difficult for most estimators to adjust to rapid drops in labor productivity.

The productivity of local labor often decreases compared to American labor when there is a language barrier between labor and their supervisors and when local labor is not experienced in dam construction and in the use of laborsaving hand tools. Production may drop when local laborers are slight in stature and lack physical stamina. The inspection team's report should be closely examined in regard to these points before direct cost is estimated.

Comparing foreign construction with construction in the United States, the unit labor cost of construction work that is done by unskilled or semiskilled labor is often less than in the United States. This results when wage rates are so low that the savings in unit cost are much more than any unit of increase caused by a reduction in labor productivity. Work in this category is hand excavation, foundation cleanup, rock drilling, moving and aligning steel forms, concrete vibration, concrete cleanup, and the placement of reinforcing steel.

In contrast, the unit labor cost of work that must be done by skilled labor often increases because the decrease in productivity of local skilled labor resulting from its inexperience causes a greater cost increase than the cost savings that results from lower wage rates. The labor cost of constructing 1 sq ft of wood forms often increases in many countries. This often occurs when local carpenters are accustomed to using pull-type handsaws resulting in low work productivity. To increase production, it may be necessary to train carpenters to use western-type handsaws and power tools. For this basic reason, other types of work requiring skilled craftsmen that may incur cost increases are piping, plumbing, and electrical installations.

The labor cost for some special items may decrease when local labor is proficient in performing these tasks, i.e., dry stonework, masonry, tile work, brick laying, applying insulation, erecting tubular scaffolding, seeding, and sod laying. Local labor may have been trained by mining or oil companies in the performance of certain tasks, which may also help decrease labor cost.

On foreign jobs, the productivity of equipment decreases if the equipment operators have not had experience in the operation of the large equipment units used in dam construction. Their inexperience will result in lower hourly productivity and less equipment availability as the equipment will need more frequent repair. If the mechanics have not had experience in repairing the heavy equipment units, repair time will increase, availability of equipment will thus drop, and the labor cost for equipment maintenance will increase.

Direct cost may also increase because of conditions unique to foreign construction. Direct cost increases in countries where rain or other weather disturbances may prevent work during some seasons of the year or will cause a definite drop in labor productivity. Direct cost increases if the work is so remotely located that communications between the project and the home office are slow, causing delays in making vital decisions. Cost also increases when inspection personnel are lacking in dam-construction experience, as they will be more inflexible and may cause work delays and the performance of unnecessary work.

XIV. ESTIMATE OF PLANT AND EQUIPMENT COST

The cost of providing construction plant and equipment is greater for foreign construction than that normal for construction in the United States. Cost of equipment, f.o.b. the jobsite, will increase due to the extra cost of providing special crates, extra transportation cost, and increased cost of erecting facilities and assembling equipment and because the equipment may be subject to harbor taxes, heavy-lift fees, import duties, or other taxes. A greater number of equipment units may be required for foreign construction than for similar work in the United States due to increased equipment downtime and lower productivity.

If local operators must be trained, equipment cost will increase since additional equipment units must be provided for this purpose.

If facilities are lacking in the surrounding areas, extensive repair shops, tire-recapping facilities, and equipment storage areas will be required at the jobsite. Since larger job inventories are necessary, larger warehousing capacity will be required. When the project is located in a remote area, special facilities not normal to United States construction may be required, such as landing fields and airplanes or helicopters.

Finally, at job completion, the disposition of the plant and equipment may present considerable difficulties and may reduce the recoverable salvage value. On some jobs it has been more economical for the contractor to return all construction equipment to the United States rather than comply with the local regulations concerning duties and other charges that are assessed by the local government when equipment is sold within the country.

XV. ESTIMATE OF INDIRECT COST

In Chap. 9, descriptions were given of the indirect expenses, which were placed in 33 Arabic-numbered accounts. Many of these descriptions can be used without change or addition in preparing indirect cost estimates for dams located in foreign countries.

In this section we will supplement the description of the indirect cost accounts that must be modified to make their use applicable for estimating the indirect cost of foreign construction projects. Descriptions of 12 additional indirect cost accounts that may be necessary for foreign construction are also presented. All these cost accounts will be required on some foreign projects, but only a few may be required on others. These additional accounts and descriptions are numbered by continuing the same numerical system used in Chap. 9, starting with account no. 34 and finishing with account no. 45.

Accounts 1—6. Indirect Labor Accounts. On construction projects located in foreign countries, the wages paid to American employees increase, the wages of local employees may decrease, and often more indirect personnel is required in comparison to domestic projects.

Americans are often required for supervisory positions, e.g., project manager, work superintendents, office manager, chief accountant, chief engineer, design engineer, chief surveyor, and personnel manager. The number of Americans that must be used in these positions varies with the project's location. Wage rates of Americans increase over stateside salaries because of foreign pay premiums. In nations where pyramiding of salaries is necessary to provide equivalent take-home pay, the wage increase may be extremely large.

Local or third-nation employees are often used to perform accounting, payroll, drafting, surveying, clerical, stenographic and other duties. Their wages vary with the nation in which the project will be located. A larger number of employees are often required since local or third-nation labor may have to be trained in these tasks and thus will not be adaptable to the performance of multiple tasks. Since often more direct labor is required on the project, more indirect labor is required to handle the work load in the personnel, payroll, and cost-accounting departments.

7. *Janitorial and Security Labor.* Often more guards are required on foreign projects than on those in the United States. In some locations it is necessary for the contractor to provide his own police force or to contract police protection to local protection agencies. In certain countries, it is a specification requirement that the contractor pay and provide living facilities for military personnel to guard and control the use of explosives.

8. *Operation of Pickups and Sedans.* More pickups and sedans for supervisory personnel are issued on foreign projects than is customary in the United States. This causes the cost of this item to increase. Since Americans often do not possess family cars on foreign projects, it may be necessary to provide transportation to the closest town for their wives and children.

11. *Telephone, Telegraph, and Cable.* The cost of these services increases drastically on foreign projects, since rapid communication between the home office and the field requires the use of intercontinental facilities.

18. *Entertainment and Travel Expense.* This expense will be high on foreign projects because more entertainment of officials and visitors is required. The traveling expenses of company executives, engineers, and consultants between the home office and the job will be a large item of cost. This travel will increase if major problems occur during construction of the project. The distribution of these traveling expenses between project cost and home-office expense varies with construction companies.

21. *Legal Cost.* The legal charges on foreign projects are high because local lawyers are often retained to advise on contract interpretation, local taxation problems, and work negotiations.

25. *Mobilization of Personnel.* This will be one of the largest indirect expenses since it covers the recruitment charges, processing charges, and transportation cost to and from the project for all American personnel. These costs run so high per employee that it is important that an accurate estimate be made of the number of employees that must be processed. The total will be the maximum number that will be employed at any period of time plus replacements for Americans who have completed their required length of foreign service and for Americans who leave before their service has been completed. The number of replacements required in this latter classification may become quite large, since some employees may become dissatisfied with local conditions and leave shortly after their arrival. The total number of Americans employed at one time on the project consists of those listed in the Indirect Cost division; plus craft supervisors; operators of specialty equipment such as cableway operators and bellboys; operators of standard equipment who may be required, until local operators become sufficiently trained, to take over operation of the equipment; instructors required to train local labor; plus any required specialty craftsmen.

On foreign work, Americans usually sign a work contract defining the rates of pay, working conditions, special compensation, length of contract, etc. The contract period is often of two years' duration. Employees who have seniority with the construction company, and other responsible employees, will abide by the provisions of the work contract. Less conscientious employees may disregard the contract and only spend a short time at the project. For this reason, recruitment of Americans for foreign work should be done with care. It costs so much to provide replacements on foreign projects that it is more economical to provide living quarters for American married personnel and thus secure stable workmen who will stay for the length of their contract than to save on camp cost by hiring single-status personnel who often will have less job stability and will leave prior to contract termination.

These recruitment processing and transportation costs are composed of the cost of recruiting the labor; the cost of providing passports, medical shots, individual medical examinations for the employee and any dependents, and living expenses during the employee's processing; the cost of providing tourist air transportation for the employee and any dependents to and from the project; and the cost of the employee's salary from the date processing starts until he arrives at the jobsite. In addition to this, transportation is allowed married employees for a reasonable amount of personal goods, and single-status personnel for a reasonable amount of excess baggage.

32. *Taxes.* The amount of taxes that must be paid varies with the project's location. Information on this subject should be included in the investigating team's report.

34. *Special Purchasing, Expediting, and Customs-clearance Expense.* This is an additional indirect expense for projects located in a foreign country. In order to expedite the purchase and delivery of equipment and materials, it may be necessary to open special purchasing and expediting offices in the United States or in other nations and to retain customs clearance firms to handle personnel and material clearance into the country.

35. *Severance Pay.* This is a large item of expense in nations where local employees receive severance payments upon completion of their employment.

36. *Labor Transportation.* In some localities, it may be either the custom or a requirement to provide truck transportation to the work site from the local employees' camp and from surrounding villages.

37. *Labor Training.* When local employees must be trained to operate equipment and to use construction tools, then the operation of these training schools becomes a large item of expense.

38. *Employees' Income Tax Payments.* At locations where the American employees are subject to the payment of high local income taxes, it may be possible for the contractor to pay any excessive charges directly to the local government. This will result in a lower cost to the contractor than the pyramiding of the employees' wages to compensate for these excessive taxes.

39. *Schooling Allowance.* When local schools are unavailable for the high school or college children of American employees, their parents are often paid a schooling allowance. This consists of reimbursement for air transportation of the children to and from suitable schools and a fixed monthly payment for each month of school attendance.

40. *Living Allowance.* When a company camp is not provided, married personnel often receive reimbursement for part of the rental cost of living facilities; single-status personnel may receive a subsistence allowance or may be reimbursed for all their living cost.

41. *Airfreight.* There will always be occurrences when critical spare parts or other materials must be shipped in by airfreight to the project. This extra transportation cost should not be placed against the inventory cost of the part or material, because it will distort these costs. This cost is a measure of the ability of the warehouse department to maintain suitable inventories.

42. *Operation of a Company Helicopter or Plane.* When the

project is remote, it may be necessary for the construction company to operate its own helicopter or plane service to the project.

43. *Construction and Operation of Hospital Facilities.* On some projects it may be specified that the contractor must construct and operate hospital facilities. At isolated locations, where there may be a lack of existing medical service, this will be a necessity. If a hospital must be constructed and operated, advice on this cost can be secured from hospital staff personnel as well as from the investigating team's report.

44. *Construction and Operation of Schools.* If the construction and operation of schools are required by the specifications or are necessary because of the project's location, then their cost must be estimated. Often American employees' wives can be hired as instructors.

45. *Language Classes.* When the local language is other than English, the contractor often conducts evening language classes in an attempt to lessen the language barrier. The cost of hiring instructors and purchasing books and supplies should be placed in this account.

XVI. ESTIMATE OF CAMP COST

When camps are required on construction projects located outside the United States, the basic facilities provided for American employees consist of married quarters for key American employees, and single rooms and a mess hall for single-status personnel. At remote locations it may be desirable to add a guest house, clubhouse, laundry facilities, swimming pool, tennis courts, and playing fields. When there is a lack of suitable local retail facilities, then it is necessary to construct and operate a company commissary.

Houses and camp buildings should be suited to the local climate. If camp facilities of a suitable design can be sold to the owner at project completion, then it is economical to construct them for permanent occupancy. When the sale of the camp cannot be arranged and when climatic conditions are not too severe, then trailer units are often imported from the United States for both the married and single-status camps. The inspection team's report should be studied to determine whether local and third-nation construction camps and schools are required, the extent and kind of facilities that are necessary, and the cost for providing these facilities. The inspection team's report should supply all the information required to estimate the cost of constructing these facilities.

The American camp facilities for foreign projects are always operated at a loss. Married couples are charged minimum monthly house rentals, resulting in an operating loss on these facilities. American single-status employees are often provided with free room and board in company camps so that the camp loss includes the entire cost of operating the American single-status camp. The camp loss may be high in countries where imported food is subject to taxation. If recreational facilities and a guesthouse are provided, there will be an additional cost of maintaining these facilities. If gardeners are required, this will be another additional cost. One camp installation that is often operated without a loss is the commissary. The prices charged the customers are usually set high enough to cover its operating cost.

The cost of operating and maintaining local and third-nation labor camp facilities must be based on the information secured by the investigation team.

XVII. ESCALATION COMPUTATIONS

Escalation costs are estimated for foreign projects in the same manner as described in Chap. 9 for domestic projects.

XVIII. ESTIMATE SUMMARY

The estimate summary form described in Chap. 9 and illustrated in Chap. 11 should be expanded so that separate listings can be made for the cost of American, local, and third-nation labor, and for expenditures in American dollars and in the local currency.

XIX. CASH FORECAST

The cash forecast is prepared in the same manner as described in Chap. 9 for domestic construction with the exception that if more than one currency is used the expenditures in each currency should be separately listed.

XX. CONTINGENCY EVALUATION

In addition to evaluating the contingency items listed in Chap. 9, consideration should also be given to the conditions peculiar to the project's location. The dam may be located in unfamiliar terrain; the labor force may be primarily untrained; weather conditions may be different; records of the river's flood stages may only cover a short period of time; the geological formations may be of types that have not been previously encountered; the work will be done under foreign laws, with foreign inspection; and the contract interpretation and negotiations will be done differently than is the standard practice in the United States.

CHAPTER ELEVEN

A Sample Estimate

This chapter contains a sample cost estimate for the construction of a fictitious concrete dam, located on a fictitious river and near a fictitious town in California. This fictitious dam has been named Denny Dam. Costs are based on the dam being constructed in a subsistence area with a contract award in 1970. The sample estimate is used to illustrate the application of the estimating tasks described in Chap. 9.

To simplify the presentation of the estimate and to eliminate unnecessary detail, typical calculations are used to illustrate how many of the estimating tasks are performed. To expedite the cross-referencing between the task descriptions in Chap. 9 and their application in this chapter, the same estimating task titling and numbering systems have been used in both chapters. A concrete-dam estimate is presented in preference to an earth- or rock-fill dam estimate, because the construction of a concrete dam is so complex that it requires the use of a more extensive construction system, the application of more types of equipment, and the performance of more construction activities than are required for the construction of either of the other types. If the reader understands the presentation of a concrete-dam construction estimate, he will have little difficulty in using the same basic estimating technique to prepare estimates for other types of dams or for any other large concrete, earth-moving, or rock-fill project.

I. SELECTION OF AN ESTIMATING FORMAT

The estimating format that will be used for the preparation of the Denny Dam estimate is as described under Estimating Task I in Chap. 9.

II. STUDY OF THE PLANS AND SPECIFICATIONS

This study was completed and the following abstract of the specifications was prepared.

Abstract of the Specifications
for the
Construction of Denny Dam

SCOPE OF WORK: The principal features involved in the construction of Denny Dam are as follows:

1. Construction of a gravity dam with a crest length of 1,656 lin ft and a crest height of 233 ft above streambed.
2. Furnishing and installing two penstock gates and hoist, and embedding two penstocks 18 ft 6 in. in diameter by 143 ft long to serve a future powerhouse.
3. Furnishing and installing six spillway tainter gates, 40 ft wide by 38 ft high.

CONTRACTING AGENCY: Scott Water District.

BID DATE: 2 P.M. PST June 10, 197_ at the Scott Water District Office, Hilldale, California.

LOCATION: On the Denny River in Hill County, California. The damsite is located approximately 20 miles from Hilldale, California.

ACCESS: A paved double-lane highway from Hilldale passes close to the dam's left abutment. The closest rail connection is at Hilldale.

DATA TO BE SUBMITTED: Sealed bid in duplicate, with unit price schedule and acknowledgment of receipt of all addenda attached.
Bid bond.
Bid envelope must be sealed, marked, and addressed as follows: "Bid for Denny Dam to be opened at 2 P.M. PST June 10, 197_"

BIDDER'S QUALIFICATIONS: Before the bid is considered for award, the district may request the contractor to submit a statement setting forth a detailed account of previous experience in performing comparable work.

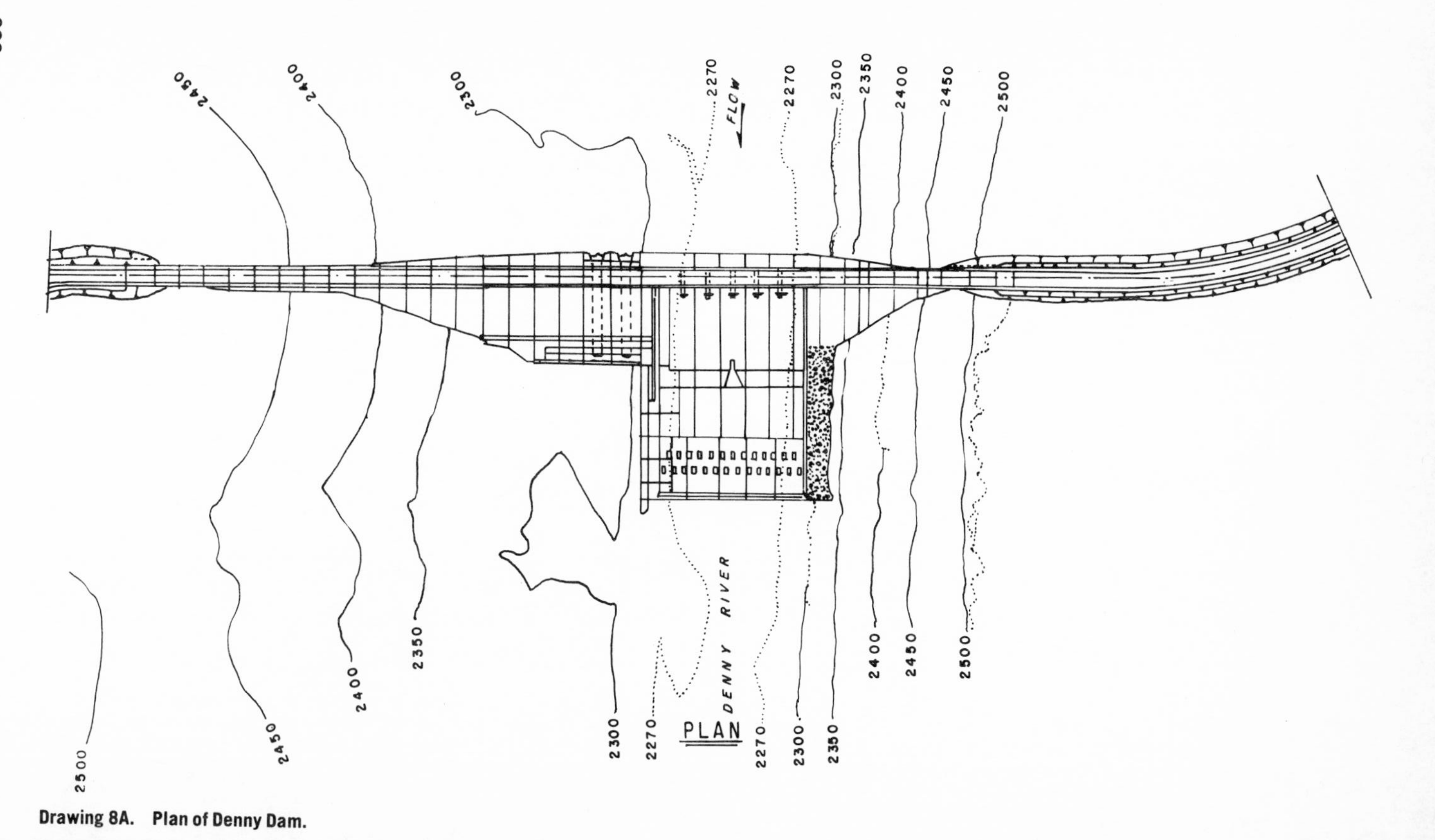

Drawing 8A. Plan of Denny Dam.

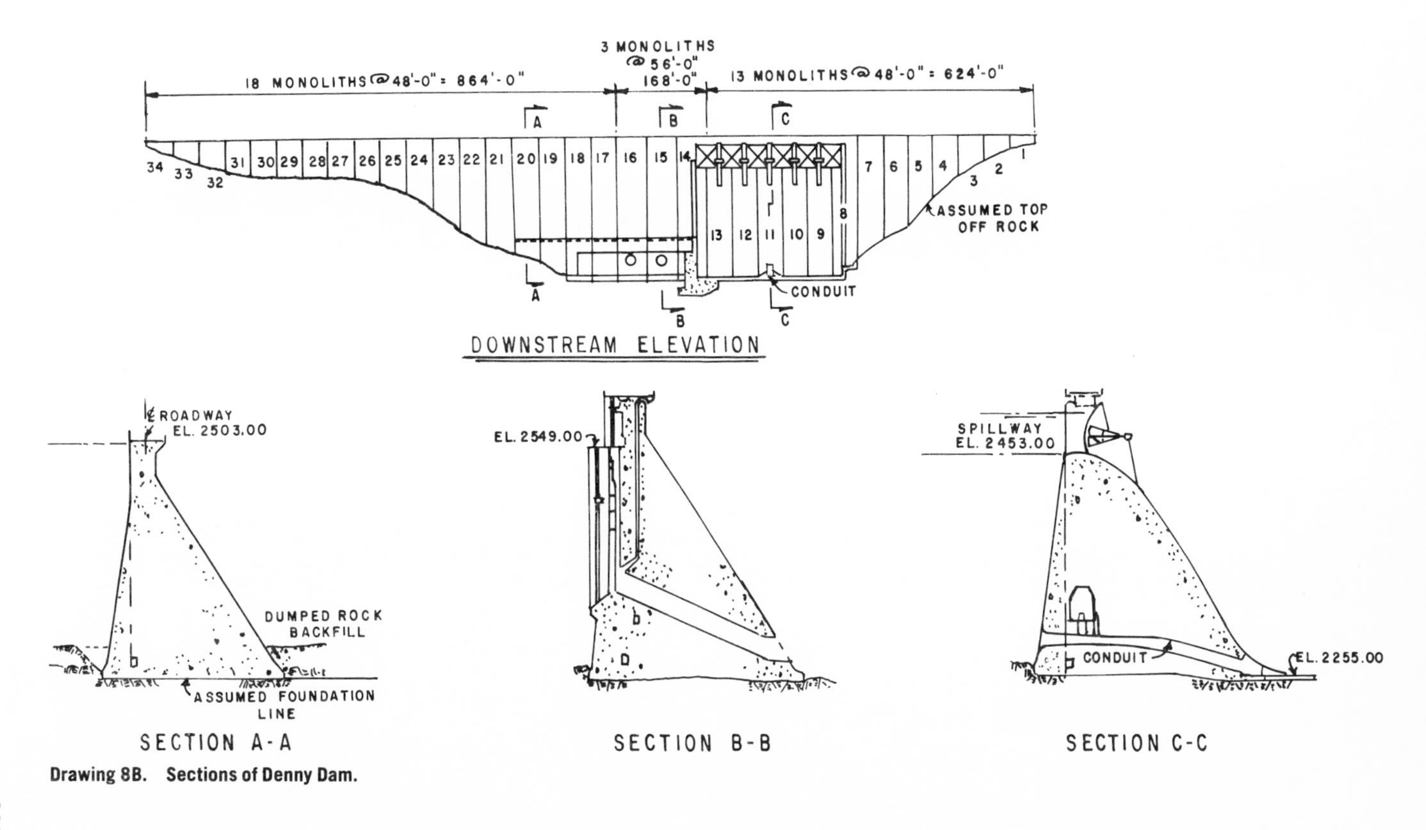

Drawing 8B. Sections of Denny Dam.

FORM OF CONTRACT: The contract form is included in the specifications.

AWARD: A written notice of acceptance of the bid will be forwarded to the successful bidder within 60 calendar days after date of bid opening.

BONDS: Bid bond—20 percent of bid price.

Payment bond—Amount shall be in accordance with the following schedule:

Amount of contract	*Payment bond*
Up to \$1 million	50%
\$1 million to \$5 million . . .	40%
Over \$5 million	\$2 1/2 million

Performance bond—50 percent of the contract amount.

TIME FOR COMPLETION: The contractor must commence work within 30 days after receipt of notice to proceed, and he will be allowed 1,250 calendar days to complete the project.

LIQUIDATED DAMAGES: The contractor shall pay the owner \$600 for each calendar day that construction time exceeds 1,250 days.

PAYMENT: As work progresses, *partial payments* will be made monthly. A *contract retention* of 10 percent of each progress payment will be retained by the owner until all work has been accepted. If work progress is satisfactory, after 50 percent of the work has been performed, the agency's engineer has authority to make the remaining payments in full.

Payments will be made for preparatory work and for the acquisition of plant and equipment. These payments will be limited to 75 percent of the contractor's cost. The amount of this prepayment will be deducted from each progress payment in accordance with a schedule presented by the contractor and meeting the approval of the district's engineer.

OWNER-FURNISHED MATERIAL: None.

RIVER DIVERSION: It is the contractor's responsibility to select the method of diversion and to divert and take care of the river. Minimum stream flow shall be maintained below the damsite. The 5 ft 8 in. by 10 ft permanent river outlet located in the spillway can be used by the contractor for diversion and for maintaining the minimum flow. The bottom of this permanent outlet is at an elevation of 2,280 ft. The approximate capacities of this outlet at different pool heights are as follows:

Pool elevation	*Capacity, cfs*
2,300	1,200
2,350	2,600
2,400	3,400
2,450	4,100
2,500	4,700

CONSTRUCTION CONSTRAINTS CONTAINED IN THE SPECIFICATIONS:

Concrete aggregates. Aggregates and sand must be manufactured from rock secured from a rock quarry located one mile from the damsite. The quarry is made available to the contractor free of charge.

Aggregate rescreening. This is a requirement. The plant must be located adjacent to the mix plant or on top of it.

Cement storage. Storage for 9,000 bbl of cement must be provided.

Allowable pour height. Contractor has the option of pouring mass concrete either in $7^1/_2$-ft lifts or in 5-ft lifts.

Concrete transportation. Concrete shall be transported in the same container from the mixing plant to the pour.

Concrete buckets. The maximum amount of concrete that can be dumped in a pour at one time is 4 cu yd.

Concrete placing temperature. Concrete shall be placed in the forms at temperatures above 40° and below 45° F.

Concrete cleanup. High-pressure air and water cutting will be acceptable.

Concrete pay lines. The concrete pay lines will be the excavated rock surface.

CLIMATOLOGICAL DATA:

Snowfall. Snowfall is limited to light falls that do not remain on the ground.

Precipitation, in inches:

Month	Maximum	Minimum	Mean
January	23.56	0.43	5.99
February	17.02	0.05	5.71
March	18.01	0.11	5.24
April	9.32	0.00	2.38
May	4.27	0.00	1.14
June	2.49	0.00	0.35
July	0.17	0.00	0.01
August	0.16	0.00	0.01
September	3.96	0.00	0.42
October	5.86	0.00	1.58
November	11.08	0.00	2.95
December	13.97	0.35	4.99
Annual	54.58	17.36	30.77

Temperature, in °F:

Month	Maximum	Minimum	Mean
January	57	35	46
February . . .	61	37	49
March	66	40	53
April	73	44	58
May	82	48	65
June	90	53	71
July	98	59	80
August	98	57	78
September .	92	53	72
October . . .	80	47	64
November . .	68	40	54
December . .	58	36	47

RIVER FLOW DATA:

Drawing 9 is of the hydrographs for the Denny River at the damsite. Recorded elevations of flows at the damsite are:

Flow, cfs	*Elevation*
Minimum flow	2,265
44,600	2,296
80,000	2,305

MAJOR QUANTITIES:

Item	Quantity
Care and diversion of river	Lump sum
Common excavation	185,000 cu yd
Rock excavation	170,000 cu yd
Preliminary rock cleanup	7,000 sq yd
Close-line drilling	30,000 sq ft
Final rock cleanup	23,000 sq yd
Drill grout, drain, and exploratory holes	55,500 lin ft
Pressure grouting	8,000 cu ft
Backfill and riprap	23,300 cu yd
Reinforcing steel	3,080,000 lb
Concrete	826,880 cu yd
Cement	63,500 bbl
Gates and guides	2,618,000 lb
Pumps, motors, and piping for gates	Lump sum
Cast-iron pipe and fittings	156,000 lb
Steel pipe and fittings	184,000 lb
Piping, water supply, and plumbing	Lump sum
Structural steel	476,500 lb
Railing and fittings	17,700 lb
Parapet railing	1,152 lin ft
Electrical system	Lump sum

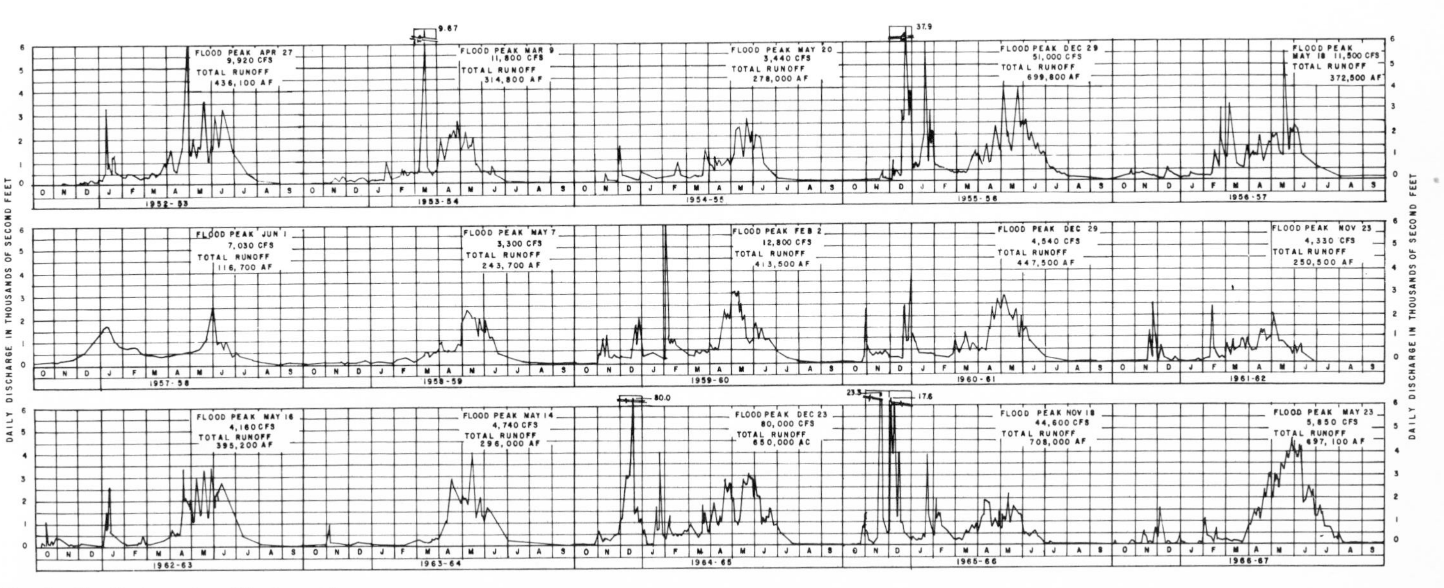

Drawing 9. Hydrographs of the Denny River.

III. PREPARATION OF A DOORKNOB ESTIMATE

A doorknob estimate will not be prepared for Denny Dam since its preparation is not difficult, and its presentation here would serve no useful purpose.

IV. PLUGGED PRICES FOR PERMANENT MATERIALS AND SUBCONTRACTS
for the Denny Dam Estimate

Bid item no.	Work description	Bid quantity	Material prices including sales tax		Subcontract prices		Painting subcontract	
			Unit	Amount	Unit	Amount	Unit	Amount
7	Drill $1\frac{1}{2}$-in.-diameter grout holes	25,000 lin ft	......	....	$ 4.00	$ 100,000	......	....
8	Drill 3-in.-diameter drain holes	17,000 lin ft	......	....	6.00	102,000	......	....
9	Core drill 3-in. exploratory holes	13,500 lin ft	......	....	6.00	81,000	......	....
10	Grout connections	800 each	......	....	10.00	8,000	......	....
11	Pressure testing	800 test	......	....	10.00	8,000	......	....
12	Grouting cement	8,000 cu ft	$ 1.40	$ 11,200	......	....	......	....
13	Pressure grouting	8,000 cu ft	......	....	4.00	32,000	......	....
18	Reinforcing steel	3,080,000 lb	......	....	0.16	492,800	......	....
25	Cement	635,000 bbl	5.00	3,175,000	......	....	......	....
26	12-in.-diameter RCCP	100 lin ft	2.50	250	......	....	......	....
27	18-in.-diameter RCCP	100 lin ft	5.00	500	......	....	......	....
28	Copper water stop	12,300 lin ft	3.00	36,900	......	....	......	....
29	6-in.-diameter porous concrete pipe	28,000 lin ft	1.25	35,000	......	....	......	....
30	12-in.-diameter half-round porous pipe	1,950 lin ft	2.00	3,900	......	....	......	....
31	Miscellaneous doors	400 sq ft	10.00	4,000	......	....	$0.70	$ 280
32	Miscellaneous metal castings	Lump sum	......	6,400	......	....	......	500
33	Slide and bulkhead gates	286,000 lb	0.50	143,000	......	....	0.01	2,860
34	Pumps, motors, piping for gates	Lump sum	......	25,000	......	....	......	250
35	Penstock gates, frames, and guides	288,000 lb	0.40	115,200	......	....	0.02	5,760
36	Penstock-gate hoist and supports	Lump sum	......	56,000	......	....	......	1,000
37	Penstock lining and frames	535,000 lb	......	....	0.40	214,000	......	....
38	Tainter gates and accessories	670,000 lb	0.25	167,500	......	....	0.02	13,400
39	Tainter-gate trunnion anchors	320,000 lb	0.30	96,000	......	....	0.001	320
40	Tainter-gate hoist and accessories	150,000 lb	1.40	210,000	......	....	0.02	3,000
41	Trashrack guides and beam	260,000 lb	0.22	57,200	......	....	0.04	10,400
42	Stop logs, protection plates, and beams	168,000 lb	0.25	42,000	......	....	0.02	3,360
43	Cast-iron pipe and fittings	156,000 lb	0.21	32,760	......	....	0.005	780
44	Steel pipe and fittings	184,000 lb	0.30	55,200	......	....	0.005	920
45	Piezometer piping and fittings	Lump sum	......	1,000	......	....	......	....
46	Sump pump and piping	Lump sum	......	6,000	......	....	......	300
47	Water supply and plumbing	Lump sum	......	12,000	......	....	......	350
48	Structural steel	66,500 lb	0.22	14,630	......	....	0.02	1,330
49	Railing and fittings	17,700 lb	0.40	7,080	......	....	0.02	354
50	Parapet railing	1,152 lin ft	12.00	13,824	......	....	......	....
51	Steel in spillway bridge	410,000 lb	0.17	69,700	......	....	0.02	8,200
52	Electrical system	Lump sum	......		......	250,000	......	
	Total	....	......	$4,397,244	......	$1,287,800	......	$53,364

V. FIELD INVESTIGATION

The field investigation was completed and is summarized as follows:

Access to the project is relatively easy since a highway passes close to the left abutment.

A single-status camp will not be required since the surrounding area will provide sufficient living accommodations. Living facilities for key married personnel can be located on a flat area adjacent to the project. Power lines pass this site and water can be secured from an existing spring. The owner of the area was contacted and he will rent the area for four years for the sum of $7,500.

Power lines pass near the project site. The power company will supply power to the project for $.01 per kwhr and will install a suitable substation for a $20,000 "up and down"* charge.

Telephone lines pass close to the damsite and the telephone company will install as many phones as required.

The rock at the damsite is a solid, medium-hard, granitic type overlain with disintegrated granite except at the streambed, where water-washed sand and gravel are encountered.

There is a level flat area on the left abutment which could be easily graded to accommodate a traveling head tower for a cableway.

The streambed is wide and the minimum water flow is so minor that low-flow diversion can be handled by small dikes and channels. The stream can be forded during the low-water season.

The quarry is composed of exposed, sound diorite not covered with vegetation or overburden and is located in a place where a quarry bench can be rapidly developed.

There is ample space on the left abutment for construction facilities, the project office, and employee parking.

Copies of all applicable labor agreements were secured. Tax rates were obtained from the county office.

*Charge for furnishing, erecting, and removing the substation.

VI. QUANTITY TAKEOFFS

Following is a summary of the quantity takeoffs for Item 24, mass concrete in the dam. This summary illustrates how quantity takeoffs should be prepared.

DENNY DAM ESTIMATE

Quantity Takeoff Summary—Bid Item 24, Mass Concrete

1. Concrete quantities:
   - Bid item quantity .... 775,000 cu yd
   - Takeoff quantity
     - Mass .... 748,150 cu yd
     - Blockout concrete .... 90 cu yd 748,240 cu yd
2. Number of lifts, pours, and size of pour:
   - Maximum number of $7^1/_2$-ft lifts 34
   - Total number of pours 693
   - Largest pour .... 3,033 cu yd
   - Smallest pour..................................... 23 cu yd
   - Average pour .... 1,080 cu yd
3. Square feet of form contact:

	Type of form			
	Built-in-place, sq ft	Lift starter, sq ft	Cantilever, sq ft	Total, sq ft
Upstream face ...	12,600	12,600	219,700	244,900
Downstream face	15,400	15,400	240,700	271,500
Bulkhead	32,000	32,000	378,900	442,900
Subtotal	60,000	60,000	839,300	959,300
Spillway splash wall				9,600
Galleries and other interior forms				83,550
Curved forms				3,250
Blockout concrete forms				3,780
Total				1,059,480
Form ratio based on takeoff yardage				1.416

4. Special forms:
   - 8-in.-diameter formed hole .... 4,410 lin ft
5. Surface finish:
   - Float .... 72,410 sq ft
   - Steel trowel .... 11,152 sq ft
6. Concrete cleanup .... 2,455,900 sq ft
7. Rock cleanup.. 23,450 sq ft

VII. LABOR COST

The following partial tabulation of labor cost for an 8-hour shift and a 5-day work week illustrates how a complete labor tabulation should be prepared.

DENNY DAM ESTIMATE
Partial Tabulation of Labor Cost

	Airtool operator	Bulldozer operator	Journeyman carpenter
Basic hourly wage	$ 5.185	$ 6.91	$ 7.025
Cost/shift:			
Basic wage for an 8-hr shift	$41.48	$55.28	$56.20
Vacation pay	4.00	3.20	4.00
Subtotal	$45.48	$58.48	$60.20
Insurance, 8.57% of basic wage	3.55	4.74	4.82
Payroll taxes, 8.90% of basic wage and vacation	4.05	5.20	5.36
Health and welfare	3.20	3.60	2.96
Pension	4.00	5.04	4.00
Apprenticeship fund	0.32	1.12	0.16
Subsistence	9.50	10.00	9.50
Total cost/8-hr shift	$70.10	$88.18	$87.00

The partial tabulation of labor cost was arranged so as to simplify its presentation in this book. When a complete tabulation of all labor classifications is prepared, the form should be changed, since its preparation is easier when the components of labor cost are listed across the page and the classifications of labor are listed down the page.

The labor cost for Denny Dam was computed using signed labor agreements for northern California, a social security tax rate of 4.8 percent, a state unemployment tax of 3.70 percent, a federal unemployment tax of 0.40 percent, a workmen's compensation insurance rate of 7.07 percent, and a public liability and property damage insurance rate of 1.50 percent.

An examination of this tabulation shows that the basic hourly wage rates are only 60 to 65 percent of the total labor cost.

VIII. EQUIPMENT OPERATING COST

The following partial tabulation of the shift cost for operating three types of construction equipment illustrates how a complete tabulation of all required equipment units should be prepared.

Cost per Eight-hr Shift

	D-8 bulldozer	$4^1/_2$-cu-yd shovel	35-ton rear-dump truck
Operating labor:			
Operator	$ 88.18	$ 92.60	
Driver	....	....	$ 81.60
Oilers	....	150.82	
Subtotal	$ 88.18	$243.42	$ 81.60
Service and maintenance labor	22.05	46.30	16.32
Total labor	$110.23	$289.72	$ 97.92
Supplies:			
Fuel, oil, grease	$ 10.78	$ 15.28	$ 10.18
Tire replacements and repair	....	....	24.90
Repair parts	33.00	60.00	25.00
Total supplies	$ 43.78	$ 75.28	$ 60.08
Total cost/8-hr shift	$154.00	$365.00	$158.00

The cost of fuel varies with the area because of local differences in its purchase price. Service and maintenance labor vary because contractors have had different cost experience. Tire replacement cost varies with anticipated tire life and with the credit given for the tires that are purchased with the truck.

IX. TRANSMITTAL OF ESTIMATING INSTRUCTIONS

It is assumed that this work will be bid by a fictitious joint venture, and that this is the sponsor's estimate. (Descriptions of joint ventures and of their formation are given in Chap. 12.) Therefore, estimating instructions will be mailed to all joint-venture partners.

A letter was mailed to all joint-venture partners informing them that a joint-venture meeting would be held at the main hotel in Hilldale, at 9:00 A.M. on June 7.

The following information was transmitted with the letter:

1. Estimating instructions—which detailed the information presented in Estimating Task I in Chap. 9.

2. Estimate comparison forms—as shown by Exhibit I in Chap. 12.

3. The plugged permanent material and subcontract prices—as prepared under Estimating Task IV in this chapter.

4. Wage-rate tabulations—as prepared for Estimating Task VII in this chapter.

5. Insurance costs—the premium for insuring the contractor's equipment is 1.8 percent of its valuation per year. The premium for insuring the construction buildings and structures is 1.2 percent of their valuation per year. Builder's risk insurance can be secured for a $62,000 total premium.

6. Taxes—the construction plant and equipment will have an assessed valuation of 25 percent of their value. The yearly tax will be 5 percent of the assessed valuation.

7. Bond rates—furnished in accordance with the tabulated rates included in the description of Estimating Task XV, Chap. 9.

8. A copy of the sponsor's concrete quantity takeoffs and a request that each partner circulate his.

9. A list of used equipment owned by the sponsor that is available for use by the joint venture.

10. A request that each partner furnish similar lists and prices of used construction equipment available for use by the joint venture.

Available Used Equipment

(The following equipment is owned by the sponsor and is available for use on the project for the listed prices, f.o.b. haul trucks at the sponsor's equipment yard.)

Equipment units	Weight, tons	Price
1 $4^1/_2$-cu-yd diesel shovel, including buckets	125	$100,000
10 35-ton rear-dump trucks, price and weight for each unit	29	32,000
1 3,000-gal water truck	...	6,000
1 Kenworth Lo-boy truck	...	15,000
1 lube truck	...	4,000
1 48- × 72-in. jaw crusher, motor, and starter	130	40,000
1 No. 12 Caterpillar grader	13	10,000
1 30-in. × 450-ft mixing-plant charging conveyor, including supports and motors	35	17,000
1 aggregate rescreening plant	80	40,000
1 mixing plant, containing 1,030 cu yd of aggregate storage and 3 4-cu-yd mixers; equipped for cold-air circulation through aggregate storage bins	240	120,000

Available Used Equipment (Cont.)

Equipment units	Weight, tons	Price
1 2,300-bbl cement-storage silo	28	4,300
1 cement screw conveyor	2	1,500
1 cement bucket elevator	9	4,800
1 100-ton per day ice machine, including screw conveyor for mixing-plant charging	5	10,000
550 tons of refrigeration equipment for producing cold air to circulate through the mix plant, and cold water to use in the mix and in cooling coils	120	150,000
2 25-ton diesel locomotives, price and weight for each unit	25	15,000
2 $12^1/_2$-ton hydraulic cranes, price and weight for each unit	17	12,000
1 lot of concrete forms for pouring 7.5-ft lifts	145	100,000
This lot includes:		
650 lin ft of upstream face forms		
650 lin ft of downstream face forms		
2,000 lin ft of bulkhead forms		
100 lin ft of upstream overhang forms		
100 lin ft of downstream overhang forms		
1,000 sq ft of wall forms		
1 ogee crest form		

X. SELECTION OF THE DIVERSION METHOD

The combination of a wide streambed at the damsite, minimum water flows during six months of each year, absence of restrictions on maximum heights between adjacent monoliths, and absence of restrictions on height differentials between the lowest and highest monoliths makes diversion through the dam the most economical method. The periods of minimum water flow, which control when succeeding diversion stages can be started, will restrict the speed of concrete placement.

An examination of the hydrographs and a review of the history of floods in northern California indicate that flood flows that exceed 13,000 cfs will only occur during the months of November or December; that the maximum recorded flood flow of 80,000 cfs occurred during the year that maximum flood flow was recorded for this area in northern California, which makes the possibility of having a flood flow exceeding 80,000 cfs extremely remote; and that the possibility of having a water flow of over 13,000 cfs in any calendar year is 1 chance in 8.

The recorded water elevation at the damsite for a flow of 80,000 cfs was elevation 2,305 ft. The cross section of the river channel at the damsite shows approximately 11,700 sq ft of channel area below this elevation.

A review of the preliminary construction schedule indicates that it will take approximately 11 months to obtain and erect the concrete plant and equipment. This establishes June 1 of the second calendar year as the earliest date that concrete placement can commence. Therefore, river diversion will be handled as shown by Drawings 10 and 11 and by the construction schedule prepared under Estimating Task XI.

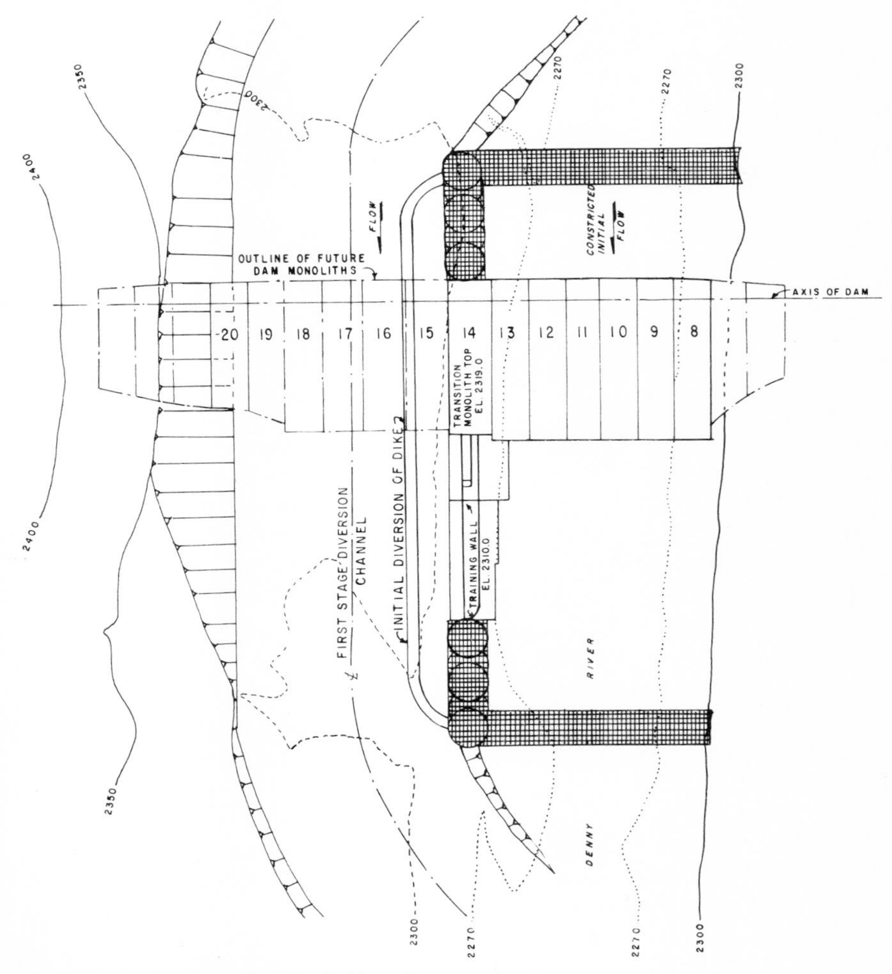

Drawing 10. First-stage diversion for Denny Dam.

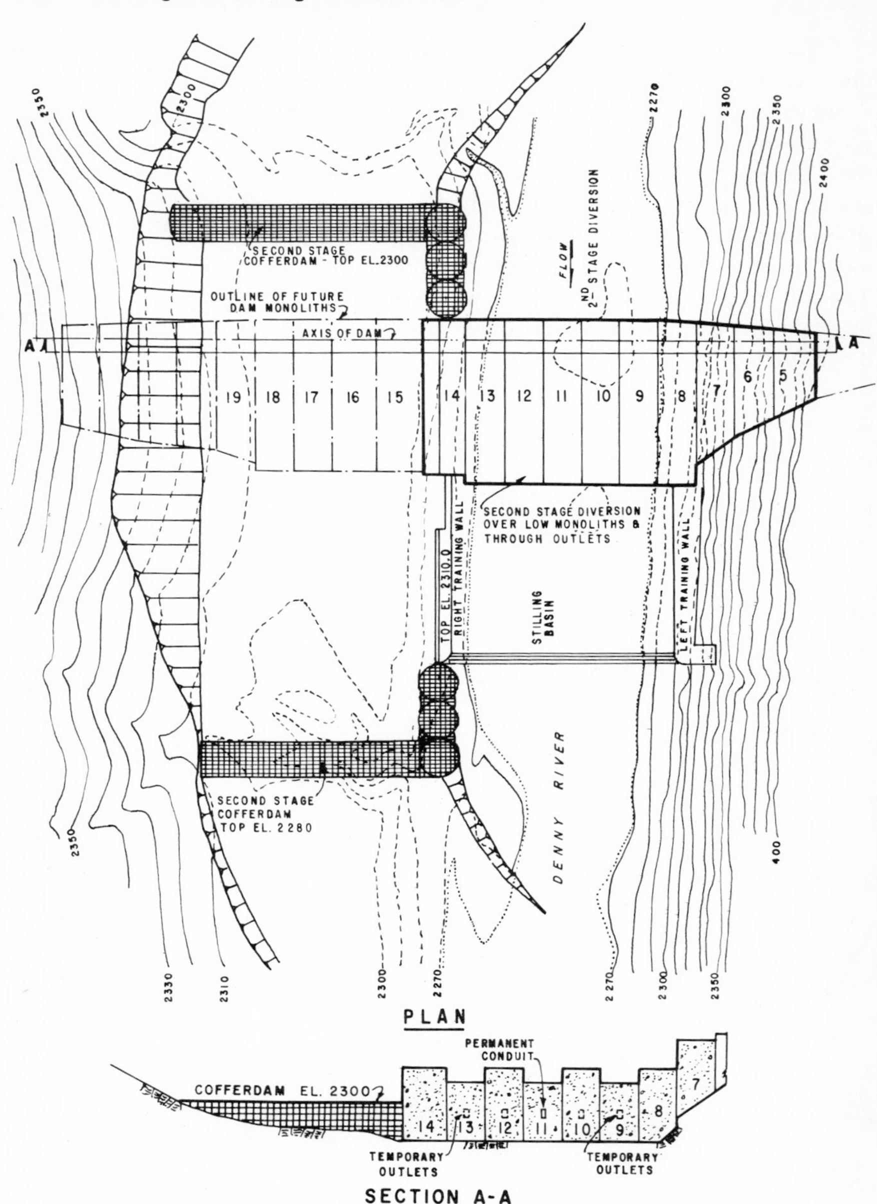

Drawing 11. Second-stage diversion for Denny Dam.

To prepare for river diversion in the fall of the first year, a diversion channel will be excavated on the right abutment passing through monoliths 15 to 19. This channel will be excavated to provide a minimum area of 10,000 sq ft below elevation 2,305 ft. Since this will be a slightly smaller area than that below this elevation in the original channel, when first-stage diversion is accomplished, the upstream elevation for a river flow of 80,000 cfs will exceed elevation 2,305 ft but should not exceed elevation 2,307 ft. Prior to construction of the diversion facilities, this plan will be checked by model studies.

The material excavated for the diversion channel from beneath future dam monoliths 15 to 19 will be paid excavation, and the cost of removing the remainder will be charged to diversion.

Diversion preparation includes the construction of three cofferdam cells upstream of monolith 14, to an elevation of 2,312 ft, and the construction of three cofferdam cells downstream of the right training wall, to an elevation of 2,310 ft.

Initial diversion of the low-water flows will be made in May of the second contract year. This initial diversion will consist of diverting the minimum flow of water through the diversion channel. To force the minimum flow into the channel, it is necessary to construct the upstream and downstream cofferdam to elevation 2,276 ft. To contain the flow in the channel, it is necessary to construct a 6-ft-high longitudinal dike along the edge of the channel connecting the upper and lower cofferdams.

During this low-flow diversion, excavation will be completed for dam monoliths 9 through 14, for the training wall, and for the spillway. Concrete placement in these structures will be started, and excavation and concrete placement will be expedited for monolith 14 and the right training wall. The concrete in the right training wall and monolith 14 will be raised to elevation 2,319 ft, prior to the start of the high-water flows, to take the place of the dirt dike as a longitudinal cofferdam.

Also, prior to November 1 of the second calendar year, the upstream cofferdam will be raised to elevation 2,312 ft, and the downstream cofferdam raised to elevation 2,310 ft. This will give a minimum of 5 ft of freeboard for a flood flow of 80,000 cfs.

During the high-water season, concrete placement will be completed in the left training wall and in the spillway apron and will continue in blocks 9 to 14 and in abutment blocks. When blocks 9, 11, and 13 reach elevation 2,326.5 ft, concrete placement in these blocks will be stopped. These blocks will remain at this elevation to pass the flood flows during the high-water season starting at the end of the third contract year. Four diversion conduits of the same size and with the same invert elevation as the permanent outlet will be formed in monoliths 9, 10, 12, and 13.

In May of the third year, the upstream and downstream cofferdams will be removed and the second-stage cofferdams will be placed upstream and downstream of blocks 15 to 19. The upstream cofferdam will be constructed to elevation 2,290 ft to pond the water so that low-water flows will pass through the diversion conduits and the permanent outlet. The downstream cofferdam will be constructed to elevation 2,270 ft to protect monoliths 15 to 19 from the tail water.

During low-flow diversion, excavation underneath monoliths 15 to 19 will be completed, and concrete placement will be expedited. By November 1 of the third year, the lowest monolith in this group will be poured to elevation 2,341.5 ft. Prior to this time, the three upstream cofferdam sheet-pile cells will be removed and the upstream cofferdam leveled. The three downstream cofferdam cells and the downstream cofferdam can be removed at any time prior to project completion.

From November 1 of the third year until all concrete is placed, low-flow diversion will be through the diversion conduits and through the permanent outlet. Monoliths 9, 11, and 13 will be kept 15 ft lower than other monoliths to handle flood flows until all floods can be handled by the combination of reservoir storage and conduit capacity. Diversion conduit and permanent-outlet capacity will be five times the capacity of the permanent outlet. This capacity for different dam elevations is:

Elevation of dam, ft	*Capacity, cfs*
2,300	6,000
2,350	13,000
2,400	17,000
2,450	20,500
2,500	23,500

Each of three low blocks will be able to handle flows of 24,500 cfs each, or the three can handle 73,500 cfs before the upstream water level reaches a high enough elevation to pass flood flows over other low blocks. These flows were computed using the broad-crested weir formula given in Chap. 2.

Finally, at the start of the low-water season in the fourth year, all the water can be carried by the permanent river outlet. Flow through the diversion conduits will be stopped by dropping concrete stop logs over their entrances and they will be filled with concrete. This will complete all the work required for river diversion.

XI. PREPARATION OF THE CONSTRUCTION SCHEDULE

The construction schedule prepared for Denny Dam is shown by a bar chart in Figure 7. A precedence diagram is not shown, since it is too large to be included in this book.

Work scheduling is controlled by the date of the receipt of the notice to proceed, the periods of minimum water flow, the date that all work must be completed, the time required for plant and equipment mobilization, the dates that each stage of diversion can be started, the time required for dam foundation excavation, and the dates that concrete placement can be started in the dam's center monoliths. Of these controls, the two crucial ones are the time required for plant and equipment mobilization and the dates that each stage of diversion can be started, since these control when the dam's foundation can be excavated and when concrete placement can start.

The excavation plant and equipment can be mobilized rapidly, since they consist basically of used equipment owned by the sponsor. This permits excavation to be started on September 1 of the first contract year. Excavation of the diversion channel can be started prior to completion of the bridge to the right abutment, since the river is fordable during the low-water season. The placement of concrete cannot start until June 1 of the second contract year, since receipt of the cableway will require nine months from the time of ordering, and two months will be required for its erection.

The diversion plan controls the sequence of placing concrete in the dam monoliths. This concrete-placing sequence was discussed in the previous section when the diversion method was selected.

Concrete placement during the first stage of diversion will be limited to the volume required to bring the low alternate monoliths of the spillway section up to elevation 2,326.5 ft and the volume of concrete that can be placed in the abutment monoliths. Concrete placement during second-stage diversion will be limited to the volume that can be placed in monoliths 15 to 19 and the concrete that can be placed in the abutment monoliths.

The concrete-placement equipment cannot be worked to full capacity when first- and second-stage diversion are being accomplished. However, the placement of concrete in the abutment monoliths during these two periods permits the concrete system to operate at a reasonable capacity and also reduces the amount of face forms that must be purchased. The total concrete placed in the dam by June 1 of the third year and by November 1 of the third year is shown on the pour diagram, Drawing 12.

After November 1 of the third year, concrete will be placed across the entire center section of the dam, leaving the alternate

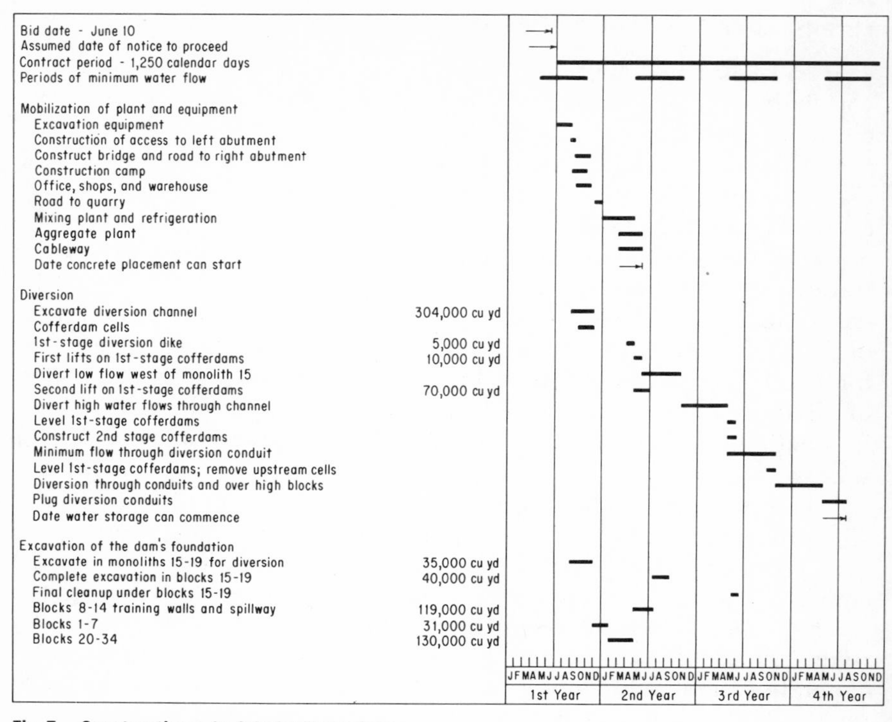

Fig. 7. Construction schedule for Denny Dam.

spillway monoliths lower than the remaining dam monoliths to provide for flood protection. The rate of concrete placement will be limited by the requirement that only one lift per week can be poured and not by the capacity of the placement equipment.

When concrete placement is started, the volume that can be placed per month will be low since only a few pouring areas will be available. As additional monoliths are started, production will increase. When the dam is being topped out, the placement rate will again decrease since all pours will be small. After the high monoliths have been topped out, it will take three weeks to top out the low monoliths. Mass-concrete placement will be completed by the

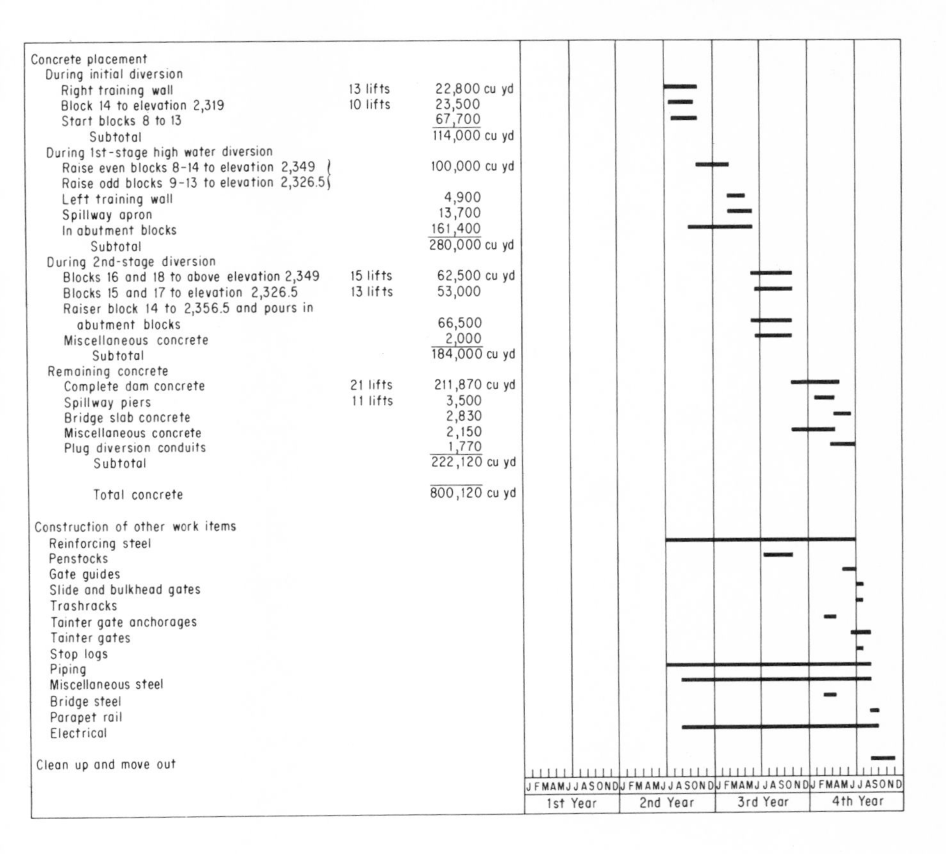

end of March of the fourth contract year, or 21 months after its start. Mass concrete will be placed during the night and swing shifts, 5 days a week, resulting in an average maximum hourly placement rate of 150 cu yd per hr, which is well within the capacity of an 8-cu-yd cableway. During the day shift, the cableway will be used for yarding. The main concrete-cleanup crew will be used on the day shift, which will make it available for completing large concrete pours whose placement extends into the day shift.

From April through July of the fourth year, the monthly concrete-placement rates are so minor that the concrete-production facilities will only be operated when required.

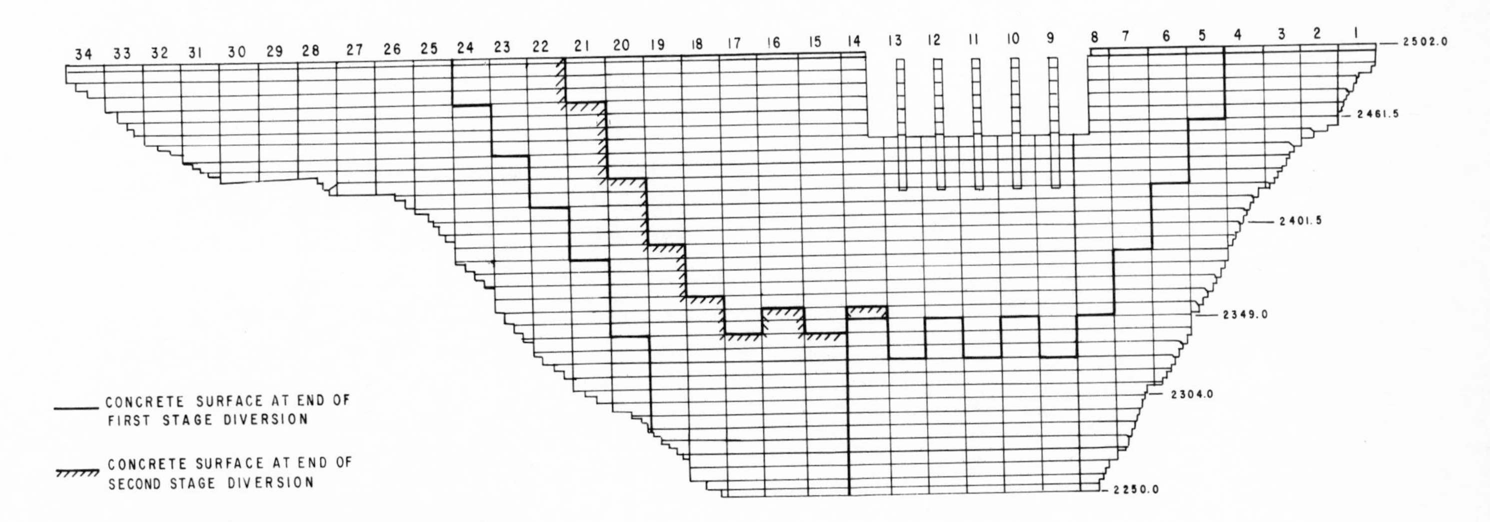

Drawing 12. Concrete pour diagram for Denny Dam.

The following is a monthly summary of the concrete-placing schedule:

Concrete-placement Schedule

	cu yd of concrete
Second contract year:	
July	11,000
August	23,000
September through December, 4 months at 40,000 cu yd/mo	160,000
Total for second year	194,000
Third contract year:	
January through May, 5 months at 40,000 cu yd/mo	200,000
June through October, 5 months at 36,800 cu yd/mo	184,000
November through December, 2 months at 44,000 cu yd/mo	88,000
Total for third year	472,000
Fourth contract year:	
January and February, 2 months at 44,000 cu yd/mo	88,000
March	25,000
April	14,120
May	3,000
June	3,000
July	1,000
Total for fourth year	134,120
Total concrete	800,120

The construction schedule and the pour diagram, Drawing 12, indicate that approximately 528 lin ft of both upstream and downstream face forms will be required for six 48-ft and three 56-ft dam monoliths. The maximum requirement for bulkhead forms will be 1,640 lin ft. These will be required when monolith 14 is at elevation 2,319 ft, requiring 320 lin ft of bulkhead forms. At this time monoliths 8, 10, and 12 will be at elevation 2,259 ft, requiring 1,100 lin ft of bulkhead forms, and an additional 130 lin ft of bulkhead forms will be required in the right training wall.

The scheduling of the construction of the remaining bid items is controlled by the concrete-placement schedule, since these items are either embedded in or supported by the concrete.

XII. SELECTION OF THE PLANT AND EQUIPMENT

The decision made in the performance of Estimating Task X, to divert the river through the dam, makes a cableway the preferable equipment to use for placing concrete in Denny Dam. Since the cableway towers are located on the dam's abutments, their erection is not dependent on completing the dam's foundation excavation. Also, a cableway will provide hook coverage of all the monoliths in the dam as soon as it is erected, and it will not interfere with the flow of the water in the natural streambed or diversion channel. Cableway coverage for the construction of Denny Dam can be secured using a radial cableway as shown by Drawing 1 in Chap. 4. A radial cableway is preferred to a parallel cableway, since its use eliminates one traveling tower and its runway, while giving adequate coverage.

In contrast to a cableway, if a trestle and cranes were used to place concrete in Denny Dam, the erection of the trestle would create scheduling problems. As shown by Drawing 2 in Chap. 4, a trestle cannot be erected across the streambed until the dam's foundation is excavated. Also, it cannot be completed until second-stage diversion is started. Its use would cause considerable delay in work scheduling and would not permit the early concrete placement in the dam monoliths located on the dam's abutment.

One 8-cu-yd cableway will have adequate placing capacity, as shown by the construction schedule. Using the rule of thumb for mixing-plant capacity (which is $1^1/_2$ times the cableway's bucket capacity), three 4-cu-yd mixers are required, or a total mixer capacity of 12 cu yd. The used mixing plant owned by the sponsor fits this requirement.

The head tower of the cableway, the concrete track, the mixing plant, and the aggregate plant will be located on the left abutment, since ample flat space is available. This location for the aggregate plant will also permit its storage piles to serve as surge piles for the mixing plant's charging conveyor.

The following calculations were made to check the capacity of the ice-making machinery and the refrigeration equipment that is securable from the sponsor. These calculations show they are adequate.

DENNY DAM ESTIMATE

Computations Made to Check the Selected Cooling Method and to Determine the Required Tonnage of Refrigeration Equipment

ASSUMED CONDITIONS

Temperature rise from mix plant to final placement will be 1°, so concrete should leave the mix plant at 39° F.

	°F
Mean dry-bulb temperature at time of pour	80
Mean wet-bulb temperature at time of pour	65
Temperature of river water	65
Temperature of cement	90
Aggregate piles will be equipped with sprinkler facilities so aggregates will be wet-bulb temperature of	65
Aggregate conveyors will be covered	

The following is an assumed mass-concrete mix, showing the percentage and weight of surface moisture and the specific heat.

	Lb of materials/ cu yd of concrete	Moisture, %	Lb of surface water	Temperature, °F	Specific heat
Cement	292	0.0	0.0	90	0.27
Large aggregate:					
6 by 3 in.	355	0.5	1.8	65	0.23
3 by 1½ in.	778	0.7	5.4	65	0.23
1½ by ¾ in.	848	1.0	8.5	65	0.23
Total large aggregate	1,981	...	15.7		
¾-by ¼-in. aggregate	810	1.5	12.2	65	0.23
Sand	740	4.0	29.6	65	0.19
Subtotal	3,823	...	57.5		
Water:					
Surface	57.5	...	....	..	1.00
Free	114.5	...	....	..	1.00
Total water	172				
Total mix	3,995				

The following heat balance table is based on the requirement of producing 39°F concrete from the mixing plant, which entails the cooling of large aggregates to 35°F and the use of 35°F water and ice in the mix.

	Lb/ cu yd of concrete	Temperature change, °F	Specific heat	Btu
Cement	292	90—39	0.27	+ 4,021
Large aggregate	1,981	35—39	0.23	− 1,823
¾-by ¼-in. aggregate	810	65—39	0.23	+ 4,844
Sand	740	65—39	0.19	+ 3,656
Water in coarse aggregate	15.7	35—39	1.00	− 63
Water in sand and fine aggregate	41.8	65—39	1.00	+ 1,087
Chilled water	31.0	35—39	1.00	− 124
Ice—change in water temperature	83.5	32—39	1.00	− 585
Ice—heat of fusion	....	−144	1.00	− 12,024
Mechanical heat of mixing	....	......	....	+ 1,000
Total	3,995	......	....	− 11

The above balance shows that concrete cooling can be accomplished using the assumed system.

Refrigeration Required for Each Cu Yd of Concrete

	Lb/ cu yd of concrete	Temperature change, °F	Specific heat	Btu
Cement	292	90—39	0.27	+ 4,021
Aggregates	2,791	65—39	0.23	+ 16,690
Sand	740	65—39	0.19	+ 3,656
Surface water	57.5	65—39	1.00	+ 1,495
Free water	114.5	65—39	1.00	+ 2,977
Mechanical heat of mixing	....	......	....	+ 1,000
Total	3,995	......	....	+ 29,839

Since one ton of refrigeration is 12,000 Btu per hr, it will require 2.5 tons of refrigeration to cool 1 cu yd of concrete per hour. With a maximum mixing-plant capacity of 240 cu yd per hr, 600 tons of refrigeration will be required.

The aggregate plant should be of a size adequate to produce as many tons of aggregates in one hour as the mixing plant consumes when it is mixing concrete at its maximum rate. This capacity is computed by multiplying the maximum production of the mix plant in cubic yards per hour by 1.85. Three 4-cu-yd mixers with a 3-minute cycle can produce 240 cu yd of mixed concrete an hour; this multiplied by 1.85 gives a maximum demand for aggregates of 444 tons per hr. This demand will be rounded to 500 tons per hr to establish aggregate plant capacity. The basic components selected for the aggregate plant are shown on Figure 5, in Chap. 5.

The quarry must be equipped to supply 500 tons per hr of broken rock to the aggregate plant. Since a bank cubic yard of diorite weighs approximately 5,000 lb, this establishes the required production in the quarry as 200 bank cu yd per hr.

In order to secure good rock breakage, it is estimated that the explosive requirements will be 1.25 lb per bank cu yd and that relatively small-diameter holes on a close drill pattern will be necessary. Five-in. diameter holes will be planned with a 10-ft burden and on a 12.5-ft drill-hole spacing. The bench height will be 30 ft, the subdrilling depth will be 3 ft, and the holes will be stemmed for 7 ft. The hole pattern requires that 0.24 lin ft of hole be drilled to break 1 bank cu yd of rock. Allowing for subdrilling and stemming, 0.19 lin ft of hole can be loaded with explosive for 1 bank cu yd of rock. The capacity of a 5-in.-diameter hole is 7.66 lb of pneumatically loaded ANFO. Therefore, the planned blast-hole diameter permits a maximum loading of 1.38 lb of ANFO per bank cubic yard of rock. For a shift production of 200 bank cu yd an hr, 48 lin ft of hole must be drilled. Since the estimated production of one drill is 30 lin ft per hr, two drills will be required. Air consumption per drill is 900 cu ft per min, so two 1,100-cu-ft-per-min, stationary, electrically driven compressors will be installed.

The $4^1/_2$-cu-yd shovel owned by the sponsor will be purchased for loading the rock. Since the aggregate plant will be located adjacent to the mixing plant, the quarried rock must be hauled one mile, requiring the use of four of the used 35-ton rear-dump trucks owned by the sponsor. To take care of equipment availability, five will be purchased. One new D-8 bulldozer will be purchased for cleanup around the shovel and for constructing the quarry road.

Since quarry operations will not start until the last of May of the second contract year, the quarry equipment can be used to excavate the dam's foundation. Additional excavation equipment will be provided by purchasing four more of the used 35-ton rear-dump trucks available from the sponsor. New excavation equipment purchases will be two light track-mounted percussion drills for drilling the foundation rock, two D-8 tractors, one 10-cu-yd articulated front-end loader, and a $^3/_4$-cu-yd backhoe.

Miscellaneous equipment to be purchased includes a 90-ton truck crane which will be used to construct the cofferdam cells, to erect the construction plant and equipment, and to have for general use. Approximately 2,200 cu ft of stationary air and 1,800 cu ft of portable air will be purchased for use during the excavation of the dam's foundation and during the concrete placement. Other equipment purchases will consist of a grader, a water wagon, service trucks, shops, a warehouse, an office, and parking facilities.

Quotations were requested on all required new equipment. A plant and equipment list was started, and this list will be completed and priced under Estimating Task XIV.

XIII. ESTIMATE OF DIRECT COST

The direct-cost section of a dam estimate is more voluminous than the remainder of the estimate since it consists of the detailing and pricing of all the work required to construct the project. To present in this book the direct cost estimating details for all the bid items for the construction of Denny Dam would be impractical. For the sake of brevity, while still providing an adequate example of the facets of the estimating procedure, only the direct-cost estimating details for Bid Item 24, mass concrete in the dam, are presented. These details will amply illustrate how direct-cost estimating is performed, since they contain representative details of all the major activities required for the construction of a concrete dam. Tabulated after these details is a summary of the estimated direct cost for all of the bid items.

Two methods were used to estimate the cost components that form the direct cost of mass concrete. The unit-price estimating method was used to estimate form cost, and the heavy-construction estimating method was used to estimate the cost of the remaining components. How these methods differ in theory and application is discussed in Chap. 1.

The presentation of the direct cost for mass concrete is illustrated by summarizing its cost components. Following this are the estimating details that produced the estimated units shown on the summary.

Direct-cost Summary of Bid Item 24
Mass Concrete in the Dam

	Cost/cu yd of concrete		
	Labor	Supplies	Total
Mixed concrete:			
Cost of aggregates in stockpiles	$1.48	$0.91	$ 2.39
Aggregate reclaiming, rescreening, and concrete mixing	0.86	0.25	1.11
Cement loss	.....	0.06	0.06
Admixture	.....	0.05	0.05
Aggregate and concrete cooling	0.28	0.16	0.44
Subtotal mixed concrete	$2.62	$1.43	$ 4.05
Place, cleanup, cure, finish:			
Concrete haul	$0.22	$0.04	$ 0.26
Cableway cost	0.52	0.32	0.84
Cableway dead time on day shift...	0.04	0.02	0.06
Placement crew	0.55	0.14	0.69
Cleanup and cure	0.90	0.08	0.98
Patch and finish	0.30	0.03	0.33
Utility service	0.11	0.05	0.16
Subtotal placement	$2.64	$0.68	$ 3.32
Form cost	2.50	0.84	3.34
Total direct cost	$7.76	$2.95	$10.71

Direct Cost of Stockpiled Aggregates

It is necessary to establish the number of shifts that the quarry and aggregate plant must be operated and the average production that will be achieved per shift before the direct cost of aggregates can be estimated. Since the aggregate plant will have 70,000 tons of aggregate storage in its finished stockpiles (enough aggregate for 37,840 cu yd of concrete), the quarry and aggregate plant can operate independently of the maximum hourly demand of the mixing plant. This amount of storage permits the quarry and aggregate plant to be operated to meet the scheduled maximum monthly rate of concrete placement. It also permits stopping the operation of the quarry and aggregate plant when the aggregate storage exceeds that required by the concrete remaining to be placed.

If the quarry and aggregate plants are worked one shift per day, 5 days per week, they will produce 4,000 tons of aggregate per shift, or 84,000 tons per month. This is sufficient aggregate for 45,400 cu yd of concrete. This monthly production slightly exceeds the scheduled concrete-placement rates, so that this shift schedule can be used to estimate aggregate cost. If for any reason actual concrete placement exceeds the amount scheduled, additional shifts can be worked. Aggregate cost can be estimated, based on suspending the operation of the quarry and aggregate plant on March 15 of the fourth year, when only 33,620 cu yd of concrete will be left to pour. The total number of shifts required is one shift per day from July 1 of the second year to March 15 of the fourth year, or 431 shifts. The concrete takeoff showed a total concrete yardage of 800,120. Using 1.85 tons of aggregate per cubic yard of concrete requires the production of 1,480,222 tons of aggregate. This tonnage, divided by 431 shifts, results in an average shift production of 3,435 tons, or 1,374 bank cu yd. With the average shift production figures and the total aggregate requirements, the direct cost of stockpiled aggregates is determined by estimating the direct cost of four operations.

1. Quarry development

	Labor	Supplies	Total
Allowance for quarry development . .	$15,000	$15,000	$30,000
Cost/ton (1,480,222 tons)	$0.01	$0.01	$0.02

2. Drill and shoot (3,435 tons/shift)

	Labor	Supplies	Total
Shift cost:			
2 drillers	$ 144.44	....	$ 144.44
2 chuck tenders	135.50	....	135.50
Drill parts and supplies	....	$ 32.00	32.00
1 compressor operator	83.86	....	83.86
Compressor parts, supplies, and maintenance	10.00	32.00	42.00
1 drill doctor and bit grinder	88.18	....	88.18
1 powder truck	95.00	16.00	111.00
1 powderman	72.22	....	72.22
Drill steel:			
Cost/set $552			
Cost/lin ft of hole $0.275			
Hole drilled/shift 330 lin ft			
Cost/shift	....	90.75	90.75
Bits:			
Cost/bit $210			
Cost/lin ft of hole $0.84			
Cost/shift	....	277.20	277.20
Powder and exploders:			
Cap and powder cost/hole $2			
No. of holes/shift 30			
Cost/shift	....	60.00	60.00
ANFO cost/lb $0.08			
Cost/cu yd $0.10			
Cost/shift	....	137.40	137.40
Small tools and supplies	....	40.00	40.00
Total/shift	$ 629.20	$685.35	$1,314.55
Cost/bank cu yd	$ 0.46	$ 0.50	$ 0.96
Cost/ton	$ 0.18	$ 0.20	$ 0.38

3. Load and haul to aggregate plant

Truck cycle:	
Load	3.5 min
Haul 1 mile at 15 mph	4.0 min
Dump in hopper	2.0 min
Return	4.0 min
Total	13.5 min
Trips per 50-min hr	3.7
Truck capacity/hr	129.5 tons
Truck capacity/shift	1,036.0 tons
Trucks required for 3,435-ton/shift	3.3 or 4

Cost/shift (3,435 tons/shift):	Labor	Supplies	Total
1 quarry foreman	$ 99.64	. . . .	$ 99.64
1 truck spotter	67.75	. . . .	67.75
1 $4^1/_2$-cu-yd shovel	289.72	$ 75.28	365.00
1 bulldozer	110.23	43.77	154.00
4 35-ton rear dumps	391.68	240.32	632.00
Road maintenance	49.00	30.00	79.00
Small tools and miscellaneous	. . . .	30.00	30.00
Total/shift	$1,008.02	$419.37	$1,427.39
Cost/ton of aggregates	$ 0.30	$ 0.12	$ 0.42

4. Operation of aggregate plant

Cost/shift (3,435 tons/shift):	Labor	Supplies	Total
1 foreman	$ 99.64	. . . .	$ 99.64
1 primary crusher operator . . .	84.70	. . . .	84.70
1 operator for other crushers .	84.70	. . . .	84.70
1 operator for screens and classifiers	84.70	. . . .	84.70
4 oilers	301.64	. . . .	301.64
2 laborers	135.50	. . . .	135.50
1 welder	88.18	. . . .	88.18
2 mechanics	176.36	. . . .	176.36
Power 700 kw per hr	. . . .	$ 44.00	44.00
Repair parts	. . . .	400.00	400.00
Oil, grease, and misc. supplies	. . . .	100.00	100.00
Total cost/shift	$1,055.42	$544.00	$1,599.42
Cost/ton	$ 0.31	$ 0.16	$ 0.47

Summary of the Direct Cost of Stockpiled Aggregates

	Labor	Supplies	Total
Quarry development	$0.01	$0.01	$0.02
Drill and shoot	0.18	0.20	0.38
Load and haul	0.30	0.12	0.42
Aggregate plant	0.31	0.16	0.47
Total cost/ton	$0.80	$0.49	$1.29
Cost/cu yd of concrete . . .	$1.48	$0.91	$2.39

Aggregate Reclaiming, Mix-plant Charging, Rescreening, Cement Handling, and Concrete Mixing

In accordance with the construction schedule, mass-concrete placement will be completed 21 months after it is started. Mixing and placing will be done two shifts per day, totaling 441 days or 882 shifts. During this period 779,000 cu yd of concrete will be poured. Thus the average production per shift will be 883 cu yd.

Cost/Shift

	Labor	Supplies	Total
1 foreman	$ 99.64	....	$ 99.64
1 belt operator	83.86	....	83.86
1 rescreen plant operator	83.86	....	83.86
1 mix-plant operator	84.70	....	84.70
1 oiler	75.41	....	75.41
1 concrete dispatcher	84.70	....	84.70
1 laborer	67.75	....	67.75
Maintenance labor	176.36	....	176.36
Power	....	$ 10.70	10.70
Repair parts and supplies	....	175.00	175.00
Miscellaneous supplies	....	40.00	40.00
Total cost/shift	$756.28	$225.70	$981.98
Total cost/cu yd	$ 0.86	$ 0.25	$ 1.11

Cement Loss and Admixture Cost

The cost of cement per cubic yard of concrete is $3.84. Cement loss will be 1½ percent of this, or .06 cu yd.

Admixture cost will be approximately $.05 per cu yd.

Aggregate and Concrete Cooling

Cost/Shift

	Labor	Supplies	Total
Operator of ice equipment	$ 77.56	....	$ 77.56
Compressor operator	83.86	....	83.86
Maintenance labor	88.18	....	88.18
Power and water	....	$ 70.00	70.00
Repair parts and supplies	....	50.00	50.00
Miscellaneous	....	20.00	20.00
Cost/shift	$249.60	$140.00	$389.60
Cost/cu yd	$ 0.28	$ 0.16	$ 0.44

Concrete Haul

Use 2 locomotive and car units

Cost/shift	$196.00	$32.00	$228.00
Cost/cu yd	$ 0.22	$ 0.04	$ 0.26

Cableway Operation

The cableway will handle approximately 800,000 cu yd of concrete, or 1,600,000 tons. The track cable should be good for the life of the project and will not require a replacement. However, approximately three travel-line replacements and approximately 14 hoisting-line replacements will be required. The cost per cubic yard of concrete for these replacements is:

Cable Replacements

	Labor	Supplies	Total
3 travel lines	$ 4,000	$ 16,000	$ 20,000
14 hoisting lines	8,000	108,000	116,000
Total	$ 12,000	$124,000	$136,000
Cost/cu yd	$ 0.015	$ 0.155	$ 0.17

Cableway operation—pouring concrete (883 cu yd/shift)

Cost/shift:			
1 operator	$ 92.60	....	$ 92.60
1 oiler	75.41	....	75.41
1 rigger	77.47	....	77.47
1 bellboy	77.47	....	77.47
Maintenance labor	120.00	....	120.00
Power	....	$ 45.00	45.00
Repair parts and supplies	....	100.00	100.00
Total/shift	$442.95	$145.00	$587.95
Cost/cu yd	$ 0.502	$ 0.164	$ 0.666
Cable replacements	0.015	0.155	0.17
Total/cu yd	$ 0.517	$ 0.319	$ 0.836
Use	$ 0.52	$ 0.32	$ 0.84

	Labor	Supplies	Total
Operation yarding, 1 shift/day, 442 shifts			
Cost/shift	$442.95	$145.00	$587.95
Cost distribution of day-shift operation:			
Reinforcing steel	$ 24,000	$ 6,000	$ 30,000
Other material items	26,000	9,000	35,000
Concrete cleanup and cure	75,000	25,000	100,000
Forms	37,500	12,500	50,000
Dead time charged to concrete	33,284	11,590	44,874
Total/442 shifts	$195,784	$ 64,090	$259,874
Cableway dead time cost/cu yd, based on 779,000 cu yd	$ 0.04	$ 0.02	$ 0.06
Placement crew (883 cu yd/shift)			
Cost/shift:			
1 foreman	$ 74.10	....	$ 74.10
5 vibrator men at $68.69	343.45	....	343.45
1 bucket dumper	68.69	....	68.69
Miscellaneous supplies	....	$ 38.00	38.00
1 carpenter (charged to forms)			
Total	$486.24	$ 38.00	$524.24
Cost/cu yd	$ 0.55	$ 0.04	$ 0.59
Furnishing and maintenance of vibrators	....	0.10	0.10
Total/cu yd	$ 0.55	$ 0.14	$ 0.69

Cleanup and Cure

This work will be estimated on a total time basis since crews are required on each of the three daily shifts and part of the cableway time on the day shift is charged to this operation. Total days are 442.

	Labor	Supplies	Total
Cost/day			
Day shift:			
1 foreman	$ 74.10	....	$ 74.10
4 air-tool operators	280.40	....	280.40
4 laborers	274.76	....	274.76
Swing shift:			
1 foreman	74.10	....	74.10
2 air-tool operators	140.20	....	140.20
2 laborers	137.38	....	137.38
Night shift:			
1 foreman	74.10	....	74.10
2 air-tool operators	140.20	....	140.20
2 laborers	137.38	....	137.38
Maintenance and weekend curing labor	90.00	....	90.00
Parts and supplies	....	$85.00	85.00
Total/day	$1,422.62	$85.00	$1,507.62
For 442 days	$ 628,798	$37,570	$ 666,368
Cableway charge	75,000	25,000	100,000
Total	$ 703,798	$62,570	$ 766,368
Cost/cu yd, based on 779,000 cu yd	$ 0.90	$ 0.08	$ 0.98

Patch and Finish

	Labor	Supplies	Total
Daily crew (1,776 cu yd/day):			
1 foreman	$ 80.62	......	$ 80.62
3 finishers (1/shift)	234.81	......	234.81
3 laborers (1/shift)	206.07	......	206.07
Supplies	....	$60.00	60.00
Cost/day	$521.50	$60.00	$581.50
Cost/cu yd	$ 0.30	$ 0.03	$ 0.33

Utility Service

	Labor	Supplies	Total
Daily cost	$200.00	$90.00	$290.00
Cost/cu yd	$ 0.11	$ 0.05	$ 0.16

Form Cost

As previously stated, most of the charges for forms are estimated by adjusting unit cost experienced on past jobs for changes in labor rates.

The cantilever forms required for the project were purchased from the sponsor and the following calculations were made to determine their write-off:

Purchased forms	$100,000
Freight and move-in	2,900
Reconditioning	20,000
	$122,900
Assumed net salvage value	25,000
Amount to be written off to project	$ 97,900
Write-off on:	
Mass concrete	748,240 cu yd
Spillway piers	3,500 cu yd
Training walls	27,700 cu yd
	779,440 cu yd
Write-off/cu yd	$ 0.125
Amount to be written off to mass concrete	$ 93,530

For the amount of forms for Bid Item 24, refer to the quantity takeoff summary in this chapter.

Cost/Sq Ft of Forms

	Labor	Supplies	Total
Purchased form write-off	....	$ 93,530	$ 93,530
Cableway charges	$ 37,500	12,500	50,000
8-in.-diameter formed hole, 4,410 lin ft at $3	8,820	4,410	13,230
Built-in-place forms, 60,000 sq ft at $6	240,000	120,000	360,000
Blockout forms, 3,780 sq ft at $6	15,120	7,560	22,680
Curved forms, 3,250 sq ft at $6	13,000	6,500	19,500
Gallery and other interior forms, 83,550 sq ft at $3.75	214,875	98,438	313,313
Shop-built lift-starter forms, 12,000 sq ft at $2.75	18,000	15,000	33,000
Set, strip, and raise panel forms, 908,900 sq ft at $1.75	1,317,905	272,670	1,590,575
Total form cost	$1,865,220	$630,608	$2,495,828
Cost/sq ft, based on 1,059,480 sq ft	$1.76	$0.60	$2.36
Cost/cu yd of mass concrete (748,240 cu yd)	$2.50	$0.84	$3.34
Man-hours/sq ft of formed contact, based on $10/hr labor cost	....	....	0.176

Summary of Direct Cost for the Denny Dam Estimate

Bid item no.	Description	Bid quantity	Labor Unit	Labor Amount	Supplies Unit	Supplies Amount	Permanent materials Unit	Permanent materials Amount	Subcontract Unit	Subcontract Amount	Total direct Unit	Total direct Amount
1	Care and diversion of river	Lump sum		$ 325,000		$ 265,900						$ 590,900
	Excavation:											
2	Common	185,000 cu yd	$ 0.87	$ 160,950	$ 0.56	$ 103,600					$ 1.43	$ 264,550
3	Rock	170,000 cu yd	2.01	341,700	1.12	190,400					3.13	532,100
4	Preliminary rock cleanup	7,000 sq yd	1.68	11,760	0.20	1,400					1.88	13,160
5	Close-line drilling	30,000 sq ft	0.46	13,800	0.46	13,800					0.92	27,600
6	Final rock cleanup	23,000 sq yd	3.36	77,280	0.40	9,200					3.76	86,480
	Total excavation			$ 605,490		$ 318,400						$ 923,890
	Drill and grout:											
7	$1\frac{1}{2}$-in.-diameter grout holes	25,000 lin ft							$ 4.00	$ 100,000	4.00	$ 100,000
8	3-in.-diameter drain holes	17,000 lin ft							6.00	102,000	6.00	102,000
9	3-in.-diameter cored exploratory holes	13,500 lin ft							6.00	81,000	6.00	81,000
10	Grout connections	800 each							10.00	8,000	10.00	8,000
11	Pressure testing	800 test							10.00	8,000	10.00	8,000
12	Grouting cement	8,000 cu ft	0.20	1,600			$ 1.40	$ 11,200			1.60	12,800
13	Pressure grouting	8,000 cu ft							4.00	32,000	4.00	32,000
	Total drill and grout			$ 1,600				$ 11,200		$ 331,000		$ 343,800
	Backfill:											
14	Uncompacted backfill	4,300 cu yd	2.11	$ 9,073	1.74	$ 7,482					3.85	$ 16,555
15	Dumped rock backfill	18,000 cu yd	1.83	32,940	1.26	22,680					3.09	55,620
16	Backfill material	560 cu yd	2.32	1,299	1.92	1,075					4.24	2,374
17	Dumped riprap	1,000 cu yd	2.21	2,210	1.33	1,330					3.54	3,540
	Total backfill			$ 45,522		$ 32,567						$ 78,089
18	Reinforcing steel	3,080,000 lb	0.008	$ 24,640	0.002	$ 6,160			0.16	$ 492,800	0.17	$ 523,600
	Concrete:											
19	Stilling basin	13,700 cu yd	10.18	$ 139,466	3.28	$ 44,936					13.46	$ 184,402
20	Roadways, sidewalks, bridge	2,830 cu yd	58.78	166,348	14.17	40,101					72.95	206,449
21	Slabs, beams, and columns	4,150 cu yd	29.90	124,085	8.32	34,528					38.22	158,613
22	Spillway piers	3,500 cu yd	33.52	117,320	9.74	34,090					43.26	151,410
23	Training walls	27,700 cu yd	11.41	316,057	3.18	88,086					14.59	404,143
24	Mass in dam	775,000 cu yd	7.76	6,014,000	2.95	2,286,250					10.71	8,300,250
	Total concrete	826,880 cu yd		$6,877,276		$2,527,991						$9,405,267

25	Cement	635,000 bbl	...	...	...	...	5.00	$3,175,000	...	...	5.00	$3,175,000
	Miscellaneous:											
26	12-in.-diameter RCCP	100 lin ft	5.10	$ 510	2.20	$ 220	2.50	$ 250	...	...	9.80	$ 980
27	18-in.-diameter RCCP	100 lin ft	7.60	760	2.40	240	5.00	500	...	...	15.00	1,500
28	Copper water stops	12,300 lin ft	2.50	30,750	0.40	4,920	3.00	36,900	...	...	5.90	72,570
29	6-in.-diameter porous concrete pipe	28,000 lin ft	2.62	73,360	0.44	12,320	1.25	35,000	...	...	4.31	120,680
30	12-in.-diameter half-round porous pipe	1,950 lin ft	3.42	6,669	0.54	1,053	2.00	3,900	...	...	5.96	11,622
31	Miscellaneous doors	400 sq ft	2.20	880	0.22	88	10.00	4,000	0.70	$ 280	13.12	5,248
	Total miscellaneous	...	...	$ 112,929	...	$ 18,841	...	$ 80,550	...	$ 280	...	$ 212,600
	Gates and guides:											
32	Miscellaneous metal castings	Lump sum	...	$ 8,000	...	$ 960	...	$ 6,400	...	$ 500	...	$ 15,860
33	Slide and bulkhead gates	286,000 lb	0.07	20,020	0.02	5,720	0.50	143,000	0.01	2,860	0.60	171,600
34	Pumps, motors, piping for gates	Lump sum	...	7,442	...	1,216	...	25,000	...	250	...	33,908
35	Penstock gates, frames, and guides	288,000 lb	0.17	48,960	0.03	8,640	0.40	115,200	0.02	5,760	0.62	178,560
36	Penstock-gate hoist and supports	Lump sum	...	20,070	...	1,021	...	56,000	...	1,000	...	78,091
37	Penstock lining and frames	535,000 lb	...	...	...	...	...	...	0.40	214,000	0.40	214,000
38	Tainter gates and accessories	670,000 lb	0.10	67,000	0.01	6,700	0.25	167,500	0.02	13,400	0.38	254,600
39	Tainter-gate trunnion anchors	320,000 lb	0.08	25,600	0.01	3,200	0.30	96,000	0.001	320	0.391	125,120
40	Tainter-gate hoist and accessories	150,000 lb	0.07	10,500	0.01	1,500	1.40	210,000	0.02	3,000	1.50	225,000
41	Trashrack guides and beam	260,000 lb	0.10	26,000	0.01	2,600	0.22	57,200	0.04	10,400	0.37	96,200
42	Stop logs, protection plates, and beams	168,000 lb	0.14	23,520	0.02	3,360	0.25	42,000	0.02	3,360	0.43	72,240
	Total gates and guides	...	...	$ 257,112	...	$ 34,917	...	$ 918,300	...	$ 254,850	...	$ 1,465,179
	Mechanical and miscellaneous metal:											
43	Cast-iron pipe and fittings	156,000 lb	0.18	$ 28,080	0.05	$ 7,800	0.21	$ 32,760	0.005	$ 780	0.445	$ 69,420
44	Steel pipe and fittings	184,000 lb	0.40	73,600	0.10	18,400	0.30	55,200	0.005	920	0.805	148,120
45	Piezometer piping and fittings	Lump sum	...	3,200	...	176	...	1,000	...	...	...	4,376
46	Sump pump and piping	Lump sum	...	1,200	...	98	...	6,000	...	300	...	7,598
47	Water supply and plumbing	Lump sum	...	11,000	...	2,000	...	12,000	...	350	...	25,350
48	Structural steel	66,500 lb	0.18	11,970	0.02	1,330	0.22	14,630	0.02	1,330	0.44	29,260
49	Railing and fittings	17,700 lb	0.60	10,620	0.07	1,239	0.40	7,080	0.02	354	1.09	19,293
50	Parapet railings	1,152 lin ft	2.20	2,534	0.50	576	12.00	13,824	...	...	14.70	16,934
51	Steel in spillway bridge	410,000 lb	0.08	32,800	0.02	8,200	0.17	69,700	0.02	8,200	0.29	118,900
	Total mechanical and miscellaneous metal	...	...	$ 175,004	...	$ 39,819	...	$ 212,194	...	$ 12,234	...	$ 439,251
52	Electrical system	Lump sum	...	...	...	...	...	...	...	$ 250,000	...	$ 250,000
	Total direct cost	...	...	$8,424,573	...	$3,244,595	...	$4,397,244	...	$1,341,164	...	$17,407,576

XIV. ESTIMATE OF PLANT AND EQUIPMENT COST for the Denny Dam Estimate

Summary of Cost

	Freight and move-in	Labor	Plant invoice	Used equipment	New equipment	Total cost	Net salvage	Net cost	Rental cost	Total
General:										
Access	......	$ 45,000	$ 55,000	....	....	$ 100,000	....	$ 100,000	.....	$ 100,000
Buildings and shops	$ 700	37,800	71,600	$ 3,500	$ 35,500	149,100	$ 19,250	129,850	.....	129,850
Service equipment and facilities	2,950	57,000	119,200	39,750	435,020	653,920	203,010	450,910	.....	450,910
Total general	$ 3,650	$139,800	$245,800	$ 43,250	$ 470,520	$ 903,020	$ 222,260	$ 680,760	.....	$ 680,760
Dewatering	$ 500	....	....	....	$ 25,000	$ 25,500	$ 10,000	$ 15,500	$5,980	$ 21,480
Excavation	$ 7,400	$ 1,000	$ 4,000	$134,400	$ 495,010	$ 641,810	$ 308,600	$ 333,210	.....	$ 333,210
Concrete:										
Quarry and haul to aggregate plant	$ 7,600	$ 24,050	$ 25,450	$276,000	$ 243,165	$ 576,265	$ 255,530	$ 320,735	.....	$ 320,735
Aggregate plant	18,590	118,750	165,445	44,000	631,119	977,904	371,100	606,804	.....	606,804
Mixing-plant charging	2,300	17,500	14,500	59,850	....	94,150	27,000	67,150	.....	67,150
Mixing plant	4,800	37,500	37,500	126,000	....	205,800	60,000	145,800	.....	145,800
Cement handling	780	11,000	12,500	11,130	....	35,410	5,800	29,610	.....	29,610
Ice manufacturer	250	8,500	18,500	10,500	....	37,750	5,000	32,750	.....	32,750
Cold air and cold water	3,600	65,500	77,500	157,500	....	304,100	75,000	229,100	.....	229,100
Concrete delivery	3,000	7,700	20,400	37,500	....	68,600	25,750	42,850	.....	42,850
Cableway	24,430	87,550	169,871	....	949,000	1,230,851	493,500	737,351	.....	737,351
Concrete placing	1,184	....	....	25,200	95,830	122,214	46,100	76,114	.....	76,114
Total concrete	$66,534	$378,050	$541,666	$747,680	$1,919,114	$3,653,044	$1,364,780	$2,288,264	.....	$2,288,264
Total plant and equipment	$78,084	$518,850	$791,466	$925,330	$2,909,644	$5,223,374	$1,905,640	$3,317,734	$5,980	$3,323,714

Details of the Plant and Equipment Cost for the Denny Dam Estimate

	Quantity	Weight/ unit	Cost/ unit	Freight and move-in	Labor	Plant invoice	Used equipment	New equipment	Total cost	Net salvage	Net cost
General:											
Access:											
Access road to left abutment	....	....	....	....	$ 2,500	$ 2,500	....	....	$ 5,000	....	$ 5,000
Road to right abutment	....	....	....	....	12,500	12,500	....	....	25,000	....	25,000
Bridge over river	....	....	....	....	30,000	40,000	....	....	70,000	....	70,000
Total access	....	....	....	....	$ 45,000	$ 55,000	....	....	$100,000	....	$100,000
Buildings and shops:											
Clearing, grading, paving	....	....	....	....	$ 7,500	$ 7,500	....	....	$ 15,000	....	$ 15,000
Fencing	....	....	....	....	3,000	7,000	....	....	10,000	....	10,000
Office trailers	3	....	$ 6,000	$ 600	....	....	....	$ 18,000	18,600	$ 9,000	9,600
Enclosure between trailers	....	....	....	....	1,000	2,000	....	....	3,000	....	3,000
First-aid trailer	1	....	3,500	100	50	100	$ 3,500	1,000	4,750	2,250	2,500
Warehouse and shelving	....	....	....	....	8,000	17,000	....	....	25,000	....	25,000
Shop	....	....	....	....	5,000	15,000	....	10,000	30,000	5,000	25,000
Tire shop	....	....	....	....	1,000	2,000	....	500	3,500	....	3,500
Pipe and rigging shop	....	....	....	....	2,000	4,000	....	2,000	8,000	1,000	7,000
Electrical shop	....	....	....	....	2,000	4,000	....	2,000	8,000	1,000	7,000
Carpenter shop	....	....	....	....	3,000	5,000	....	2,000	10,000	1,000	9,000
Form deck	....	....	....	....	750	1,500	....	....	2,250	....	2,250
Tool sheds and boxes	....	....	....	....	1,500	3,500	....	....	5,000	....	5,000
Building water and sanitation	....	....	....	....	3,000	3,000	....	....	6,000	....	6,000
Total shops and buildings	....	....	....	$ 700	$ 37,800	$ 71,600	$ 3,500	$ 35,500	$149,100	$ 19,250	$129,850

	Quantity	Weight/ unit	Cost/ unit	Freight and move-in	Labor	Plant invoice	Used equipment	New equipment	Total cost	Net salvage	Net cost
Service equipment and facilities:											
90-ton truck crane	1	31 tons		$ 930				$210,000	$210,930	$105,000	$105,930
$12^{1}/_{2}$-ton forklifts	2	16 tons	28,000	960				56,000	56,960	28,000	28,960
Grader	1	13 tons	10,500	360			$ 10,500		10,860	5,250	5,610
Water truck	1			100			6,300		6,400		6,400
Kenworth Lo-boy	1			200			15,750		15,950	7,500	8,450
Flatbeds	4							25,000	25,000	10,000	15,000
Pickups	15							36,000	36,000	11,250	24,750
Sedans	2							7,000	7,000	2,000	5,000
Ambulance							3,000		3,000	1,000	2,000
Lube truck	1			100			4,200		4,300	2,000	2,300
1,100 cu ft per min compressors	2	5 tons	22,930	300	$ 1,000	$ 1,000		45,860	48,160	22,930	25,230
Starters	2		2,080		500	500		4,160	5,160	2,080	3,080
Compressor housing					500	700			1,200		1,200
Electrical substations up and down						20,000			20,000		20,000
Power distribution and lighting					20,000	35,000		15,000	70,000		70,000
Air distribution					15,000	25,000			40,000		40,000
Water system					15,000	25,000		20,000	60,000	5,000	55,000
Welding machine and miscellaneous								4,000	4,000	1,000	3,000
Sanitation					2,000	2,000			4,000		4,000
Communication system					3,000	10,000		12,000	25,000		25,000
Total service equipment and facilities				$ 2,950	$ 57,000	$119,200	$ 39,750	$435,020	$653,920	$203,010	$450,910
Dewatering:											
Pumps				$ 500				$ 25,000	$ 25,500	$ 10,000	$ 15,500
Excavation:											
Cap and powder storage					$ 1,000	$ 4,000			5,000	1,000	4,000
Jackhammers	6		1,035					6,210	6,210		6,210
$4^{1}/_{2}$-in.-bore crawler percussion drills	2	4 tons	28,800	$ 240				57,600	57,840	28,800	29,040
900 cfm compressors	2	7 tons	38,800	420				77,600	78,020	38,800	39,220
10-cu-yd front-end loader	1	53 tons	137,800	1,590				137,800	139,390	68,900	70,490
35-ton rear-dump trucks	4	29 tons	33,600	2,120			$134,400		136,520	67,200	69,320
D-8 bulldozers	2	40 tons	84,000	2,400				168,000	170,400	84,000	86,400
$^{3}/_{4}$-cu-yd backhoe	1	21 tons	37,800	630				37,800	38,430	18,900	19,530
Farm tractor	1							5,000	5,000	1,000	4,000
Pavement breakers and miscellaneous								5,000	5,000		5,000
ANFO truck (furnished by others)											
Total excavation equipment				$ 7,400	$ 1,000	$ 4,000	$134,400	$495,010	$641,810	$308,600	$333,210

Concrete:											
Quarry and haul to aggregate plant:											
Jackhammers	3		1,035					$ 3,105	$ 3,105		$ 3,105
5½-in.-bore crawler percussion drills	2	10 tons	53,020	$ 600				106,040	106,640	$ 53,020	53,620
1,100-cfm stationary compressors	2	5 tons	22,930	300	$ 1,000	$ 1,000		45,860	48,160	22,930	25,230
Starters	2		2,080		500	500		4,160	5,160	2,080	3,080
Compressor housing					300	700			1,000		1,000
4½-cu-yd shovel	1	125 tons		2,500	750	250	$105,000		108,500	50,000	58,500
D-8 bulldozer	1	40 tons	84,000	1,200				84,000	85,200	42,000	43,200
35-ton rear dumps	5	29 tons	33,600	2,900			168,000		170,900	84,000	86,900
Tool and office trailer	1			100	500	1,000	3,000		4,600	1,500	3,100
Air and water piping					1,000	2,000			3,000		3,000
Haul-road construction					20,000	20,000			40,000		40,000
ANFO truck (furnished by others)											
Total quarry equipment				$ 7,600	$ 24,050	$ 25,450	$276,000	$243,165	$576,265	$255,530	$320,735
Cold air and cold water:											
Compressors, water coolers, cooling tower	120 tons			$ 3,600			$157,500		$161,100	$ 75,000	$ 86,100
Excavation					$ 1,500	$ 1,500			3,000		3,000
Concrete					5,000	5,000			10,000		10,000
Building					5,000	15,000			20,000		20,000
Insulation					4,000	6,000			10,000		10,000
Machinery installation					15,000	15,000			30,000		30,000
Piping and utilities					35,000	35,000			70,000		70,000
Total refrigeration plant				$ 3,600	$ 65,500	$ 77,500	$157,500		$304,100	$ 75,000	$229,100
Concrete delivery:											
Excavation and grading	800 lin ft				$ 2,500	$ 1,500			$ 4,000		$ 4,000
Ballast					1,200	1,200			2,400		2,400
Ties	1,000		2		500	2,000			2,500		2,500
Rail	100 tons		147	$ 2,000	2,000	14,700			18,700	$ 7,000	11,700
25-ton diesel locomotives	2	25 tons	15,750	1,000			$ 31,500		32,500	15,750	16,750
Delivery cars	2				1,500	1,000	6,000		8,500	3,000	5,500
Total concrete delivery				$ 3,000	$ 7,700	$ 20,400	$ 37,500		$ 68,600	$ 25,750	$ 42,850

	Quantity	Weight unit	Cost unit	Freight and move-in	Labor	Plant invoice	Used equipment	New equipment	Total cost	Net salvage	Net cost
Mixing-plant charging and rescreening:											
30 in. by 450 lin ft mixing-plant charging conveyor	....	35 tons	....	$ 700	$ 3,000	$ 3,000	$ 17,850	....	$ 24,550	$ 7,000	$ 17,550
Aggregate rescreening	....	80 tons	....	1,600	12,000	4,000	42,000	....	59,600	20,000	39,600
Allowance for used equipment repair	....	....	....	....	2,500	7,500	....	....	10,000	....	10,000
Total mixing-plant charging and rescreening	....	....	....	$ 2,300	$ 17,500	$ 14,500	$ 59,850	....	$ 94,150	$ 27,000	$ 67,150
Mixing plant:											
Plant containing 1,030-cu-yd aggregate storage, 3 4-cu-yd mixers, and outfitted for cold-air circulation through bins	....	240 tons	....	$ 4,800	$ 35,000	$ 30,000	$126,000	....	$195,800	$ 60,000	$135,800
Allowance for used equipment repair	....	....	....	....	2,500	7,500	....	....	10,000	....	10,000
Total mixing plant	....	....	....	$ 4,800	$ 37,500	$ 37,500	$126,000	....	$205,800	$ 60,000	$145,800
Cement handling and storage:											
2,300-bbl storage silo	....	28 tons	....	$ 560	....	....	$ 4,515	....	$ 5,075	$ 2,650	$ 2,425
Screw conveyor	....	2 tons	....	40	....	....	1,575	....	1,615	750	865
Bucket elevator	....	9 tons	....	180	....	....	5,040	....	5,220	2,400	2,820
Installation	....	....	....	....	$ 10,000	$ 11,000	....	....	21,000	....	21,000
Allowance for used equipment repair	....	....	....	....	1,000	1,500	....	....	2,500	....	2,500
Total cement handling	....	....	....	$ 780	$ 11,000	$ 12,500	$ 11,130	....	$ 35,410	$ 5,800	$ 29,610
Ice manufacture and storage:											
Ice manufacture and mix-plant charging equipment	....	....	....	$ 250	$ 1,000	$ 1,000	$ 10,500	....	$ 12,750	$ 5,000	$ 7,750
Storage building and reclaim	....	....	....	....	7,500	17,500	....	....	25,000	....	25,000
Total ice making	....	....	....	$ 250	$ 8,500	$ 18,500	$ 10,500	....	$ 37,750	$ 5,000	$ 32,750
Aggregate plant:											
Excavation and grading	....	....	....	....	$ 12,000	$ 10,000	....	....	$ 22,000	....	$ 22,000
Concrete	....	....	....	....	15,000	23,000	....	....	38,000	....	38,000
Structural steel supports for primary feeder, crusher, screening towers, sand tanks, and classifiers	....	55 tons	....	$ 1,650	4,500	45,695	....	....	51,845	$ 22,000	29,845
All chutes	....	20 tons	....	600	2,000	500	....	$ 20,139	23,239	10,000	13,239
Feeder for primary crusher, motor, and starter	1	13 tons	....	390	2,000	750	....	21,205	24,345	10,600	13,745
Primary crusher motor and starter	1	130 tons	....	2,600	4,000	1,000	$ 40,000	....	47,600	30,000	17,600
Gyrasphere crushers, motors, and starters	3	97 tons	....	2,910	19,400	5,000	....	193,452	220,762	97,000	123,762
Double-deck screens, motors, and starters	5	37 tons	....	1,110	7,200	2,000	....	71,515	81,825	36,000	45,825

Splitting bins	1	28 tons		840	2,200	750		21,365	25,155	10,700	14,455
Classifying tank	1	5 tons		150	850	250		8,270	9,520	4,100	5,420
Screw classifier	1	8 tons		240	1,500	400		15,015	17,155	7,500	9,655
102-in.-diameter reclaim tunnel	55 lin ft	8 tons		240	400	100		3,770	4,510	1,900	2,610
96-in.-diameter reclaim tunnel	643 lin ft	94 tons		2,820	4,200	1,500		41,884	50,404	21,000	29,404
Conveyors, motors, starters, and supports	15	155 tons		4,650	20,300	5,000		202,991	232,941	101,500	131,441
Vibrating feeders	20	13 tons		390	3,200	1,000		31,513	36,103	16,800	19,303
Air distribution					1,000	2,000			3,000		3,000
Water distribution					4,000	8,000			12,000		12,000
Electrical system					15,000	20,000			35,000		35,000
Engineering						38,500			38,500		38,500
Office trailer							4,000		4,000	2,000	2,000
Total aggregate-plant equipment				$18,590	$118,750	$165,445	$ 44,000	$631,119	$977,904	$371,100	$606,804
Cableway:											
Facility purchase:											
8-cu-yd cableway	1	495 tons		$14,850				$949,000	$ 963,850	$474,500	$489,350
Track and anchor cables		44 tons		1,760		$ 46,400			48,160		48,160
Operating cables						19,200			19,200		19,200
132-lb rail		254 tons	$149.10	7,620		37,871			45,491	19,000	26,491
Ties and rail hardware				200		5,000			5,200		5,200
Subtotal				$24,430		$108,471		$949,000	$1,081,901	$493,500	$588,401
Installation:											
Excavation and grading					$ 12,000	6,000			18,000		18,000
Install ballast, ties, and rail					5,000	2,000			7,000		7,000
Foundation concrete					2,500	7,500			10,000		10,000
Counterweight concrete	225 cu yd				3,000	8,250			11,250		11,250
Towers and running gear	315 tons		100		22,050	9,450			31,500		31,500
Machinery	180 tons		90		9,000	7,200			16,200		16,200
Cables					15,000	3,000			18,000		18,000
Electrical					15,000	10,000			25,000		25,000
Air and water					1,000	1,000			2,000		2,000
Enclosures					3,000	7,000			10,000		10,000
Subtotal					$ 87,550	$ 61,400			$148,950		$148,950
Total cableway				$24,430	$ 87,550	$169,871		$949,000	$1,230,851	$493,500	$737,351
Concrete placement:											
8-cu-yd concrete buckets each containing 2 4-cu-yd compartments	8	2 tons	7,500	$ 480				$ 60,000	$ 60,480	$ 20,000	$ 40,480
4-cu-yd concrete buckets	1	1/2 ton	5,130	15				5,130	5,145	1,000	4,145
2-cu-yd concrete buckets	2	300 lb	1,100	9				2,200	2,209		2,209
Air and water cutting								25,000	25,000	12,500	12,500
Finishers								3,500	3,500		3,500
12 1/2-ton hydraulic cranes	2	17 tons	12,600	680			$ 25,200		25,880	12,600	13,280
Total concrete placement				$ 1,184			$ 25,200	$ 95,830	$122,214	$ 46,100	$ 76,114

Details of the Plant and Equipment Cost for the Denny Dam Estimate

Rented Equipment	Quantity	Rental duration	Monthly rental	Move in and out	Rent	Total
Diesel pile hammer	1	2 months	$1,820	$100	$3,640	$3,740
Diesel pile extractor	1	2 months	1,070	100	2,140	2,240
Total outside rental	. . . .	. . . .	. . . . . .	$200	$5,780	$5,980

XV. ESTIMATE OF INDIRECT COST for the Denny Dam Estimate

Summary of Cost

Labor:	
Job management	$ 129,000
Job supervision	450,750
Engineering	618,450
Office	320,050
Purchasing and warehouse	267,940
Safety and first aid	60,000
Janitorial and security	106,400
Subtotal	$1,952,590
Burden for nonunion personnel	168,030
Total labor	$2,120,620
Other cost	250,500
Insurance premium	274,000
Taxes	110,000
Bond premium	142,000
Total indirect cost	$2,897,120

Details of the Indirect Cost for the Denny Dam Estimate

	No. of employees	No. of man months	Monthly salary	Total cost
Labor:				
Job management:				
Project manager	1	43	$2,400	$ 103,200
Secretary	1	43	600	25,800
Total	2	..	......	$ 129,000
Job supervision:				
General superintendent	1	41	1,750	$ 71,750
Swing-shift superintendent	1	22	1,500	33,000
Graveyard-shift superintendent	1	21	1,500	31,500
Excavation superintendent	1	15	1,500	22,500
Carpenter superintendent	1	40	1,500	60,000
Mechanical superintendent	1	40	1,700	68,000
Electrical superintendent	1	40	1,500	60,000
Rigging superintendent	1	40	1,500	60,000
Secretaries and clerks	2	80	550	44,000
Total	10	..	......	$ 450,750
Engineering:				
Project engineer	1	43	1,500	$ 64,500
Office engineer	1	40	1,250	50,000
Cost and progress engineer	1	40	1,000	40,000
Design engineers	2	24	1,250	30,000
Draftsmen	2	52	900	46,800
Concrete engineers	3	67	900	60,300
Chief surveyor	1	38	1,400	53,200
Surveying crews	2	50	5,000	250,000
Secretary	1	43	550	23,650
Total	14	..	......	$ 618,450
Office:				
Accountant	1	43	1,400	$ 60,200
Cost accountant	1	40	1,200	48,000
Personnel manager	1	40	1,400	56,000
Key puncher	1	40	500	20,000
Timekeeper and clerks	3	82	600	49,200
Secretaries	2	83	550	45,650
PBX operators	3	82	500	41,000
Total	12	..	......	$ 320,050
Purchasing and warehouse:				
Purchasing agent	1	40	1,250	$ 50,000
Warehousemen	3	83	1,580	131,140
Secretary	1	40	550	22,000
Truck driver	1	40	1,620	64,800
Total	6	..	......	$ 267,940
Safety and first aid:				
Safety engineer	1	40	900	$ 36,000
Nurse	1	40	600	24,000
Total	2	..	......	$ 60,000
Janitorial and security:				
Watchmen and janitors	2	80	1,330	106,400
Total labor	48	..	......	$1,952,590
Burden on nonunion labor* 12%	..	..	......	168,030
Total labor and burden	..	..	......	$2,120,620

Details of the Indirect Cost for the Denny Dam Estimate (Cont.)

	Months	Cost/ month	Total cost
Other cost:			
Operation of pickups and sedans, 520 vehicle months	..	$ 100	$ 52,000
Office supplies	40	500	20,000
Computer cost	40	700	28,000
Telephone, telegraph, and cable	40	500	20,000
Office rent	..	....	....
Heat, light, and power for office	40	100	4,000
Engineering supplies	40	300	12,000
Office equipment	..	....	15,000
Engineering equipment	..	....	5,000
First aid and safety supplies	..	....	6,000
Entertainment and travel	40	200	8,000
Blueprints, photostats, and photographs	40	200	8,000
Outside consultants	..	....	4,000
Legal cost	..	....	4,000
Audit cost	..	....	3,000
Testing laboratory expense	..	....	1,000
Job fencing	..	....	7,000
Mobilization of personnel	..	....	20,000
Licenses and fees	..	....	2,000
Sign cost	..	....	1,500
Drinking water	40	250	10,000
Miscellaneous cost—supplies and maintenance for warehouse truck	40	500	20,000
Total other cost	..	....	$250,500

Details of the Indirect Cost for the Denny Dam Estimate (Cont.)

			Total cost
Insurance premiums:			
Acquisition cost of construction equipment	$ 3,834,974	....	....
Yearly insurance premium at 1.8% per year	....	$ 69,030	....
Value of buildings	120,000	....	....
Yearly insurance premium at 1.2% per year	....	1,440	....
Total yearly premiums	....	$ 70,470	....
For three years	....	....	$211,410
Builder's risk insurance	....	....	62,000
Total insurance	....	....	$273,410
Use	....	....	$274,000
Taxes:			
Acquisition cost of equipment	$ 3,834,974	....	....
Buildings	120,000	....	....
	$ 3,954,974		
Value at completion of project	1,905,614	....	....
	$ 5,860,614	....	....
Average value	$ 2,930,307	....	....
Assessed valuation—25%	$ 732,580	....	....
Yearly tax—5%	....	$ 36,630	....
For three years	....	$109,890	....
Use	....	....	$110,000

Bond premiums:+	Contract value	Rate/ $1,000	Total
First	$ 100,000	$7.50	$ 750
Next	2,400,000	5.25	12,600
Next	2,500,000	4.50	11,250
Remainder	24,000,000	4.00	96,000
Total	$29,000,000	....	$120,600
Time premium, 18%	....	....	21,708
			$142,308
Use	....	....	$142,000

*Payroll burden for union labor has already been added to wage rate.

+Bond premiums are based on nontowner rates and on an assumed bid total of $29,000,000.

XVI. ESTIMATE OF CAMP COST

A single-status camp will not be required. In order to have key supervisory personnel live adjacent to the project, a married-quarters camp will be built. This camp will consist of 16 three-bedroom house trailers and a separate laundry building.

Construction Cost

	Labor	Invoice	Total
Grading and surfacing roads	$ 2,500	$ 2,500	$ 5,000
Water supply	2,000	2,000	4,000
Sewage disposal	4,000	6,000	10,000
Electrical distribution	2,500	5,500	8,000
16 3-bedroom, 60-ft long house trailers, at $8,500	....	136,000	136,000
Support, close in bottoms, build entrances and porches	8,000	8,000	16,000
Laundry building	4,000	4,000	8,000
Washers and dryers	800	4,000	4,800
Total construction cost	$23,800	$168,000	$191,800
Rent property	....	7,500	7,500
	$23,800	$175,500	$199,300
Salvage on trailers	....	21,760	21,760
Net camp cost	$23,800	$153,740	$177,540

Camp Operation

Rent per unit $125	
Rent for 16 units $2,000	
34 months' rental	$68,000
Maintenance, $500 month ...	17,000

XVII. ESCALATION COMPUTATIONS
for the Denny Dam Estimate

	Total labor	Labor expenditures by years			
		June 1 of first year to May 31 of second year	June 1 of second year to May 31 of third year	June 1 of third year to May 31 of fourth year	June 1 of fourth year to May 31 of fifth year
Direct labor:					
Diversion and care of river	$ 325,000	$ 205,000	$ 50,000	$ 45,000	$ 25,000
Excavation and backfill	651,012	315,000	275,000	45,000	16,012
Concrete	6,877,276	....	3,325,000	3,365,000	187,276
Remainder	571,285	70,000	185,000	200,000	116,285
Total direct labor	$ 8,424,573	$ 590,000	$3,835,000	$3,655,000	$344,573
Plant and equipment labor	518,850	518,850	....	....	....
Indirect labor	2,120,620	465,000	740,000	630,000	285,620
Camp construction labor	23,800	23,800	....	....	....
Camp operation and maintenance labor	8,500	1,750	3,000	3,000	750
Labor cost before escalation	$11,096,343	$1,599,400	$4,578,000	$4,288,000	$630,943
Estimated labor cost increase	....	....	10%	20%	30%
Labor escalation	1,504,700	....	457,800	857,600	189,300
Total labor cost	$12,601,043	$1,599,400	$5,035,800	$5,145,600	$820,243

XVIII. ESTIMATE SUMMARY
for the Denny Dam Estimate

	Labor cost	Total cost	
Direct cost:			
Labor	$ 8,424,573	$8,424,573	
Supplies	....	3,244,595	
Permanent materials	....	4,397,244	
Subcontracts	....	1,341,164	
Total direct cost	....	....	$17,407,576
Plant and equipment cost:			
Freight and move-in	....	$ 78,084	
Labor	518,850	518,850	
Plant invoice	....	791,466	
Used equipment	....	925,330	
New equipment	....	2,909,644	
Subtotal	....	$5,223,374	
Net salvage	....	1,905,640	
Net cost	....	$3,317,734	
Outside equipment rental	....	5,980	
Total job cost	....	....	$ 3,323,714
Indirect cost:			
Labor	2,120,620	$2,120,620	
Other cost	....	250,500	
Insurance premiums	....	274,000	
Taxes	....	110,000	
Bond premium	....	142,000	
Total indirect cost	....	....	$ 2,897,120
Camp cost:			
Construction cost	23,800	$ 191,800	
Operation and maintenance	8,500	17,000	
Total cost	....	$ 208,800	
Camp revenue	....	68,000	
Salvage revenue	....	21,760	
Total revenue	....	$ 89,760	
Net camp cost	....	....	$ 119,040
Cost before escalation	$11,096,343	....	$23,747,450
Escalation:			
Labor	$ 1,504,700	....	$ 1,504,700
Materials	....	....	
Total escalation	$ 1,504,700	....	$ 1,504,700
Total estimated cost	$12,601,043	....	$25,252,150

XIX. CASH FORECAST
for the Denny Dam Estimate

	Total	Cash Flow										
		First Year		Second Year			Third Year			Fourth Year		
		May–Aug.*	Sept.–Dec.	Jan.–Apr.	May–Aug.	Sept.–Dec.	Jan.–Apr.	May–Aug.	Sept.–Dec.	Jan–Apr.	May–Aug.	Sept.–Dec.
Cash revenue:												
Plant and equipment prepayment	$ 3,750,000	$ 270,000	$1,115,000	$1,125,000	$1,125,000	$ 115,000						
Bid item net revenue	25,250,000		534,000	777,000	1,416,000	3,820,000	$4,316,000	$4,187,000	$4,260,000	$3,668,000	$1,639,000	$ 633,000
Total bid (15% markup)	$29,000,000	$ 270,000	$1,649,000	$1,902,000	$2,541,000	$3,935,000	$4,316,000	$4,187,000	$4,260,000	$3,668,000	$1,639,000	$ 633,000
Retained percentage		(27,000)	(164,900)	(190,200)	(254,100)	(393,500)	(420,300)					1,450,000
Camp revenue	68,000		4,000	8,000	8,000	8,000	8,000	8,000	8,000	8,000	8,000	
Salvage of plant and equipment	1,905,640										700,000	1,205,640
Salvage of camp	21,760											21,760
Total revenue	$30,995,400	$ 243,000	$1,488,100	$1,719,800	$2,294,900	$3,549,500	$3,903,700	$4,195,000	$4,268,000	$3,676,000	$2,347,000	$3,310,400
Cash disbursements:												
Direct cost:												
Diversion	$ 590,900		$ 215,900	$ 140,000	$ 60,000	$ 20,000	$ 20,000	$ 30,000	$ 30,000	$ 15,000	$ 60,000	
Excavation	923,890		173,890	230,000	250,000	220,000		50,000				
Concrete, cement, and reinforcing	13,103,867				550,000	2,580,000	2,580,000	2,480,000	2,610,000	2,053,000	250,867	
Other items	2,788,919		100,000	180,000	250,000	320,000	320,000	320,000	320,000	320,000	400,000	$ 258,919
Plant and equipment:												
Excavation and diversion	667,310	$ 667,310										
Other	4,562,044	62,044	1,500,000	1,500,000	1,500,000							
Indirect cost:												
Labor and other	2,371,120	110,120	150,000	200,000	250,000	275,000	275,000	275,000	275,000	275,000	196,000	90,000
Bond premium	142,000	142,000										
Insurance and taxes	384,000	80,000	39,000	20,000	70,000	15,000	15,000	70,000	15,000	15,000	40,000	5,000
Camp:												
Construction	191,800	100,000	91,800									
Operating cost	17,000		1,000	2,000	2,000	2,000	2,000	2,000	2,000	2,000	2,000	
Labor escalation	1,504,700				100,000	150,000	150,000	238,000	290,000	290,000	246,000	40,700
Inventories		50,000	50,000	50,000	100,000	50,000				(100,000)	(100,000)	(100,000)
Total disbursements	$27,247,550	$1,211,474	$2,321,590	$2,322,000	$3,132,000	$3,632,000	$3,362,000	$3,465,000	$3,542,000	$2,870,000	$1,094,867	$ 294,619
Revenue less disbursements	$ 3,747,850	$ (968,474)	$ (833,490)	$ (602,200)	$ (837,100)	$ (82,500)	$ 541,700	$ 730,000	$ 726,000	$ 806,000	$1,252,133	$3,015,781
Accumulative		$ (968,474)	$(1,801,964)	$(2,404,164)	$(3,241,264)	$(3,323,764)	$(2,782,064)	$(2,052,064)	$(1,326,064)	$(520,064)	$ 732,069	$3,747,850
Working capital		631,526	598,036	595,836	758,736	676,236	717,936	747,936	773,936	579,936		
Capital from partners		$1,600,000	$2,400,000	$3,000,000	$4,000,000	$4,000,000	$3,500,000	$2,800,000	$2,100,000	$1,100,000		
Interest at 9%	$ 735,000	$ 48,000	$ 72,000	$ 90,000	$ 120,000	$ 120,000	$ 105,000	$ 84,000	$ 63,000	$ 33,000		

*4-month periods were used to conserve space. The use of monthly periods produces a more accurate cash forecast.

XX. CONTINGENCY EVALUATION

Contingency allowances are not required for the construction of Denny Dam since unusual expense should not be incurred. Unknown foundation conditions should not be encountered since the foundation rock is sound granitic material. Work access will not be a problem since there is an existing highway on the left bank of the river. A labor contingency will not be required since there is an adequate supply of construction labor in northern California. Finally, unusual diversion expenses should not be experienced, since the diversion plan provides for a flood flow of 80,000 cfs and the chance of having floods of greater magnitude is very remote.

CHAPTER TWELVE

Joint Venturing and Bid Preparation

This chapter discusses the reasons for the formation of joint ventures, explains how they operate, and describes the prebid meetings held by the partners to arrive at a mutually agreed upon bid price. The preparation of a construction bid from the estimated cost and markup established at a joint-venture prebid meeting is described and illustrated.

If a bid is submitted by a single contractor, the descriptions on joint-venture meetings will not be applicable, but the bid-preparation procedures will be similar to those described in this chapter. In this case, the contractor will often hire a consultant to prepare a check estimate. The contractor's procedure, then, in comparing the check estimate with his own estimate, will be quite similar to the bid comparison procedures used at joint-venture meetings.

JOINT VENTURES

A joint venture is the partnership of two or more contractors who join together to bid and construct a particular construction project. If the bid is unsuccessful, the joint venture is disbanded. If the bid is successful, the joint venture constructs the project and disbands after construction has been completed.

Joint ventures have the advantages and disadvantages of any other partnership. They are entered into by contractors who have mutual trust in each other's integrity and construction ability.

Joint venturing is advantageous to contractors because it provides them with a more reliable bid estimate, increases the economic stability of each joint-venture partner, and permits contractors to participate in larger construction projects.

A joint-venture bid estimate is more reliable than an estimate prepared by a single contractor since it consists of estimated cost developed during the comparison and discussions of the estimates presented by each joint-venture partner. By comparing estimates, each partner is able to adjust his estimate in the event of any errors or oversights or for the use of preferable construction systems developed by other partners. After each partner has adjusted his estimate, the best solution presented for each phase of the dam's construction is discussed, and if agreeable to all partners, it is used as the agreed estimate. Therefore, an agreed estimate is the result of the combined thinking of all the partners' estimators.

Joint venturing increases each partner's economic stability by permitting him more financial diversification. Each contractor can only bid and construct a limited volume of construction work. These limits are established by his available bonding capacity, his working capital, and the number of experienced supervisory personnel that he has available. If he bids and constructs large projects without partners, he may only be able to bond, finance, and provide supervision for a few. If one of these construction projects operates at a loss, then he must absorb the total loss, which may be financially disastrous. However, if he bids large construction projects as a member of different joint ventures, he will be able to spread his bonding capacity and working capital over a large number of projects, and he needs only to supply supervisory personnel to the projects he sponsors. If one of these projects operates at a loss, he must absorb only his share of that loss and will enjoy the profits from a large number of other, more lucrative projects to counterbalance this loss.

Joint venturing permits contractors to submit bids on contracts too large for them to handle individually. Some dams are so massive and their construction so time-consuming that the largest heavy-construction contractors do not wish to undertake the risk of bidding the work by themselves. However, they are willing to form joint ventures to bid these large jobs since this spreads the total risk among the partners in accordance with their percentage of participation in the joint venture. Many competent contractors are so small that they cannot secure enough bonding capacity to individually submit bids on large construction projects. By forming joint ventures, they are able to submit bids since they then need furnish only their share of the bonding and financing requirements.

The formation of joint ventures is to the owners' advantage as well, for it enables them to receive competitive bids on the largest of construction projects. By enabling more contractors to participate in the bidding of large projects, the competition between contractors increases. Many large construction contracts are bid by ten or more joint ventures. Assuming that each joint venture is composed of four contractors, 40 or more contractors will be bidding for the construction of the project. This increases bidding competition.

The formation of a joint venture is usually initiated when one contractor decides to bid on a certain project and contacts other contractors with whom he has been associated in the past, or with whom associations are desired, to determine if they are interested in participating in the joint venture. The initiating contractor is the sponsor; the other participating companies are partners in the joint venture. The sponsor inquires what percentage each contractor desires in the joint venture. When enough contractors indicate interest to subscribe for 100 percent of the work, the joint venture is formed.

The sponsor takes the largest percentage of participation in the joint venture. In the majority of cases, the sponsor is in charge of the work and receives only advice on major decisions and financial help from the other partners. In some cases a committee may be formed, composed of a member from each of the major partners, which meets and decides on major decisions. In other joint ventures, the work is divided among the partners with each partner responsible for certain phases of construction, while each shares in the common profit. Others are formed where one partner is responsible for one phase of the work and takes the profit on this phase, and other partners are responsible for other phases and take the corresponding profits from their phases. It is the practice with some organizations that the sponsoring partner gets an additional percentage of profits for sponsoring and administering the work, but this is not commonly done. Usually, each partner puts up working capital in accordance with his subscribed percentage and shares in the profits accordingly; the sponsor does not receive any additional compensation for manning and directing the work.

A prebid joint-venture agreement is signed by all the partners prior to the submission of the bid. This agreement generally:

Dates the agreement.

Lists the names of each partner and states their agreement to form a joint venture.

States that the partners are agreeing to submit a construction bid and perform construction on a certain specified contract.

States that the bid shall be satisfactory to all partners.

Delegates authority to individuals to sign the bid, bid bond, and other bid documents.

Defines the percentage of participation of each partner and states that each partner agrees to furnish his share of deposits or other security for bidding.

Designates the sponsoring company.

States that the partners agree that if the bid is successful, they will enter into a joint-venture agreement covering construction of the contract.

Is executed by each partner.

When the joint venture submits a successful bid, the partners then execute a joint-venture agreement and the power of attorney. This generally:

Designates the sponsoring partner's personnel who have been nominated as attorneys-in-fact for the joint venture and states the authority that has been delegated to them.

Dates the agreement.

Lists the partners.

Lists what particular work is to be accomplished by the joint venture.

States the name of the joint ventrue and its legal position.

Designates the sponsor.

Details the percentage of participation of each partner.

States that partners will put up their proportional share of working capital.

Provides for a separate accounting system for the joint venture.

Gives the joint venture power to purchase equipment, open bank accounts, etc.

Lists the reasons for default by any partner and the consequences in case of default.

Specifies how profits and losses will be shared.

Provides for audits.

ESTIMATORS' MEETING

All or several of the joint-venture partners prepare independent cost estimates for construction of the project. To arrive at a mutually agreed upon estimate on which to base the bid, a joint-venture estimators' meeting is held prior to bid submittal. The sponsoring partners' chief estimator informs each partner of the date, time, and location of the meeting and sends out estimating instructions. These estimating instructions were described under Estimating Task IX in Chap. 9.

The sponsoring company's chief estimator presides over the estimators' meeting and may also present his company's estimate. At this meeting each partner reads his estimated cost. All the meeting members tabulate these costs on the estimate comparison forms. Exhibit 1 shows how the estimates of three joint-venture partners at the fictitious estimators' meeting for Denny Dam were tabulated and adjusted.

After all estimates are tabulated, the differences between the estimates are resolved. This is done by each partner describing his construction system and its advantages and by subsequent group comparison and discussion of each item of estimated cost. As the details of the estimates are reviewed, it is also determined whether any partner has overlooked an element of cost.

Estimating differences can seldom be resolved without reviewing items in greater detail than that shown on the estimate comparison forms. The additional items reviewed may be the quantities of work, the construction schedule and system, the construction equipment, the supervisory and indirect personnel, and the estimated direct cost of many individual bid items. To expedite this detailed comparison, all estimates should be arranged so that unit cost and total cost of each operation are readily available. Furthermore, the partners' estimators should use the estimating format illustrated by the sponsoring partner's instructions to facilitate subsequent discussions.

The estimate comparisons and the accompanying descriptions and discussions help to reveal any mistake or oversight in an estimate, to indicate which partner has used the most efficient construction systems, and to place emphasis on obtaining the most accurate estimated cost for all phases of the dam's construction. During the progress of the meeting, the partners often correct their estimates to account for mistakes, omissions, or improvements suggested during the meeting. Improvements may include the use of more economical construction systems or more accurately estimated cost suggested by other partners. The columns headed "Revised" in Exhibit 1 show how the joint-venture partners for Denny Dam made revisions of their estimates.

After each partner has completed the adjustments to his estimate, the estimators arrive at a mutually satisfactory, agreed estimate that is approved by each joint-venture partner. Often agreement is reached by arriving at an agreed cost for each item listed on the bid comparison forms.

Exhibit 1 ESTIMATE COMPARISON FORMS FOR THE DENNY DAM JOINT VENTURE MEETING

	Sponsoring partner		First participating partner	
	Original	Revised	Original	Revised
Direct cost:				
Labor	$ 8,424,573	$ 8,464,573	$ 9,284,500	$ 8,599,000
Supplies	3,244,595	3,261,454	3,979,092	3,685,392
Permanent material	4,397,244	4,397,244	4,397,244	4,397,244
Subcontracts	1,341,164	1,341,164	1,341,164	1,341,164
Total direct cost	$17,407,576	$17,464,435	$19,002,000	$18,022,800
Plant and equipment cost:				
Freight and move-in	$ 78,084	....	$ 60,000	$ 60,000
Labor	518,850	....	610,000	610,000
Plant invoice	791,466	....	625,000	625,000
Used equipment	925,330	....	....	925,330
New equipment	2,909,644	....	4,760,000	2,960,000
Subtotal	$ 5,223,374	....	$ 6,055,000	$ 5,180,330
Net salvage	1,905,640	....	2,320,000	1,870,000
Net cost	$ 3,317,734	....	$ 3,735,000	$ 3,310,330
Outside equipment rental	5,980	....	....	....
Total job cost	$ 3,323,714	$ 3,323,714	$ 3,735,000	$ 3,310,330
Indirect cost:				
Labor	$ 2,120,620	....	$ 2,450,000	....
Other cost	250,500	....	350,000	....
Insurance premiums	274,000	....	310,000	....
Taxes	110,000	....	150,000	....
Bond premium	142,000	....	160,000	....
Total indirect cost	$ 2,897,120	$ 2,897,120	$ 3,420,000	$ 2,897,120
Camp cost:				
Construction cost	$ 191,800	....	....	$ 191,800
Operating and maintenance	17,000	....	....	17,000
Total cost	$ 208,800	....	....	$ 208,800
Camp revenue	$ 68,000	....	....	$ 68,000
Camp salvage	21,760	....	....	21,760
Total revenue	$ 89,760	....	....	$ 89,760
Net camp cost	$ 119,040	$ 119,040	....	$ 119,040
Total cost before escalation	$23,747,450	$23,804,309	$26,157,000	$24,349,290
Escalation:				
Labor	$ 1,504,700	$ 1,509,000	$ 1,700,000	$ 1,564,000
Materials	....	....	....	....
Total escalation	$ 1,504,700	$ 1,509,000	$ 1,700,000	$ 1,564,000
Total estimated cost	$25,252,150	$25,313,309	$27,857,000	$25,913,290

The order of review of these listed items varies with the sponsoring company. Some sponsors start their review with the direct cost breakdown, while others start with indirect cost. One advantage to starting with indirect cost is that it quickly establishes what cost should be included in indirect and what in direct cost. After agreement is reached on each listed cost, they are added to arrive at the total agreed cost. The agreed cost for Denny Dam is shown in the last column in Exhibit 1.

Second participating partner		
Original	Revised	Agreed estimate
$ 8,671,000	$ 8,754,000	$ 8,495,000
2,890,392	2,918,192	3,275,159
4,397,244	4,397,244	4,397,244
1,341,164	1,341,164	1,341,164
$17,299,800	$17,410,600	$17,508,567
$ 90,000	....	$ 78,084
450,000	....	518,850
510,000	....	791,466
890,000	....	925,330
3,210,000	....	2,909,644
$ 5,150,000	....	$ 5,223,374
1,750,000	....	1,905,640
$ 3,400,000	....	$ 3,317,734
....	....	5,980
$ 3,400,000	$ 3,400,000	$ 3,323,714
$ 2,600,000	....	$ 2,120,620
500,000	....	250,500
200,000	....	274,000
90,000	....	110,000
120,000	....	142,000
$ 3,510,000	$ 2,897,120	$ 2,897,120
$ 210,000	....	$ 191,800
26,000	....	17,000
$ 236,000	....	$ 208,800
$ 50,000	....	$ 68,000
40,000	....	21,760
$ 90,000	....	$ 89,760
$ 146,000	$ 146,000	$ 119,040
$24,355,800	$23,853,720	$23,848,441
$ 1,000,000	$ 1,500,000	$ 1,514,000
....	....	....
$ 1,000,000	$ 1,500,000	$ 1,514,000
$25,355,800	$25,353,720	$25,362,441

Exhibit 1 (continued) ESTIMATE COMPARISON FORMS FOR THE DENNY DAM JOINT VENTURE MEETING

Bid item no.	Description	Sponsoring partner		First participating partner		Second participating partner		Agreed estimate
		Original	Revised	Original	Revised	Original	Revised	
	Comparison of direct cost:							
1	Care and diversion of river	$ 590,900	$ 590,900	$ 750,000	$ 750,000	$ 500,000	$ 500,000	$ 590,900
2–6	Excavation	923,890	960,000	1,250,000	960,000	850,000	900,000	940,000
7–13	Drill and grout	343,800	343,800	344,200	344,200	346,000	346,000	343,800
14–17	Backfill	78,089	78,089	110,000	110,000	120,000	120,000	90,000
18	Steel reinforcement	523,600	523,600	492,800	523,600	492,800	523,600	523,600
19–24	Concrete	9,405,267	9,405,267	10,120,000	9,600,000	9,050,000	9,300,000	9,405,267
25	Cement	3,175,000	3,175,000	3,175,000	3,175,000	3,175,000	3,175,000	3,175,000
26–31	Miscellaneous	212,600	212,600	250,000	250,000	276,000	276,000	240,000
32–42	Gates and guides	1,465,179	1,465,179	1,750,000	1,600,000	1,820,000	1,600,000	1,500,000
43–51	Mechanical	439,251	460,000	510,000	460,000	420,000	420,000	450,000
52	Electrical system	250,000	250,000	250,000	250,000	250,000	250,000	250,000
	Total direct cost	$17,407,576	$17,464,435	$19,002,000	$18,022,800	$17,299,800	$17,410,600	$17,508,567
	Labor summary:							
	Direct labor	$ 8,424,573	$ 8,464,573	$ 9,284,500	$ 8,599,000	$ 8,671,000	$ 8,754,000	$ 8,495,000
	Plant and equipment labor	518,850	518,850	610,000	610,000	450,000	450,000	518,850
	Indirect labor	2,120,620	2,120,620	2,450,000	2,120,620	2,600,000	2,120,620	2,120,620
	Camp construction labor	23,800	23,800	....	23,820	30,000	30,000	23,800
	Camp operation and maintenance labor	8,500	8,500	....	8,500	20,000	20,000	8,500
	Subtotal	$11,096,343	$11,136,343	$12,344,500	$11,361,940	$11,771,000	$11,374,620	$11,166,770
	Labor escalation	1,504,700	1,509,000	1,700,000	1,564,000	1,000,000	1,500,000	1,514,000
	Total labor	$12,601,043	$12,645,343	$14,044,500	$12,925,940	$12,771,000	$12,874,620	$12,680,770
	Maximum capital requirements	$ 4,000,000	....	....	....	....	$ 3,000,000	$ 4,000,000
	Interest expense	$ 735,000	....	....	....	....	$ 460,000	$ 735,000

In some cases, the estimators cannot agree on an estimated cost. When this happens, a tabulation of the various estimates is presented to the principals, who then must establish a bid price without having the advantage of an agreed estimate.

The agreed estimate is something of a compromise, but special consideration is given to the sponsor's estimate and the estimates prepared by partners who have had construction experience on similar projects. Individual estimators should relinquish any pride of authorship and be willing to adjust their estimate to take into account any method or cost better than what they originally had in their estimates. In the majority of cases, the final agreed estimate will reflect the use of the best construction system and the closest approximation of the actual cost of doing the work. The final agreed estimate is, of course, only an estimate, and the final cost of doing the work will vary to some degree from this estimate.

After agreement has been reached on the estimated cost, the estimators discuss the amount of capital required, interest expense, and project contingency. The capital requirements and interest expense are mathematically determinable as explained under Estimating Task XIX in Chap. 9 and as illustrated under the same estimating task in Chap. 11. Project contingency is more difficult to establish since it is mathematically indeterminate and can only be evaluated by exercising engineering judgment. To establish the amount of contingency, the estimators review the overall concept of the construction work, consider whether the specifications contain adequate force majeure or unforeseen condition clauses, and discuss the risk that may be incurred during the project's construction. Detailed descriptions of contingency items were given under Estimating Task XX, in Chap. 9 and 10.

At the conclusion of the estimators' meeting, each estimator presents to his principal: the agreed estimate, the capital requirements, the interest expense, and the results of the contingency evaluation.

PRINCIPALS' MEETING

After the estimators have concluded their meeting, the results of the meeting are submitted by the estimators to their respective principals. The principals then hold a meeting to determine the amount of markup to be added to the agreed estimate to arrive at the bid price.

Principals should be well-informed about the various aspects of the job. This job information is available from discussions of the work with their estimators, review of the results of the estimators' meeting, and a study of the bidding documents and job description. Many principals attend the estimators' meeting since exposure to the estimate comparisons, descriptions, and discussions assist them in evaluating the proposed project.

The markup, as determined by the principals, is composed of the following elements:

1. A proportional share of each participating company's home office expense.
2. Interest cost of the capital required to finance the work.
3. Contingency—in setting the contingency, the principals start with the job risk information that is presented to them by their estimators and with knowledge they have acquired by job review or attendance at the estimators' meeting. They must then determine the amount of contingency to place in the markup.
4. Profit—in setting the job profit, the principals are influenced by the following job requirements: the amount of the contractor's capital required to finance the job; how many of the contractor's key personnel will be required to supervise the job, and for how long; and the estimated salvage value of the plant and equipment. This salvage value is a matter of judgment since it must be predetermined for a future date when the equipment may or may not be marketable. Sometimes specialized dam equipment, i.e., cableways, gantry cranes, mixing plants, and aggregate plants, cannot be sold until several years after they have been placed on the market. During these years, costs are incurred for storage and care and for property taxes assessed by the county in which the equipment is stored.

If the bid documents have not been circulated and signed prior to the prebid meeting, they are signed at the principals' meeting. At the meeting's conclusion, the markup and other bidding instructions are given to the chief estimator of the sponsoring company, who has the responsibility for the bid preparation.

For the Denny Dam bid, it is assumed that the principals agreed to a 15 percent markup on the agreed cost, composed of the following elements:

Interest on capital required	$ 735,000
$1^1/_2$ percent of the total cost as home office expense allocation	380,438
Contingency and markup represented by the salvage value of the equipment	1,905,640
Contingency and markup represented by a cash profit	783,288
Total markup—15 percent	$3,804,366

BID PREPARATION

It is the responsibility of the sponsoring company's estimator to prepare the final bid. The period immediately before bid submittal is a very hectic one. In the short time available, the estimator must adjust his estimated direct cost to equal the agreed total direct cost, spread the remainder of the agreed cost and the markup to the bid items, check all quotations and make quotation adjustments, prepare the bidding papers in final form, check all extensions and additions, and submit the bid to the bid-opening agency.

Spreading Cost

There often are discussions by owners and engineers about *balanced* and *unbalanced* bids, without a good understanding of what constitutes a balanced bid. In preparing a dam estimate, the direct cost is estimated for each bid item, but the other divisions of cost, namely, plant and equipment cost, camp cost, indirect cost, and escalation, are estimated for the total job and must then be divided or spread among the individual bid items. A balanced bid is one in which these other divisions of cost and profit are spread to the individual bid items which cause this cost and profit to occur. An unbalanced bid occurs when these costs and profits are spread to bid items that do not cause the cost and profit to occur. To arrive at a truly balanced bid, the spread of these costs to the bid items would require a very detailed study which would take more time than is available between the prebid meeting and bid submission time. Because of this time limitation, it is impossible to submit a perfectly balanced bid on heavy-construction work. Cost spreading is therefore done by various approximations, with reliance placed on the experience of the estimator. The methods of spreading vary among companies, and since the human element is involved, no two estimators with similar cost to spread would spread it in the same manner.

The easiest method of cost spreading (but the one which takes the least recognition of where cost occurs) is to divide the total direct costs into the total bid price in order to obtain a factor by which each direct unit cost is multiplied to arrive at the unit bid price. With this method, plant and equipment cost, indirect cost, labor escalation, etc., are spread against many items which would not incur this cost, resulting in unit costs that are not representative of the work required. If time becomes critical, every contractor must resort to this simplified method in order to submit the bid by the deadline. Some contractors, in order to simplify bid submission, use this method on every bid, but this is not recommended.

The preferred cost-spreading method is to use experience and judgment in arbitrarily placing cost against the proper items so that as good a unit cost is produced as is possible in the time available. This will not be a truly balanced bid but only an approximation of one. It is important that the contractor develop a true unit bid price, because, if he does not do so and quantities vary from the estimated quantities shown on the bid-comparison forms, he may not receive proper compensation for the work he performs. For instance, if the foundation excavation for a dam is extended deeper than originally planned, the rock-excavation and mass-concrete quantities will overrun the bid quantities. If insufficient amounts for plant and equipment, indirect, camp, escalation, and profit, were spread against these two bid items, the contractor would not be adequately reimbursed for performing this extra work.

Cost should be spread to produce unit costs that are as accurate as possible. The cost-spreading method shown in Exhibit 2 for the Denny Dam estimate accomplishes this. The spread was started by adjusting the sponsor's estimated direct cost to that agreed to at the prebid meeting.

Plant and equipment cost was spread by first dividing the net cost of the general plant and equipment cost between the net cost of the excavation, diversion, and concrete plant and equipment. This resultant cost was then spread to the excavation, diversion, and concrete bid items. This spreading placed most of the plant and equipment cost against the concrete items, which was correct since they constitute the bulk of the prime contractor's work.

Indirect cost and camp cost were added together and spread to the bid items at two different rates. The spread to bid items whose costs were primarily composed of purchased materials or subcontracts was done at 5 percent of their direct cost. This left the major portion of this cost to be spread to the remaining bid items, whose cost included large amounts of direct labor. This remaining cost was spread at the rate of 24 percent of the bid items' direct cost.

Labor escalation was spread against the bid items that were scheduled to be constructed in the years that the escalation occurred.

All costs were then added together to produce the agreed total cost.

The agreed total cost was then transferred to Exhibit 3. This transfer of the agreed cost from one spread sheet to another is not done in actual practice. Instead, a spread sheet is used that contains enough columns to contain all spread calculations and quotation adjustments.

Exhibit 3 illustrates how the markup was spread for Denny Dam, how the initial bid prices were computed, and how they were adjusted to reflect the quotations received prior to bid submittal. Markup was spread using the same technique as used to spread indirect and camp cost. The spread to bid items whose cost was primarily permanent materials and subcontracts was done at the rate of 5 percent of their total cost. The spread to remaining bid items whose cost included large amounts of direct labor was done at a rate of 18.57 percent of their total cost. The spread of markup should also be made to round out the unit bid prices.

Markup was also spread to round out bid prices so that their extensions would result in even-dollar amounts, thereby eliminating the necessity for tabulating cents on the bid schedule. This also produces unit bid prices that are easily extendable during the construction of the project.

The foregoing described methods of spreading cost and markup represent the preferred method. Often there is insufficient time to complete all these steps, so that short cuts must be taken and reliance must be placed on the ability of the estimator who makes or directs the spread.

After the initial bid prices are established, the estimator should review the unit bid prices and adjust the price of items whose construction might result in extra expense to the contractor. Since items of this nature were not present in the Denny Dam bid, these types of adjustments are not shown.

To illustrate the use of this type of unit bid price adjustment, if the dam had been located on incompetent rock and the foundation exploration had not been adequately performed, the estimator might decide to increase the unit bid price for preliminary rock cleanup. Justification for such an increase would be that if preliminary rock cleanup overruns in quantity, it will be because the owner's engineers are uncertain about the dam's foundation. To prove the foundation material, they may force the contractor to excavate the dam's foundation in shallow layers and call for preliminary rock cleanup after each layer is removed. This procedure is repeated until suitable foundation has been uncovered. This procedure will increase excavation cost; delay the completion of excavation, which in turn will delay the completion of the project; and increase indirect cost and the depreciation charge for plant and equipment. The best way for the contractor to be sure that he will receive adequate compensation for overruns in preliminary rock cleanup quantities is to increase its unit bid price, since in preparing his bid the cost placed against this item assumed that excavation in shallow layers would not be

Exhibit 2 SPREAD OF PLANT AND EQUIPMENT, INDIRECT COST, CAMP COST, AND ESCALATION FOR THE DENNY DAM BID

Bid item no.	Description	Bid quantity	Agreed direct cost		Agreed plant and equipment		Agreed indirect and camp cost		Agreed escalation		Agreed total cost	
			Unit	Amount	Unit	Amount	Unit	Amount	Unit	Amount	Unit	Amount
1	Care and diversion of river	Lump sum	......	$ 590,900	.....	$ 166,834	......	$ 140,157	......	$ 27,819	....	$ 925,710
	Excavation:											
2	Common	185,000 cu yd	$ 1.45	$ 268,250	$0.40	$ 74,000	$ 0.35	$ 64,750	......	....	$ 2.20	$ 407,000
3	Rock	170,000 cu yd	3.15	535,500	0.90	153,000	0.76	129,200	$ 0.20	$ 34,000	5.01	851,700
4	Preliminary rock cleanup	7,000 sq yd	2.00	14,000	0.50	3,500	0.48	3,360	0.12	840	3.10	21,700
5	Close-line drilling	30,000 sq ft	1.00	30,000	0.20	6,000	0.24	7,200	0.06	1,800	1.50	45,000
6	Final rock cleanup	23,000 sq yd	4.00	92,000	1.00	23,000	0.96	22,080	0.24	5,520	6.20	142,600
	Total excavation	....	......	$ 939,750	.....	$ 259,500	......	$ 226,590	......	$ 42,160	....	$ 1,468,000
	Drill and grout:											
7	1½-in.-diameter grout holes	25,000 lin ft	4.00	$ 100,000	.....	....	0.20	$ 5,000	......	....	4.20	$ 105,000
8	3-in.-diameter drain holes	17,000 lin ft	6.00	102,000	.....	....	0.30	5,100	......	....	6.30	107,100
9	3-in.-diameter cored exploratory holes	13,500 lin ft	6.00	81,000	.....	....	0.30	4,050	......	....	6.30	85,050
10	Grout connections	800 each	10.00	8,000	.....	....	0.50	400	......	....	10.50	8,400
11	Pressure testing	800 test	10.00	8,000	.....	....	0.50	400	......	....	10.50	8,400
12	Grouting cement	8,000 cu ft	1.60	12,800	.....	....	0.08	640	......	....	1.68	13,440
13	Pressure grouting	8,000 cu ft	4.00	32,000	.....	....	0.20	1,600	......	....	4.20	33,600
	Total drill and grout	....	......	$ 343,800	.....	....	......	$ 17,190	......	....	....	$ 360,990
	Backfill:											
14	Uncompacted backfill	4,300 cu yd	5.00	$ 21,500	1.00	$ 4,300	1.20	$ 5,160	0.30	$ 1,290	7.50	$ 32,250
15	Dumped rock backfill	18,000 cu yd	3.43	61,740	1.00	18,000	0.82	14,760	0.20	3,600	5.45	98,100
16	Backfill material	560 cu yd	5.00	2,800	1.00	560	1.20	672	0.30	168	7.50	4,200
17	Dumped riprap	1,000 cu yd	4.00	4,000	1.00	1,000	0.96	960	0.24	240	6.20	6,200
	Total backfill	....	......	$ 90,040	.....	$ 23,860	......	$ 21,552	......	$ 5,298	....	$ 140,750
18	Reinforcing steel	3,080,000 lb	0.17	$ 523,600	.....	....	0.01	$ 30,800	......	....	0.18	$ 554,400
	Concrete:											
19	Stilling basin	13,700 cu yd	13.46	$ 184,402	4.00	$ 54,800	3.23	$ 44,251	......	....	20.69	$ 283,453
20	Roadways, sidewalks, bridges	2,830 cu yd	72.95	206,449	4.00	11,320	17.50	49,525	10.94	$ 30,960	105.39	298,254
21	Slabs, beams, and columns	4,150 cu yd	38.22	158,613	4.00	16,600	9.17	38,055	5.73	23,780	57.12	237,048
22	Spillway piers	3,500 cu yd	43.26	151,410	4.00	14,000	10.38	36,330	6.49	22,715	64.13	224,455
23	Training walls	27,700 cu yd	14.59	404,143	4.00	110,800	3.50	96,950	......	....	22.09	611,893
24	Mass in dam	775,000 cu yd	10.71	8,300,250	3.44	2,666,000	2.57	1,991,750	1.61	1,247,750	18.33	14,205,750
	Total concrete	826,880 cu yd	......	$ 9,405,267	.....	$2,873,520	......	$2,256,861	......	$1,325,205	....	$15,860,853

25	Cement	635,000 bbl	5.00	$ 3,175,000	....	....	0.25	$ 158,750	....	....	5.25	$ 3,333,750
	Miscellaneous:											
26	12-in.-diameter RCCP	100 lin ft	10.00	$ 1,000	....	....	2.40	$ 240	....	....	12.40	$ 1,240
27	18-in.-diameter RCCP	100 lin ft	15.00	1,500	....	....	3.60	360	....	....	18.60	1,860
28	Copper water stop	12,300 lin ft	6.00	73,800	....	....	1.44	17,712	....	....	7.44	91,512
29	6-in.-diameter porous concrete pipe	28,000 lin ft	5.21	145,880	....	....	1.25	35,000	....	....	6.46	180,880
30	12-in.-diameter half-round porous pipe	1,950 lin ft	6.00	11,700	....	....	1.44	2,808	....	....	7.44	14,508
31	Miscellaneous doors	400 sq ft	15.30	6,120	....	....	3.67	1,468	....	....	18.97	7,588
	Total miscellaneous	....	....	$ 240,000	....	....	....	$ 57,588	....	....	....	$ 297,588
	Gates and guides:											
32	Miscellaneous metal castings	Lump sum	....	$ 16,000	....	....	....	$ 800	....	$ 960	....	$ 17,760
33	Slide and bulkhead gates	286,000 lb	0.60	171,600	....	....	0.03	8,580	0.04	11,440	0.67	191,620
34	Pumps, motors, piping for gates	Lump sum	....	34,000	....	....	....	1,700	....	2,000	....	37,700
35	Penstock gates, frames, and guides	288,000 lb	0.63	181,440	....	....	0.03	8,640	0.04	11,520	0.70	201,600
36	Penstock-gate hoist and supports	Lump sum	....	79,190	....	....	....	3,960	....	4,800	....	87,950
37	Penstock lining and frames	535,000 lb	0.41	219,350	....	....	0.02	10,700	0.02	10,700	0.45	240,750
38	Tainter gates and accessories	670,000 lb	0.40	268,000	....	....	0.02	13,400	0.02	13,400	0.44	294,800
39	Tainter-gate trunnion anchors	320,000 lb	0.41	131,200	....	....	0.02	6,400	0.02	6,400	0.45	144,000
40	Tainter-gate hoist and accessories	150,000 lb	1.51	226,500	....	....	0.08	12,000	0.09	13,500	1.68	252,000
41	Trashrack guides and beam	260,000 lb	0.38	98,800	....	....	0.02	5,200	0.02	5,200	0.42	109,200
42	Stop logs, protection plate, and beam	168,000 lb	0.44	73,920	....	....	0.02	3,360	0.03	5,040	0.49	82,320
	Total gates and guides	....	....	$ 1,500,000	....	....	....	$ 74,740	....	$ 84,960	....	$ 1,659,700
	Mechanical and miscellaneous metal:											
43	Cast-iron pipe and fittings	156,000 lb	0.45	$ 70,200	....	....	0.02	$ 3,120	0.03	$ 4,680	0.50	$ 78,000
44	Steel pipe and fittings	184,000 lb	0.81	149,040	....	....	0.04	7,360	0.05	9,200	0.90	165,600
45	Piezometer piping and fittings	Lump sum	....	6,000	....	....	....	250	....	300	....	6,550
46	Sump pump and piping	Lump sum	....	9,000	....	....	....	300	....	360	....	9,660
47	Water supply and plumbing	Lump sum	....	27,483	....	....	....	1,280	....	1,547	....	30,310
48	Structural steel	66,500 lb	0.44	29,260	....	....	0.02	1,330	0.03	1,995	0.49	32,585
49	Railing and fittings	17,700 lb	1.09	19,293	....	....	0.05	885	0.07	1,239	1.21	21,417
50	Parapet railing	1,152 lin ft	14.70	16,934	....	....	0.70	807	0.90	1,037	16.30	18,778
51	Steel in spillway bridge	410,000 lb	0.30	123,000	....	....	0.01	4,100	0.02	8,200	0.33	135,300
	Total mechanical and miscellaneous metal	....	....	$ 450,210	....	....	....	$ 19,432	....	$ 28,558	....	$ 498,200
52	Electrical system	Lump sum	....	$ 250,000	....	....	....	$ 12,500	....	....	....	$ 262,500
	Total agreed cost	....	....	$17,508,567	....	**$3,323,714**	....	$3,016,160	....	$1,514,000	....	**$25,362,441**

Exhibit 3 SPREAD OF MARKUP AND QUOTATION ADJUSTMENTS FOR THE DENNY DAM BID

Bid item no.	Description	Bid quantity	Agreed total cost		Spread of markup		Initial bid price open items		Quotation adjustments		Final bid price	
			Unit	Amount	Unit	Amount	Unit	Amount	Unit	Amount	Unit	Amount
1	Care and diversion of river	Lump sum	...	$ 925,710	...	$ 174,720	...	$ 1,100,430	...	$ (1,985)	...	$ 1,098,445
	Excavation:											
2	Common	185,000 cu yd	$ 2.20	$ 407,000	$ 0.41	$ 75,850	...	...	...	...	$ 2.61	$ 482,850
3	Rock	170,000 cu yd	5.01	851,700	0.93	158,100	...	...	...	...	5.94	1,009,800
4	Preliminary rock cleanup	7,000 sq yd	3.10	21,700	0.58	4,060	...	...	...	...	3.68	25,760
5	Close-line drilling	30,000 sq ft	1.50	45,000	0.28	8,400	...	...	...	...	1.78	53,400
6	Final rock cleanup	23,000 sq yd	6.20	142,600	1.15	26,450	...	...	...	...	7.35	169,050
	Total excavation	...	...	$ 1,468,000	...	$ 272,860	...	...	...	...	...	$ 1,740,860
	Drill and grout:											
7	$1^{1}/_{2}$-in.-diameter grout holes	25,000 lin ft	4.20	$ 105,000	0.21	$ 5,250	...	...	...	...	4.41	$ 110,250
8	3-in.-diameter drain holes	17,000 lin ft	6.30	107,100	0.32	5,440	...	...	...	...	6.62	112,540
9	3-in.-diameter cored exploratory holes	13,500 lin ft	6.30	85,050	0.32	4,320	...	...	...	...	6.62	89,370
10	Grout connections	800 each	10.50	8,400	0.50	400	...	...	...	...	11.00	8,800
11	Pressure testing	800 test	10.50	8,400	0.50	400	...	...	...	...	11.00	8,800
12	Grouting cement	8,000 cu ft	1.68	13,440	0.08	640	...	...	...	...	1.76	14,080
13	Pressure grouting	8,000 cu ft	4.20	33,600	0.21	1,680	...	...	...	...	4.41	35,280
	Total drill and grout	...	...	$ 360,990	...	$ 18,130	...	...	...	...	...	$ 379,120
	Backfill:											
14	Uncompacted backfill	4,300 cu yd	7.50	$ 32,250	1.39	$ 5,977	...	...	...	...	8.89	$ 38,227
15	Dumped rock backfill	18,000 cu yd	5.45	98,100	1.01	18,180	...	...	...	...	6.46	116,280
16	Backfill material	560 cu yd	7.50	4,200	1.40	784	...	...	...	...	8.90	4,984
17	Dumped riprap	1,000 cu yd	6.20	6,200	1.15	1,150	...	...	...	...	7.35	7,350
	Total backfill	...	...	$ 140,750	...	$ 26,091	...	...	...	...	...	$ 166,841
18	Reinforcing steel	3,080,000 lb	0.18	$ 554,400	0.01	$ 30,800	...	...	...	...	0.19	$ 585,200
	Concrete:											
19	Stilling basin	13,700 cu yd	20.69	$ 283,453	3.84	$ 52,608	...	...	...	...	24.53	$ 336,061
20	Roadway, sidewalks, and bridges	2,830 cu yd	105.39	298,254	19.61	55,496	...	...	...	...	125.00	353,750
21	Slabs, beams, columns	4,150 cu yd	57.12	237,048	10.60	43,990	...	...	...	...	67.72	281,038
22	Spillway piers	3,500 cu yd	64.13	224,455	11.91	41,685	...	...	...	...	76.04	266,140
23	Training walls	27,700 cu yd	22.09	611,893	4.10	113,570	...	...	...	...	26.19	725,463
24	Mass in dam	775,000 cu yd	18.33	14,205,750	3.40	2,635,000	$ 21.73	$16,840,750	$(0.30)	$(232,500)	21.43	16,608,250
	Total concrete	826,800 cu yd	...	$15,860,853	...	$2,942,349	...	...	...	$(232,500)	...	$18,570,702

25	Cement	635,000 bbl	5.25	$ 3,333,750	0.26	$ 165,100					5.51	$ 3,498,850
	Miscellaneous:											
26	12-in.-diameter RCCP	100 lin ft	12.40	$ 1,240	2.30	$ 230					14.70	$ 1,470
27	18-in.-diameter RCCP	100 lin ft	18.60	1,860	3.45	345					22.05	2,205
28	Copper water stop	12,300 lin ft	7.44	91,512	1.38	16,974					8.82	108,486
29	6-in.-diameter porous concrete pipe	28,000 lin ft	6.46	180,880	1.20	33,600					7.66	214,480
30	12-in.-diameter half-round porous pipe	1,950 lin ft	7.44	14,508	1.38	2,691					8.82	17,199
31	Miscellaneous doors	400 sq ft	18.97	7,588	3.52	1,408					22.49	8,996
	Total miscellaneous			$ 297,588		$ 55,248						$ 352,836
	Gates and guides:											
32	Miscellaneous metal castings	Lump sum		$ 17,760		$ 890						$ 18,650
33	Slide and bulkhead gates	286,000 lb	0.67	191,620	0.03	8,580					0.70	200,200
34	Pumps, motors, and piping for gates	Lump sum		37,700		1,890						39,590
35	Penstock gates, frames, and guides	288,000 lb	0.70	201,600	0.04	11,520					0.74	213,120
36	Penstock-gate hoist and supports	Lump sum		87,950		4,400						92,350
37	Penstock lining and frames	535,000 lb	0.45	240,750	0.02	10,700					0.47	251,450
38	Tainter gates and accessories	670,000 lb	0.44	294,800	0.02	13,400					0.46	308,200
39	Tainter-gate trunnion anchors	320,000 lb	0.45	144,000	0.02	6,400					0.47	150,400
40	Tainter-gate hoist and accessories	150,000 lb	1.68	252,000	0.08	12,000					1.76	264,000
41	Trashrack guides and beam	260,000 lb	0.42	109,200	0.02	5,200					0.44	114,400
42	Stop logs, protection plate, and beam	168,000 lb	0.49	82,320	0.02	3,360					0.51	85,680
	Total gates and guides			$ 1,659,700		$ 78,340						$ 1,738,040
	Mechanical and miscellaneous metal:											
43	Cast-iron pipe and fittings	156,000 lb	0.50	$ 78,000	0.03	$ 4,680					0.53	$ 82,680
44	Steel pipe and fittings	184,000 lb	0.90	165,600	0.05	9,200					0.95	174,800
45	Piezometer piping and fittings	Lump sum		6,550		330						6,880
46	Sump pump and piping	Lump sum		9,660		480						10,140
47	Water supply and plumbing	Lump sum		30,310		1,515						31,825
48	Structural steel	66,500 lb	0.49	32,585	0.02	1,330					0.51	33,915
49	Railing and fittings	17,700 lb	1.21	21,417	0.06	1,062					1.27	22,479
50	Parapet railing	1,152 lin ft	16.30	18,778	0.70	806					17.00	19,584
51	Steel in spillway bridge	410,000 lb	0.33	135,300	0.02	8,200					0.35	143,500
	Total mechanical and miscellaneous metal			$ 498,200		$ 27,603						$ 525,803
52	Electrical System			$ 262,500		$ 13,125						$ 275,625
	Total bid			$25,362,441		$3,804,366				$(234,485)		$28,932,322

required. A more detailed description of how construction cost may increase if quantities of preliminary rock cleanup overrun is included in Chap. 3.

When the estimator increases one unit bid price, he must decrease the unit bid price on one or more other items since he is preparing a bid that totals to a sum agreed to at the principals' meeting. Therefore an increase in one extension must be balanced by an equal decrease in other extensions. This makes it imperative for the estimator to use proper judgment in adjusting tentative bid prices, since if he increases the bid price on a quantity that underruns, the contractor will lose a portion of his anticipated profit. Unit bid prices should never be adjusted lower than their unit direct cost, since if this is done and the quantity overruns, the contractor will not recover his direct cost when he performs the extra work.

To be competitive in the bidding and to give the owner a lower price for the work, some contractors increase unit bid prices for bid-item work that will be constructed at the start of the project. To balance this increase in the bid, the unit bid prices for bid-item work that is to be constructed toward the end of the contract must be reduced. This type of bid unbalancing gives the contractor early money, reduces his financing costs which in turn reduces his interest expense, and permits a lower markup. This type of spreading may work to the advantage of both the owner and the contractor.

The owner often recognizes that the contractor must invest a large sum of money to finance the work. Thus he may include a provision for payments for preparatory work in the specifications, or he may include a pay item for mobilization in the bid schedule in order to relieve the contractor of a part of this financing load. Such payment provisions often specify that the contractor will receive reimbursement for 75 percent of his equipment cost, when the equipment arrives at the jobsite and is paid for by the contractor, and for 75 percent of any plant construction work performed at the jobsite.

Plant and equipment prepayment provisions in the specifications are advantageous both to the owner and to the contractor. They are advantageous to the owner since they reduce the capital the contractor must advance to the project, thus reducing his interest charges and allowing him to reduce his markup and submit a lower bid. They are advantageous to the contractor because they allow him to reduce his capital requirements and still submit a balanced unit cost in his bid. Then if quantities overrun, he will receive proper compensation for the extra work involved.

In contrast to this, the inclusion of a mobilization bid item in the bid schedule helps the contractor finance the work but often it is to his disadvantage if work quantities overrun. A major overrun in work quantities will increase the time required for construction and increase the hours worked by the equipment. This results in the equipment being older and having less useful life at job completion, which decreases its sale value and increases the depreciation charges to the construction project. When a contractor has placed equipment depreciation charges in the mobilization bid item, the unit bid prices for the work quantities do not include equipment depreciation, and he does not recover increased depreciation costs when he receives payment for overruns in quantities. By placing equipment depreciation against the mobilization bid item, the contractor established it as a fixed sum based on the original construction schedule and the original bid quantities and therefore he unbalanced the unit bid prices.

Cost spreading becomes more difficult when the estimated bid quantities shown on the bid schedule are inaccurately listed. If the bid quantities are too large, then underruns will occur when the final pay quantities are determined, which forces the contractor to increase the unit prices in order to properly recover his cost, since a large percentage of the cost spread over these items is fixed and varies only slightly with the amount of work performed. Such fixed costs will not be recoverable on the quantities of work bid but not done or paid for, unless the contractor increases the amount spread to these items in proportion to the amount each item will underrun. This increase in unit prices will increase the bid total and often will make a bid noncompetitive. If a contractor does not catch this quantity error, he may be the low bidder and then will not receive adequate compensation for his work.

Owner's engineers too often do not understand this principle of bidding and do not pay sufficient attention to the accuracy of the bid quantities. For example, when a certain dam was bid, the engineers threw out all the bids and completely redesigned the dam, reducing the quantities of work, but they readvertised the work with the original bid quantities. This made it very difficult for the contractor to submit a bid that would add up to a proper total and still allow the contractor proper reimbursement. When such flagrant errors in bid quantities occur, it forces the contractor to completely unbalance the bid by placing as much cost as possible against lump-sum items or items where quantities are correctly stated.

Spreading of cost is such an important part of bidding that it should be better understood by estimators, owners, and engineers than it is today.

BID CLOSING AND QUOTATION ADJUSTMENTS

As soon as the spread sheets are finished and bid prices computed, most of these units and extensions should be written into the bidding schedule. The Denny Dam bid schedule contains only 53 bid items, and this posting procedure would be comparatively short and simple. Many bid schedules are quite complicated with hundreds of bid items. Posting a bid of this type may take hours and, in some cases, days. Most of the plugged permanent material prices and subcontract prices will require adjustments when quotations are received immediately prior to bid submittal. Therefore, the bid total, the unit price, and the bid price extensions for some of the large bid items are not written into the bid, in order that all final quotation adjustments can be made by adjusting the price of these open items. For the Denny Dam bid, it is assumed that all bid prices and extensions are written in, except for the bid total, the unit bid prices and extensions for Bid Item 1, diversion and care of river, and for Bid Item 24, mass concrete in the dam. The written-in extensions should be totaled so that the bid total can be rapidly adjusted and checked when adjustments are completed shortly before the bid closing time.

It is very important that the bid prices be checked and that the extensions and additions be checked on the bid schedule. When bid totals do not agree with the total found by extending bid quantities by unit bid prices and adding the extensions, then the bid total is disregarded and the total of the unit price extensions is used to determine the low bidder. Therefore, the low bidder on a construction project is not definitely determined until the owner's representative has checked the extensions and additions of every bid submitted. Occasionally, because of sloppy extensions or because erroneous unit bid prices have been written in the bid, the apparent low bidder at bid reading time was not the successful bidder. Therefore, the need for accuracy in this work cannot be overstressed.

During bid preparation, replacing plugged prices with actual quotations, changing the bid total to reflect these quotations, and keeping a running total of the bid adjustments resulting from quotations is of prime importance, since the receipt of a final low quote may make the difference between having the low bid or being just another bidder. The use of plugged prices in a contractor's estimate was explained under Estimating Task IV in Chap. 9.

There are many ways of keeping track of quotations so that adjustments of the bid can be readily made. The estimator should decide on the method that suits him best and be familiar with its procedure. Exhibit 4 shows an over-and-under sheet which illustrates how quotation adjustments were tabulated for the Denny Dam

Exhibit 4 OVER-AND-UNDER QUOTATION ADJUSTMENTS FOR THE DENNY DAM BID

Bid item no.	Description	Bid quantity	Cost in estimate			Lowest quote		Under −	Over +	Name of supplier or subcontractor
			Unit	Materials total	Subcontracts total	Unit	Total			
7	Drill $1^1/_2$-in.-diameter grout holes	25,000 lin ft	$ 4.00		$ 100,000	$ 3.90	$ 97,500	$ 2,500		Hardrock Drilling Co.
8	Drill 3-in.-diameter drain holes	17,000 lin ft	6.00		102,000	5.50	93,500	8,500		Hardrock Drilling Co.
9	Core drill 3-in. exploratory holes	13,500 lin ft	6.00		81,000	6.25	84,375		$ 3,375	Hardrock Drilling Co.
10	Grout connections	800 each	10.00		8,000	10.00	8,000			Hardrock Drilling Co.
11	Pressure testing	800 test	10.00		8,000	15.00	12,000		4,000	Hardrock Drilling Co.
12	Grouting cement	8,000 cu ft	1.40	$ 11,200		1.30	10,400	800		Kilm Cement Co.
13	Pressure grouting	8,000 cu ft	4.00		32,000	3.50	28,000	4,000		Hardrock Drilling Co.
18	Reinforcing steel	3,080,000 lb	0.16		492,800	0.15	462,000	30,800		Hill Steel Co.
25	Cement	635,000 bbl	5.00	3,175,000		4.80	3,048,000	127,000		Kilm Cement Co.
26	12-in.-diameter RCCP	100 lin ft	2.50	250		2.50	250			Sutter Pipe Co.
27	18-in.-diameter RCCP	100 lin ft	5.00	500		4.50	450	50		Sutter Pipe Co.
28	Copper water stop	12,300 lin ft	3.00	36,900		3.25	39,975		3,075	Kidder Supply Co.
29	6-in.-diameter porous concrete pipe	28,000 lin ft	1.25	35,000		1.50	42,000		7,000	Sutter Pipe Co.
30	12-in.-diameter half-round porous pipe	1,950 lin ft	2.00	3,900		2.25	4,388		488	Sutter Pipe Co.
31	Miscellaneous doors	400 sq ft	10.00	4,000		8.00	3,200	800		Kidder Supply Co.
32	Miscellaneous metal castings	Lump sum		6,400			7,600		1,200	Steel Fabrication Co.
33	Slide and bulkhead gates	286,000 lb	0.50	143,000		0.52	148,720		5,720	Kidder Supply Co.
34	Pumps, motors, and piping for gates	Lump sum		25,000			20,000	5,000		True Steel Co.
35	Penstock gates, frames, and guides	288,000 lb	0.40	115,200		0.43	123,840		8,640	True Steel Co.
36	Penstock-gate hoist and supports	Lump sum		56,000			49,000	7,000		Steel Fabrication Co.
37	Penstock lining and frames	535,000 lb	0.40		214,000	0.38	203,300	10,700		Steel Fabrication Co.
38	Tainter gates and accessories	670,000 lb	0.25	167,500		0.23	154,100	13,400		Steel Fabrication Co.
39	Tainter-gate trunnion anchors	320,000 lb	0.30	96,000		0.28	89,600	6,400		Steel Fabrication Co.
40	Tainter-gate hoist and accessories	150,000 lb	1.40	210,000		1.52	228,000		18,000	Steel Fabrication Co.
41	Trashrack guides and beam	260,000 lb	0.22	57,200		0.22	57,200			True Steel Co.
42	Stop logs, protection plate, and beam	168,000 lb	0.25	42,000		0.23	38,640	3,360		True Steel Co.
43	Cast-iron pipe and fittings	156,000 lb	0.21	32,760		0.18	28,080	4,680		Sutter Pipe Co.
44	Steel pipe and fittings	184,000 lb	0.30	55,200		0.31	57,040		1,840	Sutter Pipe Co.
45	Piezometer piping and fittings	Lump sum		1,000			800	200		Sutter Pipe Co.
46	Sump pump and piping	Lump sum		6,000			4,000	2,000		Sutter Pipe Co.
47	Water supply and plumbing	Lump sum		12,000			15,000		3,000	Sutter Pipe Co.
48	Structural steel	66,500 lb	0.22	14,630		0.21	13,965	665		Steel Fabrication Co.
49	Railing and fittings	17,700 lb	0.40	7,080		0.45	7,965		885	Steel Fabrication Co.
50	Parapet railing	1,152 lin ft	12.00	13,824		10.00	11,520	2,304		Kidder Supply Co.
51	Steel in spillway bridge	410,000 lb	0.17	69,700		0.16	65,600	4,100		True Steel Co.
52	Electrical system	Lump sum			250,000		227,500	22,500		Hill Electric Co.
	Painting for all items				53,364		48,000	4,364		White Painting Co.
	Subtotal			$4,397,244	$1,341,164		$5,534,508	$261,123	$57,223	
					4,397,244			57,223		
	Total adjustment				$5,738,408		$5,534,508	$203,900		

Note: All prices include sales tax.

estimate. This is a very simple and effective method of tabulating quotations and keeping track of bid price adjustments.

On the Denny Dam over-and-under sheet, the painting subcontract prices that were plugged into the estimate are shown as a line item to shorten the time required to compare the plugged painting prices with the painting quotation. Comparing plugged and quoted painting prices for each bid item would be so slow and tedious that it should not be attempted just prior to bid submittal.

In using an over-and-under sheet, the lowest quote for any item is posted; it is compared to the amount in the estimate, and the adjustment is tabulated. Any new quotation received is compared to the one posted; if it is higher, it is put aside; if it is lower, the one on the over-and-under sheet is erased and the new one posted, extensions made, and the new adjustment made. This is a foolproof method since each quote needs to be handled only once.

The receipt of quotes must be halted sufficiently ahead of bid closing time to allow time for closing the bid and for transporting the bidding documents to the point of submission. This work involves the totaling of the adjustments on the over-and-under sheet. The bid items left open on the bidding papers and the bid total then must be adjusted to reflect this adjustment; the adjusted bid total must be written in; these computations should be checked; the bid envelope containing all the bidding documents must be sealed; and the bid must be carried to and deposited at the location stated in the bidding instructions. In order to allow the bid to remain open as long as possible, a temporary office can be located near where the bid must be submitted. Hotel rooms are commonly used for this purpose.

Bid closing for Denny Dam was done by totaling the adjustments on the over-and-under sheet, Exhibit 4, which amounted to a bid reduction of $203,900. This quotation reduction also results in a reduction in markup. The new bid total was then adjusted in the following way for these reductions:

Agreed bid price		$29,166,807
Quotation reduction	$203,900	
Reduction in markup	30,585	
Bond premium adjustment	none	
Total bid reduction		234,485
		$28,932,322

For the Denny Dam bid, it was assumed that all bid prices and extensions were written in the bid schedule except for items 1 and 24; therefore, the final quotation adjustment had to be made to these items, as shown on Exhibit 3.

After bid closing time, bids are opened by the owner's representative, read, and the apparent low bidder announced. The owner's representative checks all bids for any errors in extensions or additions and then determines the low bidder. As previously explained, the unit prices control and the apparent low bidder may not be low when the extensions are verified and the additions are checked on all the bids.

BUDGET ESTIMATE

If the bid is successful, then the sponsoring company's estimator prepares a budget estimate. This budget estimate is used by the job management to compare the production and cost achieved during construction with the production and cost that was estimated for the work. The estimator should also make these comparisons so that he can utilize this work experience in his next estimate.

To prepare the budget estimate, the adjustments made at the joint-venture estimators' meeting and the quotation adjustments must be carried back into all the estimating details. The cash forecast should also be revised to show the agreed cost, the actual quotations for permanent materials and subcontracts, and the amount of markup used in the bid. It then can be used to control job financing.

Glossary of Dam Terminology

AIR GUN A steel cylindrical-shaped vessel with gates to allow the receipt of concrete and with air connections and discharge-pipe connections. After the gun is charged, the concrete is forced through the discharge pipe by air pressure.

ANFO An explosive consisting of a mixture of ammonium nitrate prills and a carbonizing agent. Originally the carbonizing agent was fuel oil, hence the initials.

ARROW DIAGRAMING A method of making a construction schedule using arrows to indicate the sequence of operations or events.

BALONEY CABLE Heavy insulated electric drag cable often used to supply power to movable cableway towers, cranes, shovels, and similar types of equipment.

BANK YARD A cubic yard of material in its natural state in a borrow pit or quarry.

BARRAGES Diversion dams.

BARRING AND WEDGING Procedure used to remove the last portion of unsuitable rock from the foundation surface of a dam.

BATHTUBS The concrete containers on dumpcrete trucks.

BELLBOY Man who transmits voice and signal directions to the operator of a cableway or crane to control the spotting of a concrete bucket.

BENTONITE SLURRY TRENCH Bentonite slurry is often used to maintain the sides of the excavation when deep trenches are excavated through common material.

BID QUANTITIES Quantities published on the bid schedule for unit-price items. Bid quantities are extended by unit prices to arrive at the total bid price for each unit-price bid item.

BIFURCATION The dividing of one penstock into two or more smaller penstocks, each of which is then connected to a turbine scroll case.

BLOCK HOLING Secondary breakage of blasted rock accomplished by drilling holes in the rock and then using explosives in these holes.

BLOCK-OUT CONCRETE Concrete surrounding gate guides and other similar members that is not placed when the main structure is poured. After completion of the main structure, the steel gate guides are placed and aligned, and the block-out concrete is then placed.

BOGIE PRICES See Plugged Prices.

BUCKET CAR A rail car fabricated to carry as many as four 8-yd buckets, usually equipped with a catwalk at the proper elevation to expedite hooking and unhooking the concrete buckets from a cableway. If an automatic hook is used on the cableway, no catwalk is needed.

BUCKET DOCK A landing platform for concrete buckets with a length that permits the cableway to land a bucket when its traveling tower or towers are at any travel position. The bucket dock is located adjacent to and slightly below the transfer track, permitting the transfer car to discharge concrete directly into the bucket.

BUDGET ESTIMATE A bid estimate adjusted for late quotes and other bid adjustments, used as a reference for control of the cost of constructing a project.

BUILT-IN-PLACE FORMS Forms with such irregular shape that they must be built in place and hand-tailored to fit the opening. An example is the first lift forms for dam concrete which are placed against the irregular rock surface forming the dam's foundation.

BULKHEAD FORMS Dam forms which separate the concrete in a dam into a series of individual blocks or monoliths.

BUTTON LINE A line used on older-type cableways to control the spacing along the track cable of the slack-line carriers.

BUY-OUT PRICES Prices actually paid for equipment and materials required during the construction of a dam project.

CABLE SAG Amount of track-cable sag at the center of the span of a cableway. Six percent of the span is a rough approximation of its extent.

CANDY WAGON Service truck.

CARRIAGE The frame and wheel sheaves that are pulled along the track cable of a cableway by the conveying line. Framed beneath the wheels are sheaves over which pass the hoisting line which suspends and controls the elevation of the cableway hook.

COMPARATIVE ESTIMATES Rough estimates comparing the cost of doing one operation in two different manners or the utilization of two different types of equipment. Comparative estimates assist in the selection of the most economical method or type of equipment.

CONCRETE CLEANUP The removal of laitance and the roughing and preparation of the top of a previously poured concrete lift prior to placing the next concrete lift.

CONCRETE PUMP A pump that mechanically forces concrete through a pipeline into a formed area.

CONFINED EXCAVATION Excavation performed in a narrow slot for a dam's foundation. The narrow working area prevents efficient utilization of equipment.

CONSTRUCTION SYSTEM An integrated unit consisting of construction time, manpower, construction equipment, and supporting facilities required for the economical construction of a project.

CONTRACT RETENTION Percentage withheld by the owner from each progress payment made to a contractor until the work is completed.

CONVEYING LINE Cable that moves the cableway carriage along the track cable.

COOLING COILS Small-diameter pipe placed in coils on top of each concrete lift. After being covered by the next concrete lift, cold or refrigerated water is pumped through the coils to remove the heat generated in the concrete by the hydration of the cement.

CORRECTIVE CONCRETE Concrete placed beneath any rock overhang occurring on the rock surface of the foundation for the impervious core of an earth- or rock-fill dam.

COYOTE Small-sized tunnel drift excavated into a quarry face and loaded with an explosive to break large quantities of rock.

CRESCENT SCRAPER A large open-bottomed scraper shaped like a crescent used to drag material off a slope.

CRITICAL PATH A phrase used to describe the sequence of those construction operations which are critical in determining the time required for completion of a construction project.

CURING COMPOUND Liquid sprayed on a freshly poured concrete surface which prevents water from evaporating from the surface, thus retaining sufficient water in the concrete to assure complete hydration of the cement.

DAY SHIFT Shift that extends from 8:00 A.M. to 4:00 P.M. See Swing Shift, Graveyard Shift.

DENTAL EXCAVATION Excavation performed in pockets, shafts, or drifts in the foundation of a dam to remove unsuitable foundation rock.

DISPOSAL AREAS Areas where excavated materials can be wasted.

DOORKNOB ESTIMATE A rough approximation of the construction cost of a construction contract prepared by a few man-hours of work.

DOWN-THE-HOLE DRILLS Percussion drills that follow the bit down the drilled hole.

DREDGER TAILINGS Piles of gravel left by gold dredges as they dredged river channels or old river channels.

DRIFTER-TYPE DRILL Percussion drills with bore diameters of from $3^1/_2$ to 5 in.

DUMMY ACTIVITIES Nonevent arrows required by PERT or CPM scheduling to enable the schedule to show proper sequence of activities.

DUMPCRETE Concrete hauled from the mixer to the pour in open-top tank-type truck bodies. These tanks may or may not be equipped with paddles for agitating the concrete.

ELASTIC FRACTIONATION Method of removing lightweight materials from concrete aggregates, based on the principle that the denser aggregates will rebound a greater distance when they are dropped on a steel plate.

ELEPHANT TRUNK Steel pipe or rubber hose used to drop concrete vertically into a pour.

EMBANKMENT YARDS Volume of fill material in an earth- or rock-fill embankment in its compacted state.

ENDLESS LINE Another name for the conveying line of a cableway. See Conveying Line.

EQUIPMENT AVAILABILITY Percentage of time that construction equipment will be available for use. The remainder of the time it will be out of service for repair or servicing.

EQUIPMENT LEVELING Scheduling the construction of noncritical work items for periods when construction equipment will be available.

ESCALATION The amount that labor, material, and equipment costs will increase during the life of a construction project.

EXPLORATORY DRILLING Drilling performed to secure information about dam foundations, borrow pits, quarries, tunnels, etc.

FINENESS MODULUS Measure of the fineness of sand computed from its screen analysis.

FIRST-STAGE CONCRETE The principal structural concrete. The term is used only when second-stage concrete is to be poured at a later stage of construction.

FLY ASH Flue dust reclaimed from the stacks of coal-fired steam plants. It is used as a pozzolanic material in dam concrete.

FORCE MAJEURE Clause in the specifications that relieves the contractor of financial responsibility for certain events not under his control. Examples include acts of God, war, earthquakes, floods, and strikes.

FORM CONTACT Square feet of concrete surface area that must be formed.

FORM HARDWARE Form ties, she-bolts, snap ties, etc., used to hold forms in position.

FORM OIL Oil painted on the contact surface of concrete forms to expedite form stripping and protect the formed surface.

FORM RATIO Square feet of formed surface required per cubic yard of concrete.

FOUNDATION CLEANUP The removal of all loose material and water from a rock surface before it is covered with concrete.

FOUNDATION SLOPE CORRECTION Excavation that must be performed to comply with some specifications that may require that the rock surface under the impervious core of earth- or rock-fill dams must have an upstream slope in relation to the dam axis. This excavation must be done in areas where the rock slopes downstream to change its slope.

GABION A group of medium-sized (8 to 24 in.) rock particles contained in a wire netting.

GETTING-OFF-THE-ROCK The completion of the first concrete lift on a dam foundation.

GOB HOPPER Hopper located under the mixers in a concrete-mixing plant into which the mixers dump the concrete. Concrete buckets, cars, or trucks can then be filled with concrete from the gob hopper, freeing the mixer for charging and mixing.

GRAVEYARD SHIFT Shift that extends from midnight to 8:00 A.M. See Day Shift, Swing Shift.

GREEN-SHEET ESTIMATE Another name for a doorknob estimate. See Doorknob Estimate.

GRIZZLY A heavy screen with large openings used to scalp off large rocks from quarried rock or from concrete aggregates. Often grizzlies consist of spaced rails.

GROUT CURTAIN A continuous zone of grouted rock, formed by pumping grout into a regular pattern of drill holes.

GUT A section of unsuitable foundation material that passes through the foundation of a dam.

HEAD TOWER The tower of a cableway that contains or is near the drums and hoist that operate the cableway.

HEAVY-MEDIA SEPARATION A method of eliminating lightweight particles in concrete aggregates by flotation in a mixture of water, magnetite, and ferrosilicon.

HIGH LINE Slang terminology for a cableway.

HOIST LINE A line that controls the elevation of the cableway hook.

HOLE BURDEN The horizontal distance from the rock face of an excavated area or quarry to the blast holes.

HOLE PATTERN Horizontal spacing of drill holes.

HOLE SPACING The horizontal distance between drill holes measured parallel to the rock face.

HYDRAULIC MONITOR A nozzle that discharges a jet of high-pressure water used as a means of removing common materials from a dam's foundation and as a means of improving rock consolidation in a rock-fill dam.

IMPERVIOUS ZONE The clay or silt zone of an earth- or rock-fill dam which provides a water barrier.

INCLINED DRILLING Drill holes inclined so that they are parallel to the vertical face of a quarry. Also used to describe nonvertical exploratory, grouting, or drainage holes.

JOE MAGEE CABLEWAY A construction cableway used during the start of a construction project to provide temporary access to one dam abutment. Often it consists of a cable rigged across the canyon from a logging winch on a tractor.

LAITANCE Dehydrated cement mortar that forms on the top of a concrete lift.

LEFT-ON-THE-TABLE The difference between the low and second bid.

LIFT The vertical thickness of concrete placed during one pouring operation.

LIFT DIFFERENTIAL The vertical distance between the top of the lowest and highest blocks (monoliths) of a concrete dam.

LIFT-STARTER FORMS Noncantilever forms used after the first pour has been made on a dam's foundation, and until sufficient height has been poured in each monolith to provide clearance for cantilever forms.

LOOSE YARDS Space occupied by material that has been excavated and deposited in a haulage vehicle.

LUFFING CABLEWAY A cableway with two fixed towers which provide transverse hook coverage by drifting the tops of the towers as the lengths of the side guys are adjusted.

MAIN GUT Track cable of a cableway.

MASS CONCRETE Concrete in massive structures which has a low form ratio.

MISCELLANEOUS CONCRETE Concrete in slender structures that has a high form ratio.

MONKEY SLIDE Inclined trackage equipped with a winch-controlled car.

MONOLITH A block of concrete in a dam separated from the others by bulkhead forms.

MUD A slang term for concrete.

MUDCAPPING A method of performing secondary rock breakage. The explosive is placed on the surface of large rocks and covered with mud.

MULTIPLE-SHEAVE DRIVE A double set of sheaves controlling the travel of a conveying line of a cableway.

MULTIPLE SHIFT The operation of a construction project for two or more shifts per day.

OFF-HIGHWAY Trucks with heavy bodies and with such a payload capacity that the weight on each tire will exceed that permitted for highway usage.

OLD-MAN SLACK-LINE CARRIERS The two slack-line carriers at each end of a cableway that are very seldom moved except when the carriage is brought into a tower for servicing.

ON-HIGHWAY Trucks constructed with light bodies and with payloads limited to weights that will permit the load on each tire to conform to highway vehicle codes.

ON-SITE COST Cost generated or money expended at the site of a construction project.

OPTIMUM BANK HEIGHT The height that a bank of rock or common material must be to secure maximum production from a power shovel.

OVER-AND-UNDER SHEET Method of tabulating prebid quotes for materials and subcontracts.

OVERBREAK Excavation that extends past pay lines.

PANTS LEGS A slang term for bifurcation. See Bifurcation.

PARALLEL CABLEWAY A cableway that has movable head and tail towers which maintain the same relative position on parallel or concentric runways.

PAY QUANTITIES Quantities of work performed for each unit bid item. Pay quantities times unit bid price will equal the final payment for each unit-price bid item.

PAYLOAD The weight of the material that can be hauled by each haulage unit.

PENETRATION RATE The length of blast hole produced by a drill in a unit time interval.

PICKUP POINT The position under a cableway where the cableway transfers concrete buckets or has a concrete bucket filled by a transfer car.

PLUGGED PRICES Approximate prices used in an estimate for permanent materials and subcontracts until quotes are received from suppliers and from subcontractors.

POWDER FACTOR The pounds of explosives required to break one bank cu yd of rock.

POZZOLAN Finely ground shale, pumice, volcanic material, or reclaimed flue dust produced by the combustion of coal (fly ash), used to disperse cement in concrete and in some instances to react with the free alkali in the cement.

PRECEDENCE DIAGRAMING One method of preparing a construction schedule, using arrows to show the sequence of work.

PREPACK CONCRETE Concrete formed by first placing the coarse aggregates and then pumping sand, cement, grout, and a dispersing agent into the preplaced aggregates.

PROPORTIONAL SLACK-LINE CARRIERS Wheeled carriers traveling on the track cable of a cableway that support the operating ropes. Proportional carriers space themselves across the span by differences in travel-speed gearing. They are actuated by the movement of the endless line.

PUMPED STORAGE POWER PROJECTS Power-generating installations that can function as hydrogenerating units and as motor-driven pumps. Such plants are located between two reservoirs that are at different elevations. During periods when excess power is available from thermal plants, they pump water from the lower reservoir to the higher. During periods of maximum power demands, the water flow is reversed and they act as generators.

PUNCH LIST A list of deficiencies in an almost completed construction unit or project.

PUSHER A power unit used to assist in the loading of a scraper.

QUARRY FACE The nearly vertical surface of unbroken rock in a rock quarry.

RADIAL CABLEWAY A cableway with a fixed tower and a movable tail tower. The movable tower travels on a track which is a constant distance from the fixed tower.

RANDOM FILL Fill placed in an embankment which may consist of a variety of materials.

REACTIVE AGGREGATES Concrete aggregates that contain siliceous ingredients that combine with the free alkali in the cement to form an alkali silica gel. Over a period of time, the concrete may be broken by the osmotic swelling of this gel.

RECLAIM TUNNEL A tunnel under stockpiled material, usually containing a conveyor belt. The purpose of the tunnel is to facilitate loading out of the stockpiled material.

REFRACTION SEISMOGRAPH An instrument for determining the characteristics of rock by measuring the velocity at which seismic waves travel through the material.

RESCREENING PLANT An aggregate-screening plant located adjacent to or on top of the concrete-mixing plant. The rescreening plant is used to correct any size degradation caused by rehandling the various sizes of aggregates.

RESTEEL A slang term for reinforcing steel.

ROCK CLEANUP The cleanup of a rock surface on which concrete is to be placed.

ROCK LADDERS Fabricated steel framework placed under the discharges of aggregate stockpiling conveyor belts to restrict the free fall of the aggregate to between 2 and 3 ft.

ROCK NECKLACES Large rocks that have been drilled and strung on wire ropes so that when they are used to close a water-diversion opening, they can better resist the erosive force of the water.

ROD BUSTERS Men who install reinforcing steel.

RUNNER Operator of a shovel or crane.

S.I. A surveyor.

SALAMANDERS Oil heaters.

SCALPING The removal of oversized rock or boulder particles from the remainder of the material.

SECOND-STAGE CONCRETE Concrete placed within a partially completed structure to embed turbine scroll cases or similar items.

SECONDARY BREAKAGE The breaking of oversized rocks into smaller particles in quarry operations.

SELF-RAISING FORMS Forms equipped with powered jacks. After concrete has been poured and set within the forms, the jacks raise the forms to the proper position to form the next concrete lift.

SHE-BOLTS Form hardware used to anchor vertical forms. A she-bolt is threaded externally to accept clamps which bear upon the forms and internally to accept the threads of spacer rods. All parts are salvageable after a pour is completed except the spacer rods.

SHELL The zones of an earth- or rock-fill dam which support the impervious zones.

SINK-FLOAT A method for removing lightweight portions from concrete aggregates. The lightweight particles are floated off using a mixture of water, magnetite, and ferrosilicon. See Heavy-media.

SKELETON BAYS Future generating bays of a hydro powerhouse in which only the first-stage concrete has been poured.

SKINNER Operator of a tractor or similar type of equipment.

SKIPS Metal boxlike containers used to contain debris cleaned from the rock surface of a dam's foundation, from concrete surface areas, and for similar uses.

SLACK-LINE CARRIERS Wheeled frames riding on the track cable of a cableway that support the operating cables of a cableway.

SLIP FORMS Forms that are raised as a concrete pour progresses so that there is a continuing formed surface within which concrete can be poured.

SLURRY An explosive consisting of a mixture of water, ammonium nitrate prills, a sensitizer, and other ingredients. Also a bentonite slurry to control ground water as an aid to excavation.

SNAP TIES Form hardware consisting of small-diameter rods spanning between two wall forms to hold them at a constant distance from each other. When the concrete has set they can be broken at two deformed points located back from the surface of the concrete.

SOFFIT FORMS Forms supporting the bottom of a concrete beam or other suspended concrete member.

SONOTUBE FORMS Circular concrete forms made from treated paper.

SPIDERS Temporary supports which maintain a penstock in circular shape while it is being transported and erected.

SPOTTER The individual who instructs a truck driver where to spot his truck for loading or for dumping.

SPREAD SHEETS Estimate forms used to distribute the various general construction costs to individual bid items.

STEMMING Inert material placed on top of the explosives in a blast hole to confine the force of the explosion.

STEP EXCAVATION Excavation done on the abutments of a dam to provide level foundation surfaces that form steps up the abutments.

SUBDRILLING The distance drilled beneath the desired surface of an excavated area or quarry floor to ensure that all rock will be broken to the required elevation.

SUBSISTENCE The daily living allowance paid to workers in remote areas.

SUBSISTENCE AREAS Areas defined by labor agreements in which the employer must pay living allowances to his employees.

SURFACE GROUTING Grouting the rock surface of the dam foundation to improve its characteristics.

SURFACE MOISTURE The amount of water on the surface of concrete aggregates.

SWELL FACTOR The percentage that one bank yard will increase in volume after it has been excavated and before it is compacted.

SWING SHIFT The shift that extends from 4:00 P.M. to midnight. See Day Shift and Graveyard Shift.

TAIL TOWER The tower of a cableway that does not contain the operating machinery. See Head Tower.

TAKEOFF QUANTITIES Quantities of work taken off the contract drawings by the contractor.

TETRAHEDRON A concrete block with four sides used to close off a diversion opening. The shape of the block allows block interlocking which permits it to resist the transporting effect of flowing water.

TETRAPOD A concrete block with four arms that interlock when used to close a diversion opening. See Tetrahedron.

TRACK CABLE The main cable of a cableway on which the cableway carriage travels. It is made with a track lay that presents flat surfaces of the wire strands of the cable to resist the wear of the wheels of the carriage. The general range in cable diameter is from $1^1/_2$ to 4 in., depending on the capacity of the cableway.

TRANSFER CAR A railroad car which may be self-propelled or may be pulled by a locomotive, containing concrete compartments of the same size as the concrete buckets used on a cableway. Used to transfer concrete from the mix plant to the bucket dock of a cableway. At the bucket dock the car dumps each compartment into a concrete bucket hooked to the cableway.

TRANSFER TRACK The railroad track running between the mixing plant and a cableway. See Transfer Car.

TRAVEL LINE Another name for conveying line. See Conveying Line.

UNCLASSIFIED EXCAVATION Bid item terminology used when both rock and common excavation are undifferentiated for payment purposes and are paid for under one bid item.

UP-AND-DOWN CHARGES Charges made by a utility company to furnish, erect, and dismantle a power substation.

VACUUM PROCESS The process used to cool concrete aggregates by causing surface water to evaporate from the aggregates in a partial vacuum.

VIBROFLOTATION A patented process of compacting sandy material.

WELL POINTS A series of pipes installed in shallow wells around an area to be dewatered. The pipes are connected to headers and pumps to remove the water.

WOOD BUTCHER A carpenter.

WORK CONSTRAINTS Specification or labor contract requirements that specify how construction work must be performed, or the type of labor and working rules that must be used during its construction.

Index